Seismic Analysis and Design of Building Structures

Seismic Analysis and Design of Building Structures

Guest Editors

Bo Fu
Bo Wang
Xinxin Wei
Qing Lv

MDPI

Basel • Beijing • Wuhan • Barcelona • Belgrade • Novi Sad • Cluj • Manchester

Guest Editors

Bo Fu
School of Civil Engineering
Chang'an University
Xi'an
China

Bo Wang
School of Civil Engineering
Chang'an University
Xi'an
China

Xinxin Wei
Faculty of Civil and
Environmental Engineering
Ruhr-Universität Bochum
Bochum
Germany

Qing Lv
School of Civil and
Transportation Engineering
Ningbo University of
Technology
Ningbo
China

Editorial Office
MDPI AG
Grosspeteranlage 5
4052 Basel, Switzerland

This is a reprint of the Special Issue, published open access by the journal *Buildings* (ISSN 2075-5309), freely accessible at: https://www.mdpi.com/journal/buildings/special_issues/T2ELV3GUAV.

For citation purposes, cite each article independently as indicated on the article page online and as indicated below:

Lastname, A.A.; Lastname, B.B. Article Title. *Journal Name* **Year**, *Volume Number*, Page Range.

ISBN 978-3-7258-6606-9 (Hbk)
ISBN 978-3-7258-6607-6 (PDF)
https://doi.org/10.3390/books978-3-7258-6607-6

Contents

buildings

MDPI

Article

Two-Stage Optimal Design Method for Asymmetric Base-Isolated Structures Subject to Pulse-Type Earthquakes

Jiayu Zhang [1], Ai Qi [1,*] and Mianyue Yang [2]

[1] College of Civil Engineering, Fuzhou University, Fuzhou 350116, China; n190510007@fzu.edu.cn
[2] Faculty of Architecture and Civil Engineering, Huaiyin Institute of Technology, Huaian 223003, China; mianyueyang@hyit.edu.cn
* Correspondence: qikai@fzu.edu.cn

Abstract: Asymmetric base-isolated structures subjected to severe torsion may suffer further aggravation of their torsional and translational responses under pulse-type earthquakes. To counteract these detrimental impacts, this study introduces a two-stage optimal design method. The first stage involved the application of the NSGA-II algorithm for determining an optimal isolator arrangement—namely, position and category—with the objective of reducing both the maximum interstory rotation of the superstructure and the isolation layer. In the second stage, the inclusion of viscous dampers served to minimize the excessive translational response triggered by pulse-type earthquakes. The influence of these dampers' positions on the structural response was carefully evaluated. The final application of this optimal design method was demonstrated on an asymmetric base-isolated structure. The results indicated a significant reduction in the translational and torsional responses of the asymmetric base-isolated structure when the two-stage optimal design method was utilized, compared to those of structures designed using traditional conceptual methods. It was found that by installing viscous dampers in the isolation layer along both the x and the y directions—specifically, underneath the mass center of the superstructure (CMS)—the effectiveness of the torsional resistance from the first stage could be effectively maintained.

Keywords: asymmetric base-isolated structures; pulse-type earthquakes; NSGA-II algorithm; two-stage optimal design method; isolator arrangement; viscous dampers; structural response

1. Introduction

The growing popularity of the seismic isolation technology is primarily attributed to its ability to isolate the majority of the energy from earthquakes by extending buildings' natural period [1–3]. However, recent research suggests that the efficacy of isolators diminishes under long-period pulse-type earthquakes [4–7]. Increased deformation is a common outcome for base-isolated structures exposed to pulse-type earthquakes [8–10]. Furthermore, the challenge is amplified for asymmetric base-isolated structures that may experience severe torsion due to the eccentricity of the isolation system (e_b) (i.e., the distance between the CMS and the stiffness center of the isolation system (CSI)) [11–13]. Such structures may suffer an exacerbated adverse impact from pulse-type earthquakes [14]. Thus, this work proposes an optimization design method aiming to effectively mitigate the detrimental impacts of pulse-type earthquakes on asymmetric base-isolated structures.

Several studies have already proposed approaches that lessen the significant torsion of asymmetric base-isolated structures and decrease the excessive translational response caused by pulse-type earthquakes. The torsion of asymmetric base-isolated structures is predominantly influenced by structural parameters such as e_b, the ratio of torsional to lateral frequency, and the effective period of base-isolated structures [15–19]. One of the most efficient approaches that lessen the torsional response involves altering the arrangement of isolators in the isolation system [20,21], enabling the adjustment of structural parameters to favor the torsional resistance. Additionally, it was found that adding viscous dampers to the isolation systems can successfully lessen the excessive structural response of base-isolated structures induced by pulse-type earthquakes. For instance, Kelly [22] used a two-degree-of-freedom linear model to simulate a base-isolated structure, demonstrating that damper installation in the isolation layer can lessen the isolation layer displacement while elevating the interstory drift. Politopolous [23] found that as the viscous damping ratio increases, the acceleration and shear of isolated structures decrease, using a model similar to Kelly's. He and Agrawal [24] argued that viscous damping is most effective when the pulse period closely aligns with the system period, as shown through base-isolated building structural analysis. Additionally, Mazza and Vulcano [25] conducted a dynamic response analysis of a five-story isolated frame structure with viscous dampers under pulse-type earthquakes to assess the effect of additional damping of the isolation layer. Their findings suggest that viscous dampers can effectively manage the displacement of isolators.

However, when an asymmetric base-isolated structure with severe torsional response is adversely affected by pulse-type earthquakes, it becomes imperative to explore whether these two aforementioned measures retain their effectiveness in reducing the structural response and how to synergize these measures to optimize their advantages. Consequently, this work suggests a two-stage optimal design method to leverage both measures. This optimization approach considers the impacts of both structural parameters and pulse-type characteristics on the structural response to more effectively mitigate the translational and torsional responses of asymmetric base-isolated structures under pulse-type earthquakes.

The first stage of the optimization design method aims to adjust the isolator arrangement to minimize the torsional response as much as possible. The traditional conceptual design of torsional resistance for asymmetric base-isolated structures usually employs a trial-and-error method to modify the isolator arrangement in order to lessen the structural torsional response. However, this method can be labor-intensive and inefficient. Moreover, the final isolator arrangement scheme obtained may not always be effective, even if it can somewhat reduce the torsional response. To enhance the efficiency of the torsional resistance, this paper introduces a multi-objective optimization algorithm in the first stage. Some scholars have conducted research on multi-objective optimal design methods [26,27]. It is worth noting that metaheuristic algorithms are increasingly popular for solving multi-objective optimization problems. For example, Ali et al. [28,29] achieved the optimal isolation period and isolation damping ratio of isolated structures through the multi-objective water cycle (MOWCA), crow search (CSA), and whale optimization (WOA) algorithms and the grey wolf optimizer (GWO). Kaveh and Zakianan [30] obtained minimal structural weight and cost through an improved grey wolf optimizer (IGWO) algorithm in truss systems. However, the decision variables of these metaheuristic algorithms are structural parameters with continuous values. When referring to a set of isolated points comprising parameters for different isolator categories in the first stage of the optimization design method, metaheuristic algorithms have certain limitations. Except for the metaheuristic algorithms, the fast non-dominated sorting genetic algorithm (NSGA-II), which offers advantages such as quick operation, excellent solution set convergence, high optimization accuracy, and guaranteed population diversity, is superior to general genetic algorithms [31–35]. Further, the decision variables of the NSGA-II algorithm are suitable for both continuous values and discrete values. Owing to these benefits, the NSGA-II method can rapidly generate a balanced set of Pareto optimal solutions, enabling decision-makers to select an optimal isolator arrangement scheme that maximizes the torsional response reduction while keeping the superstructure in an elastic state. The effectiveness and advantages of the NSGA-II algorithm have also been thoroughly demonstrated in research, with significant work by Song et al. [36], Xu et al. [37], Li et al. [38], and Zhong et al. [39]. Therefore, this paper applied the NSGA-II algorithm to achieve the optimal isolator arrangement.

The second stage of the optimization design method aims to install viscous dampers in the isolation system to decrease the excessive translational displacement induced by pulse-type earthquakes. The traditional conceptual design for enhancing the torsional resistance, which was described in JGJ297-2013 [40] and DBJ53/T-125-2021 [41], suggests that installing viscous dampers around a structure's perimeter can increase its torsional stiffness due to the dampers' additional stiffness. However, for asymmetric base-isolated structures that have undergone the first stage of torsional resistance, the isolation layer already possesses adequate torsional stiffness, and e_b is nearly zero. In this scenario, even though the installation of viscous dampers in the outer layer of isolation could provide more torsional stiffness for the optimized structure, the additional stiffness from the viscous dampers would shift the CSI significantly away from the CMS, severely compromising the structural torsional resistance. To prevent this negative impact, it is crucial to identify the most suitable position for the viscous dampers, enabling a reduction in the excessive translational displacement caused by pulse-type earthquakes while also preserving the efficiency achieved in the first stage.

This study first elaborates on the principle of the two-stage optimal design method before applying it to a five-story L-shaped planar asymmetric base-isolated structure exposed to pulse-type earthquakes. To assess the benefits of this optimal design method, the structural response of the optimized asymmetric isolated structure is compared to that of an asymmetric base-isolated structure designed using traditional conceptual methods. Moreover, we examine the effect of the viscous dampers' position on the structural response. The results suggest that in order to maintain the efficiency of the torsional resistance achieved in the first stage, it is advisable to install the viscous dampers in the isolation layer along the x and y directions, directly below the CMS, in the second stage. Further, the asymmetric base-isolated structure designed using this optimal method was shown to better mitigate the translational and torsional responses when subjected to pulse-type earthquakes compared to structures designed using traditional methods.

2. Specific Implementation of the Two-Stage Optimal Design Method

This section outlines the detailed implementation of the proposed two-stage optimal design method. The first stage seeks to eliminate the torsional response by optimizing the isolator arrangement using the NSGA-II algorithm [32–35]. Subsequently, the second stage aims to decrease the excessive translational displacement of this optimized structure, caused by pulse-type earthquakes, by installing viscous dampers in the isolation layer.

2.1. *First Stage of the Optimization Design Method*

2.1.1. Problem Formulation for the Optimization of Isolation Systems

The optimization objective of the first stage is to identify an optimal isolator arrangement, including positions and categories, as described in Equations (1) and (2), that simultaneously minimizes the maximum interstory rotation of the superstructure ($\theta_{\mathrm{s,max}}$), the maximum rotation of the isolation layer ($\theta_{\mathrm{b,max}}$), and the maximum interstory displacement of the superstructure ($u_{\mathrm{s,max}}$) (see Equation (3)) without exceeding the displacement limit of the isolators ($d_{\max}$) (see Equation (4)). The optimization problem for the first stage can be stated as follows:

$$\text{find} \qquad \mathbf{X} = [x_1, x_2, \cdots, x_N] \tag{1}$$

$$x_i \in R^d \tag{2}$$

$$\text{minimize} \qquad f(\mathbf{X}) = [u_{\mathrm{s,max}}, \theta_{\mathrm{s,max}}, \theta_{\mathrm{b,max}}] \tag{3}$$

$$\text{subject to} \qquad d_0 \leq d_{\max} \tag{4}$$

where $\boldsymbol{X}$ is the isolator arrangement scheme encompassing positions and categories; $x_i(i = 1, 2, \cdots, N)$ represents the isolator category installed in the isolation layer in position i; N indicates the total number of isolators; R^{d} stands for the selection range of the isolator categories, including lead rubber bearings (LRBs) and laminated rubber bearings (LNRs); and d_0 denotes the maximum calculated displacement of the isolator.

2.1.2. Specific Implementation for the Optimization of the Isolation System

The three optimization objectives in Equation (3) have differing optimization weights. Both the rotation of the isolation layer and the interstory rotation of the superstructure should be reduced as much as possible, while the interstory translational displacement of the superstructure only needs to be reduced within its elastic limits. Achieving these specific objectives is difficult using general genetic algorithms. However, owing to the advantages of the NSGA-II algorithm [32–35], it is possible to quickly generate a set of well-balanced Pareto optimal solutions, enabling decision-makers to choose an optimal isolator arrangement scheme that fulfills this specific demand effectively. Therefore, the NSGA-II algorithm was applied to obtain a set of well-balanced Pareto optimal solutions in the first stage of the optimization design. From these selected solutions, the one that concurrently minimized both $\theta_{\mathrm{b,max}}$ and $\theta_{\mathrm{s,max}}$ while maintaining $u_{\mathrm{s,max}}$ in its elastic limits was ultimately selected.

Figure 1 presents a detailed illustration of the first stage of the optimization design method. During this optimization stage, several factors must be considered:

1. To determine the range of the isolator categories, R^{d}, the isolator's diameter should be estimated, adhering to the vertical compressive stress limit requirements of the isolator under gravity load.
2. To maintain high-quality individuals within the population, the objective functions of those individuals whose isolator displacements do not meet the constraint $d_0 \leq d_{\max}$ should be assigned high values during the non-dominated sorting process.
3. To improve the optimization efficiency, large-diameter LRBs should be used as peripheral isolators to enhance base-isolated structures' torsional resistance, as defined in Equation (1).
4. The structural analysis of asymmetric base-isolated structures and the first-stage NSGA-II algorithm are executed using ETABS v21.1.0 [42] and MATLAB [43], respectively. ETABS API v21.1.0 allows for the integration of a structural analysis in ETABS with the NSGA-II algorithm in MATLAB, courtesy of its function library. Consequently, these software tools can effectively implement the first stage of the optimal design of asymmetric base-isolated structures using the NSGA-II algorithm to enhance the torsional resistance.
5. The principle of adjusting the isolator arrangement to minimize the torsional response is predicated on the understanding that altering the isolator placement allows for the adjustment of structural parameters that determine the structural properties of torsional resistance for asymmetric base-isolated structures. This adjustment can notably enhance the torsional resistance. As the time history analysis method is time-consuming and these structural properties remain unaffected by different analysis methods and ground motion spectrum characteristics, this paper adopted the response spectrum analysis method for the first optimization stage to derive the optimal isolator arrangement.

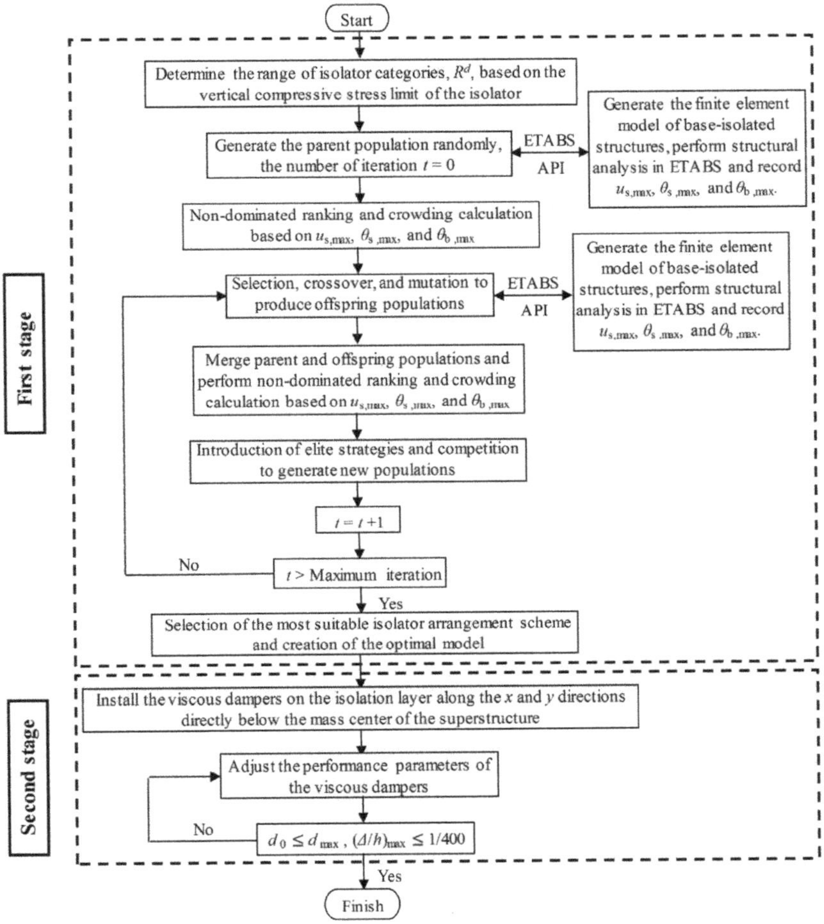

Figure 1. Flow chart of the two-stage optimal design method.

2.2. Second Stage of the Optimization Design Method

Following the first stage of optimization, the asymmetric base-isolated structure's structural parameters are all favorable for the torsional resistance. For instance, the torsional stiffness is ample, and the isolation system's eccentricity is nearly zero. At this point, the optimized structure's torsion could be contained within a limited range, even under pulse-type earthquakes. The second stage aims to reduce the excessive translational displacement caused by pulse-type earthquakes by strategically placing viscous dampers in the isolation layer. It is worth noting that viscous dampers and the parts that link them may build up fatigue as a result of repeated pulse-type earthquakes. Thus, it is important to monitor how well they function after prolonged usage. In order to preserve their performance, viscous dampers need to have their damping fluids and other components replaced on a regular basis. Figure 1 depicts the detailed process of the second stage of the optimization. During this optimization stage, several considerations should be made:

1. Energy dissipators in structures designed for energy dissipation and vibration reduction provide additional dynamic stiffness to the structure, as stipulated in the standards JGJ297-2013 [40] and DBJ 53/T-125-2021 [41]. Hence, the added stiffness given by energy dissipators to the structural response should be considered while

installing viscous dampers in the isolation layer. Following the first stage of optimization, the isolation layer possesses sufficient torsional stiffness, and e_b is nearly zero. At this point, installing viscous dampers in the outer isolation layer can cause the CSI to move significantly away from the CMS due to the dampers' additional stiffness, which would be detrimental to the structure's torsional resistance. Therefore, to preserve the first stage's torsional resistance efficiency, it may be advisable to install the viscous dampers in the isolation layer along the x and y directions, directly beneath the CMS. This method can ensure ample torsional stiffness and maintain e_b close to zero.

2. The time history analysis method is employed to evaluate the potential negative impacts of the pulse-type ground motion's pulse features on the base-isolated structures' translational displacement.
3. The performance parameters of the viscous dampers are iteratively adjusted until the maximum isolator displacement ($x_{b,max}$) and the maximum interstory drift ratio $(\Delta/h)_{max}$ meet the specified limits. $(\Delta/h)_{max}$ is the drift (Δ) divided by the story height (h). According to the drift ratio threshold defined in the GB/T51408-2021 code [44], the status of the superstructure is elastic when the $(\Delta/h)_{max}$ is less than 1/400. A limited number of dampers and their predetermined locations allow designers to modify only the performance parameters of the viscous dampers to reduce the base-isolated structures' translational response. The second stage, compared to the first stage of the optimization design for improving the torsional resistance, is straightforward to execute and does not require optimization algorithms.

3. Description of the Scheme Study

To assess the two-stage optimal design method's effectiveness, a typical five-story L-plan asymmetric base-isolated structure containing a reinforced concrete frame superstructure was selected. The structure was subjected to a seismic fortification intensity of eight, with a basic design acceleration of 0.2 g, a class II site category, and a group II seismic design according to the GB50011-2010 code [45]. The structure's original isolator arrangement was designed using traditional conceptual design methods. One of the most crucial torsional design principles is that large-diameter LRBs should be used as peripheral isolators to enhance base-isolated structures' torsional resistance [21]. Figures 2 and 3 illustrate the original object's finite element model and isolator arrangement, respectively, while Table 1 presents the isolators' property parameters.

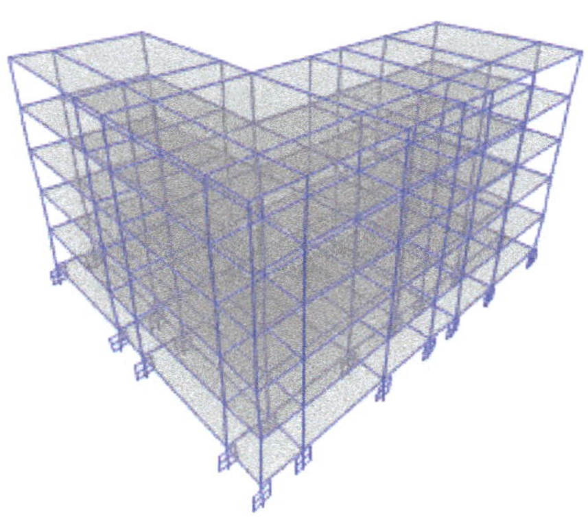

Figure 2. The finite element model of the original asymmetric base-isolated structure.

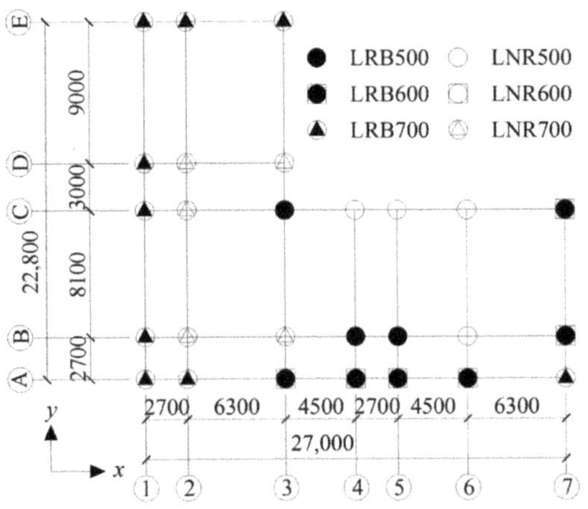

Figure 3. The arrangement of the isolation layer of the original asymmetric base-isolated structure without viscous dampers (unit: mm).

Table 1. The property parameters of isolators.

No	Name	Type	K (kN·m^{-1})	γ = 100% (kN·m^{-1})	γ = 250% (kN·m^{-1})	K_v (kN·m^{-1})	Q_d (kN)	*Ratio*	d_{max} (mm)
1	LRB500	LRB	6411	1374	950	1,721,000	62.8	0.1	275
2	LRB600	LRB	7918	1697	1174	2,321,000	90.4	0.1	330
3	LRB700	LRB	8741	1873	1296	2,784,000	123.1	0.1	385
4	LNR500	LNR	798	798	798	1,592,000	0	0	275
5	LNR600	LNR	987	987	987	2,186,000	0	0	330
6	LNR700	LNR	1090	1090	1090	2,632,000	0	0	385

Note: K symbolizes the pre-yield stiffness of the isolator; γ indicates the effective stiffness of the isolator at a shear strain equal to 100% and 250%; K_v represents the vertical stiffness of the isolator; Q_d denotes the yield force of the isolator; and *Ratio* is the ratio of the post-yield stiffness to the pre-yield stiffness of the isolator.

Viscous dampers can notably decrease the excessive translational displacement resulting from pulse-type earthquakes. Compared to the two-stage optimal design method suggested herein, the original object's viscous dampers were installed in the isolation layer using traditional conceptual design methods, as illustrated in Figure 4.

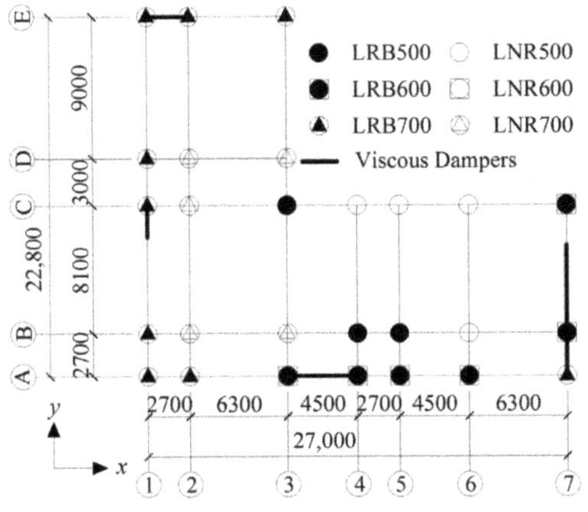

Figure 4. The arrangement of the isolation layer of the original asymmetric base-isolated structure with viscous dampers (unit: mm).

4. Selecting the Earthquakes

This section details the earthquakes used for the comparison of the structural responses to both non-pulse-type and pulse-type earthquakes.

4.1. Non-Pulse-Type Earthquakes

Seven non-pulse-type earthquakes were chosen depending on the site condition of the selected structure. The specifics about the characteristics and selection principles of the earthquakes were suggested by Reference [46]. Figure 5 displays the response spectra of these seven non-pulse-type earthquakes and the average spectrum that aligns with the target spectrum of the standard GB/T51408-2021 [44].

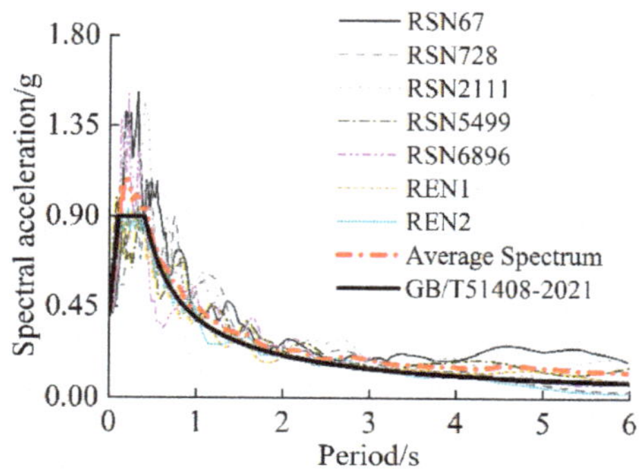

Figure 5. The response spectra of the selected non-pulse-type earthquakes [44].

4.2. Pulse-Type Earthquakes

The near-field coefficient, as per standard GB50011-2010 [45] and in accordance with code UBC97 [47], was utilized to augment the design response spectrum of near-fault earthquakes. The theory dictates that the near-field coefficient should be 1.5 within a 5 km radius of the fault and more than 1.25 within a 5–10 km radius. This strategy accounts for the effects of pulse-type earthquakes by amplifying non-pulse-type earthquakes' acceleration.

To better align with the standard GB50011-2010 [45], the design response spectrum—as per the standard GB/T51408-2021 [44] for a seismic fortification intensity of eight and a class II site category—was first multiplied by the near-field coefficient values of 1.5 and 1.25. This accounted for pulse-type earthquakes' amplification. The resulting amplified design response spectra were then compared with the average response spectrum of 29 near-field pulse-type earthquakes for a class II site category in the Pacific Earthquake Engineering Research Center (PEER) [48] Ground Motion Database.

It was found that the design response spectrum with a near-field coefficient of 1.5 for long periods (greater than the characteristic period T_g) better fitted the average response spectrum of these 29 near-field pulse-type earthquakes for this site condition. To further improve the match with the average response spectrum, T_g of the design response spectrum for pulse-type earthquakes was set to 0.8 s. The detailed process can be found in Reference [49]. This design response spectrum of pulse-type earthquakes was then applied herein, as depicted in Figure 6.

For the structural analysis, seven pulse-type earthquakes with long pulse periods (T_p) were selected from the PEER Ground Motion Database. These motions were chosen as their response spectra perfectly aligned with the pulse-type design response spectrum during primary vibration periods. When the building's seismic fortification intensity was eight under rare earthquakes, the peak ground acceleration of the seven selected earthquakes had to be adjusted to 0.4 g, according to the GB50011-2010 [45]. Figure 6 presents the response spectra of the selected pulse-type earthquakes. Table 2 presents the characteristics of the selected pulse-type earthquakes. These pulse-type earthquakes were input in the x and y

directions. The structural response was evaluated based on the average value computed from the seven chosen pulse-type earthquakes.

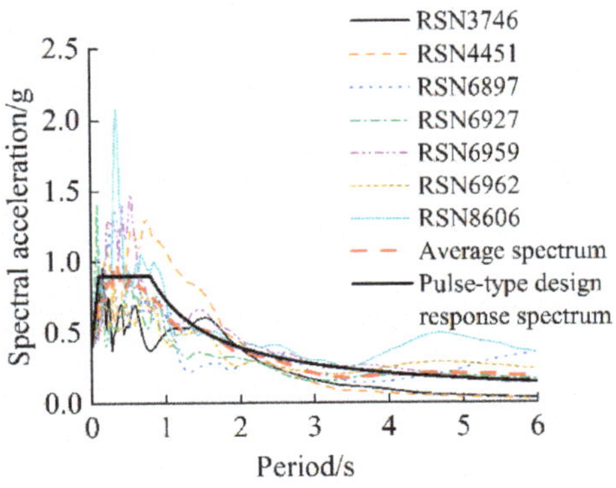

Figure 6. The response spectra of the selected pulse-type earthquakes.

Table 2. The characteristics of the selected pulse-type earthquakes.

Earthquake ID	Event	Station	M_w	T_p (s)	V_{S30} (m/s)	S_{90} (s)
RSN3746	Cape Mendocino	Centerville Beach_ Naval Fac	7.01	1.97	459.04	10.6
RSN4451	Montenegro_ Yugoslavia	Bar-Skupstina Opstine	7.10	1.44	462.23	21.3
RSN6897	Darfield_ New Zealand	DSLC	7.00	7.83	295.74	19.6
RSN6927	Darfield_ New Zealand	LINC	7.00	7.37	263.20	12.7
RSN6959	Darfield_ New Zealand	Christchurch Resthaven	7.00	12.02	141.00	30.5
RSN6962	Darfield_ New Zealand	ROLC	7.00	7.14	295.74	11.4
RSN8606	El Mayor-Cucapah_ Mexico	Westside Elementary School	7.20	7.08	242.00	25.3

5. First Stage of the Optimization Design of the Selected Scheme

In this section, we employed the first stage of the optimization design method to optimize the isolator arrangement of the original asymmetric base-isolated structure, as detailed in Section 3. The effectiveness of this stage was then assessed by contrasting the structural responses of the optimal and original asymmetric base-isolated structures.

5.1. Optimization Process and Results of the First Stage

Following the optimization process of the first stage described in Section 2.1, the isolator range was initially determined to be LRB500, LNR500, LRB600, LNR600, LRB700, and LNR700. This was based on the gravity load at the base of each column and the isolators' ultimate vertical compressive stress, which was set to 12 MPa for the chosen scheme in the standard GB/T51408-2021 [44]. The characteristics of these isolators are detailed in Table 1.

In this study, the parameters of the NSGA-II algorithm were first designed with a focus on establishing a balance between computational precision and time efficiency, as suggested by Deb [32,35]. Throughout the iterative process, these parameters were adjusted in response to emerging solution trends. Finally, the population size was set to 50, with a crossover probability of 0.8, a mutation probability of 0.05, and an iteration number of 100. The values of the three optimization objectives remained stable when the number of iterations exceeded 85, indicating the convergence of the set of solutions. Thus, 100 iterations were sufficient for the NSGA-II algorithm to demonstrate effective convergence characteristics for this optimization problem's first stage.

After establishing the range of isolators and parameters for multi-objective optimization, the iteration for the first stage was successfully executed, as depicted in Figure 1.

A set of Pareto optimal solutions was swiftly obtained, which achieved a good balance among the three optimization objectives, as displayed in Figure 7. Following the optimization objectives of the first stage, the solutions exhibiting minimal maximum interstory displacement of the superstructure were first chosen. From these selected solutions, the one that concurrently minimized both $\theta_{s,max}$ and $\theta_{b,max}$ was ultimately selected as the optimal isolator arrangement scheme (Figure 8).

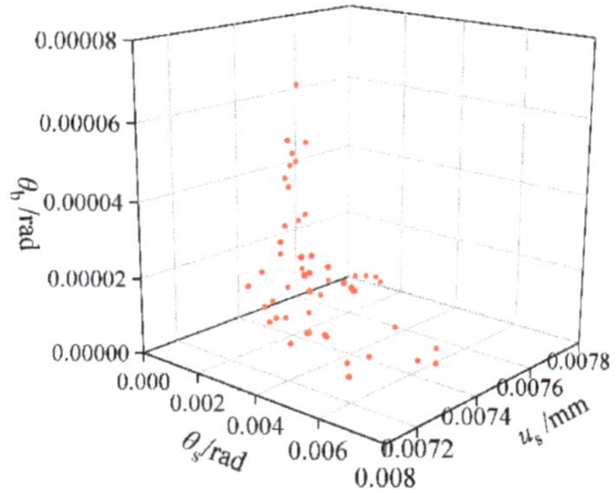

Figure 7. Optimal points of the Pareto optimal solutions.

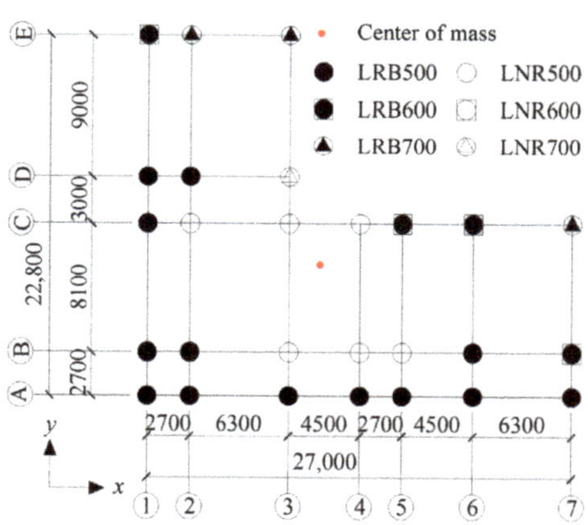

Figure 8. The optimal isolator arrangement scheme of the asymmetric base-isolated structure (unit: mm).

5.2. Discussion of the Results of the First Stage

The efficiency of the first stage was evaluated by comparing the structural responses of schemes 1 and 2. The effects of pulse-type earthquakes on the structural response of asymmetric base-isolated structures were evaluated by comparing the structural responses of schemes 2 and 3. Detailed descriptions of these three schemes are tabulated in Table 3. The isolator arrangement of scheme 1 is presented in Figure 3, while those of schemes 2 and 3 are shown in Figure 8. The primary structural responses are represented by $(\Delta/h)_{max}$, $(\Delta/h)_{max}$ of the superstructure $(\Delta/h)_{max,1}$, the maximum displacement of the isolation layer $(x_{b,max})$, the maximum interstory rotation (θ_{max}), and the maximum displacement ratio $(\Delta_{max}/\Delta_{ave})_{max}$.

Table 3. The description of the different schemes in this paper.

Scheme	Type of Earthquakes	Design Methods of the Isolation Layer	
		First Stage	**Second Stage**
1	Non-pulse-type	Traditional conceptual design	-
2	Non-pulse-type	Optimization design	-
3	Pulse-type	Optimization design	-
4	Pulse-type	Optimization design	Optimization design
5	Pulse-type	Traditional conceptual design	Traditional conceptual design
6 and 7	Pulse-type	Optimization design	Traditional conceptual design

Table 4 presents the CMS, the CSI, and e_b of schemes 1–3. It is clear that e_b for the optimal asymmetric base-isolated structure was significantly less than for the original asymmetric base-isolated structure and very close to zero. This indicates that the first stage of the optimization design method effectively minimized e_b for the asymmetric base-isolated structure.

Table 4. The CMS and CSI of the structures.

Structures	Coordinates of CMS (m)		Coordinates of CSI (m)		e_b (m)	
	X	Y	X	Y	X	Y
Original asymmetric base-isolated structure (Scheme 1)	10.9346	8.2445	10.4576	7.7271	0.4770	0.5174
Optimal asymmetric base-isolated structure (Scheme 2 and Scheme 3)	10.9346	8.2445	11.0181	8.2865	0.0835	0.0420

Note: e_b = | Mass center − stiffness center |.

Figure 9 displays $(\Delta/h)_{max}$ for all the superstructure stories in schemes 1–3. Notably, $(\Delta/h)_{max}$ for scheme 3 was substantially larger than for schemes 1 and 2. $(\Delta/h)_{max}$ for schemes 1–3 peaked on the first story, with these maximum values summarized in Table 5 as $(\Delta/h)_{max,1}$. Both Figure 9 and Table 5 reveal that the values of $(\Delta/h)_{max}$ for schemes 1 and 2 were less than the elastic limit of 1/400, while $(\Delta/h)_{max,1}$ for scheme 3 was 1/369 in the x direction and 1/319 in the y direction. Table 5 also reports the values of $x_{b,max}$. $x_{b,max}$ for scheme 3 was significantly larger than for schemes 1 and 2, leading to $x_{b,max}$ for scheme 3 exceeding the isolator's displacement limit (d_{max}). These results highlight that pulse-type earthquakes substantially increased the translational displacements in both the superstructure and the isolation layer of the selected scheme, causing both the interstory drift ratio and the isolators' displacement to exceed their respective limits.

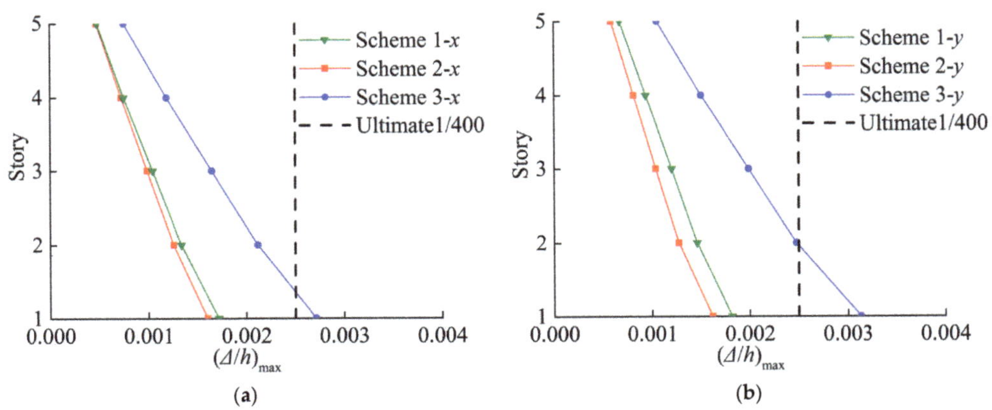

Figure 9. The values of $(\Delta/h)_{max}$ for schemes 1–3 in the (**a**) x direction and (**b**) y direction.

Table 5. The values of $(\Delta/h)_{max,1}$ and $x_{b,max}$ for schemes 1–3.

Scheme	$(\Delta/h)_{max,1}$		$x_{b,max}$ (mm)
	x	y	
1	1/582	1/550	222.96
2	1/624	1/619	191.91
3	1/369	1/319	317.14

Table 6 and Figure 10 display the torsional response of schemes 1–3, which includes θ_{max} and $(\Delta_{max}/\Delta_{ave})_{max}$. $(\Delta_{max}/\Delta_{ave})_{max}$ is the ratio between the maximum or inter-story translation and the average value of the floor's vertical elements' translation. The code GB/T51408-2021 [44] defines particular displacement ratio thresholds linked to different levels of structural torsion-related injury to reinforced concrete elements. The structural elements have the potential to suffer significant torsional damage if the value of $(\Delta_{max}/\Delta_{ave})_{max}$ surpasses 1.2, indicating a severe torsional effect. If the value of $(\Delta_{max}/\Delta_{ave})_{max}$ is between 1 and 1.2, the damage to elements caused by structural torsion is slight. Nevertheless, if the value of $(\Delta_{max}/\Delta_{ave})_{max}$ is close to 1.0, the structural torsion effect is minimal, and the elements almost suffer no damage caused by structural torsion.

Table 6. The values of θ_{max} for schemes 1–3.

Story	θ_{max} (rad)		
	Scheme 1	Scheme 2	Scheme 3
5	-3.14×10^{-6}	-8.57×10^{-7}	4.14×10^{-6}
4	-5.14×10^{-6}	-3.43×10^{-6}	4.86×10^{-6}
3	-7.43×10^{-6}	-5.43×10^{-6}	5.00×10^{-6}
2	-9.71×10^{-6}	-7.43×10^{-6}	3.86×10^{-6}
1	-1.27×10^{-5}	-8.14×10^{-6}	4.29×10^{-7}
0	4.90×10^{-4}	3.02×10^{-4}	3.05×10^{-4}

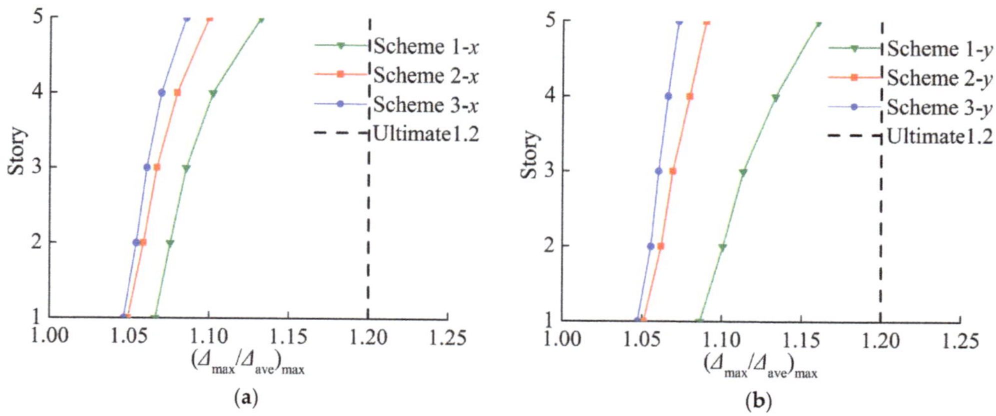

Figure 10. The values of $(\Delta_{max}/\Delta_{ave})_{max}$ for schemes 1–3 in the (**a**) x direction and (**b**) y direction.

Upon completion of the first stage of optimization, both θ_{max} and $(\Delta_{max}/\Delta_{ave})_{max}$ for scheme 2 were noticeably lower than for scheme 1. This reduction signified the first stage's effectiveness in enhancing the structure's torsional resistance. While θ_{max} for scheme 3 was slightly higher than for scheme 2 due to pulse-type earthquakes, $(\Delta_{max}/\Delta_{ave})_{max}$ for schemes 2 and 3 was near 1.0.

In conclusion, the first stage of the optimization considerably reduced the torsional response of the optimal asymmetric base-isolated structure. Moreover, pulse-type earthquakes had a little impact on the torsion of this structure, given that all structural parameters were optimized for torsional resistance. However, the translational displacements in the superstructure and the isolation layer of the optimal asymmetric base-isolated structure increased dramatically under pulse-type earthquakes, thereby exceeding the maximum limits for interstory drift ratio and isolator displacement. This finding suggests the need for further design enhancements to improve the lateral resistance under pulse-type earthquakes.

6. Second Stage of the Optimization Design of the Selected Scheme

This section discusses the second stage of the optimization design. The optimal asymmetric base-isolated structure was fitted with viscous dampers to mitigate the excessive translational response induced by pulse-type earthquakes, using the optimization method proposed in Section 2.2. The optimal structure's primary responses were then compared with those of a structure designed using traditional methods. Lastly, the influence of the viscous dampers' positions on the structural response was evaluated.

6.1. Optimization Process and Results of the Second Stage

The Maxwell model [50], a simplified mechanical model for viscous dampers, was employed in this study. Figure 11 depicts the model's diagram, while Equation (5) describes the damping force of a viscous damper

$$F = CV^{\alpha} + KD \tag{5}$$

where C is the damping coefficient of the viscous damper; V indicates the relative speed of the two opposite ends of the viscous damper; $\alpha \in [0.2, 0.4]$ represents the damping exponent of the viscous damper and its range, as specified by the internal construction of the viscous damper; K denotes the stiffness of the viscous damper; and D is the relative displacement of the two opposite ends of the viscous damper.

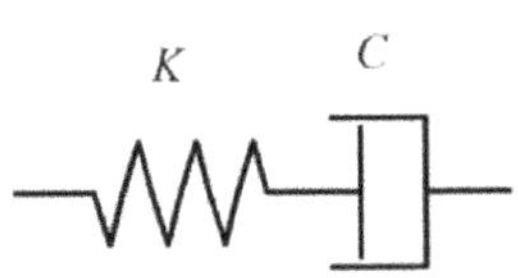

Figure 11. The diagram of the simplified mechanical model for viscous dampers.

In order to maintain the isolation system's eccentricity close to zero, the viscous dampers were initially applied to the isolation layer, directly beneath the CMS, along the x and y directions. Subsequently, the structural response was evaluated using the average value derived from the seven pulse-type earthquakes specified in Section 4.2. The length and the parameters of the viscous dampers were continually adjusted according to the structural response. This adjustment ensured that $x_{b,\max}$ and $(\Delta/h)_{\max}$ remained within the specification limits. The final damping coefficient for the viscous dampers was set at C = 700 kN·s/m, with a damping exponent $\alpha = 0.3$. The dampers, featuring ball hinges at the bottom, were installed at an oblique angle. As a construction precaution, it was necessary to place secondary beams under the floor slab at the damper positions. A stirrup reinforcement was employed to ensure an adequate bearing capacity at the joint where the dampers intersected the beams. Consequently, the two-stage optimal design method resulted in the optimal asymmetric base-isolated structure with viscous dampers, referred to as scheme 4 in Table 3. Figure 12 details the isolation layer arrangement in scheme 4.

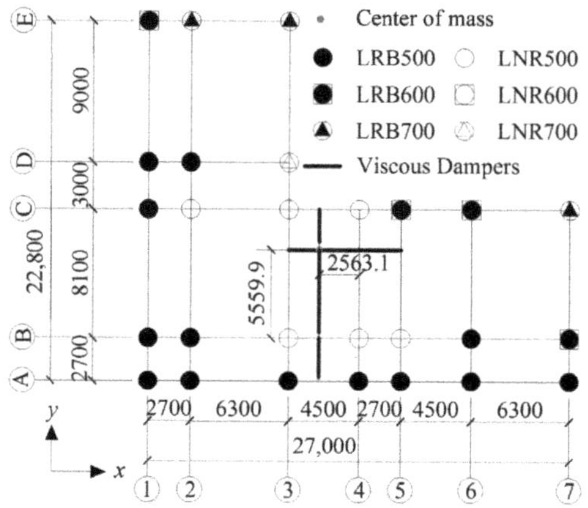

Figure 12. The isolation layer arrangement scheme of the optimal asymmetric base-isolated structure with viscous dampers (unit: mm).

6.2. Discussion of the Results of the Second Stage

This section first compares the primary structural responses of schemes 3 and 4 to assess the impact of the viscous dampers in alleviating the excessive translational displacements caused by pulse-type earthquakes. Following that, the primary structural response of the optimal scheme 4 is compared with that of scheme 5 (as seen in Table 3), which employed traditional design methods for arranging isolators and viscous dampers. Figure 4 illustrates the arrangement of the isolators and viscous dampers in scheme 5.

In order to preserve the torsional resistance efficiency achieved in the first stage, the dampers were applied to the isolation layer directly beneath the CMS, along the x and y directions. This installation contradicts the traditional damper arrangement principle, which recommends dispersing viscous dampers around the isolation layer's perimeter. Consequently, schemes 6 and 7 were designed with dampers distributed around the isolation layer's perimeter, and the response of scheme 4 was compared with those of schemes 6 and 7. To enhance the analysis of the damper position effects on the structural response, the structural arrangements of schemes 6 and 7 mirrored that of scheme 4, except for damper positioning. Figure 13 illustrates the arrangement of the isolators and viscous dampers in schemes 6 and 7.

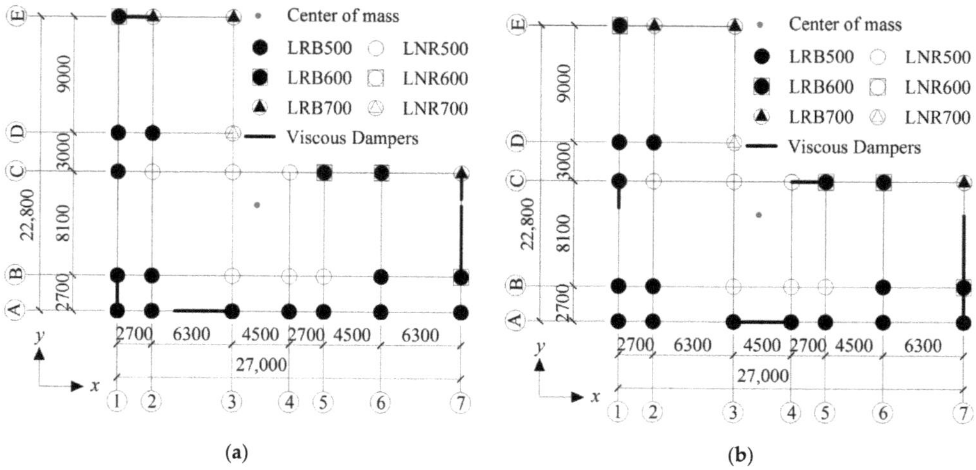

Figure 13. The arrangement of isolators and viscous dampers in schemes (**a**) 6 and (**b**) 7. (unit: mm).

Figure 14 presents $(\Delta/h)_{max}$ for schemes 3–7 in both the x and y directions. Each scheme's maximum value occurred on the first story, and these values are summarized as $(\Delta/h)_{max,1}$ in Table 7. Table 7 also displays $x_{b,max}$ for schemes 3–7. Figure 14 shows that $(\Delta/h)_{max}$ for scheme 4 was less than that for scheme 3. Moreover, it is evident that $(\Delta/h)_{max,1}$ and $x_{b,max}$ for scheme 4 were significantly reduced, staying within acceptable limits, due to the addition of the viscous dampers. This stands in contrast to what observed for scheme 3, as in that case, both exceeded their respective limits. These results suggest that viscous dampers can notably decrease both $(\Delta/h)_{max}$ and $x_{b,max}$ in a base-isolated structure under the influence of pulse-type earthquakes.

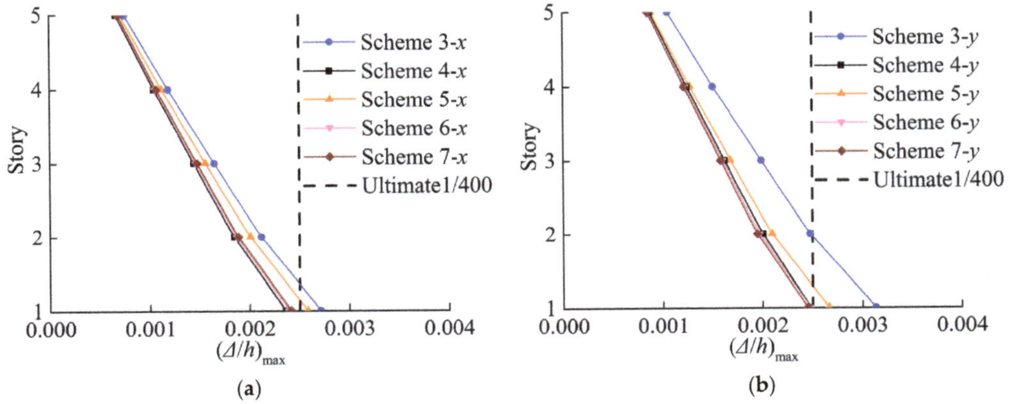

Figure 14. The values of $(\Delta/h)_{max}$ for schemes 3–7 in the (**a**) x direction and (**b**) y direction.

Table 7. The values of $(\Delta/h)_{max,1}$ and $x_{b,max}$ for schemes 3–7.

Scheme	$(\Delta/h)_{max,1}$		$x_{b,max}$ (mm)
	x	y	
3	1/369	1/319	317.14
4	1/424	1/403	233.71
5	1/388	1/376	215.46
6	1/418	1/404	274.16
7	1/415	1/408	277.66

Furthermore, for scheme 5, designed using traditional conceptual design methods, there was a 9.29% increase in $(\Delta/h)_{max,1}$ over scheme 4, and the elastic threshold of 1/400 in terms of $(\Delta/h)_{max,1}$ was exceeded, even though $x_{b,max}$ remained within its limits. This revealed that the two-stage optimal design method outperformed the traditional conceptual design methods, particularly in reducing the superstructure's maximum interstory drift.

Additionally, it is clear that $(\Delta/h)_{max}$ and $(\Delta/h)_{max,1}$ for schemes 4, 6, and 7 share similar values and did not exceed their respective limits, as seen from Figure 14 and Table 7. This suggests that the viscous dampers' position had a little adverse impact on the structure's interstory drift in the second stage. However, schemes 6 and 7 showed a 17.31% increase in $x_{b,max}$ over scheme 4. The reason for the increase might be that the traditional arrangement method resulted in a larger θ_{max} of the isolation layer in schemes 6 and 7, which increased the corner isolator displacements with respect to that observed for scheme 4. The following provides a comparative analysis of the torsional response of several schemes under various design methods in order to validate this issue.

Table 8 and Figure 15 show the θ_{max} and $(\Delta_{max}/\Delta_{ave})_{max}$ values for schemes 3–7. Notably, θ_{max} for scheme 4 was significantly less than for scheme 5, as seen in Table 8. Specifically, θ_{max} of the isolation layer in scheme 4 was merely 14.16% of that in scheme

5, and θ_{max} of the superstructure in scheme 4 was reduced to a maximum of 51.28% of that in scheme 5. Moreover, for scheme 5, $(\Delta_{max}/\Delta_{ave})_{max}$ greatly exceeded the limit of 1.2, whereas for scheme 4, it was notably lower. This revealed that the two-stage optimal design method outperformed traditional conceptual design methods, particularly in reducing the interstory rotation and the displacement ratio.

Table 8. The maximum interstory rotation (θ_{max}) for schemes 3–7.

Story	θ_{max} (rad)				
	Scheme 3	Scheme 4	Scheme 5	Scheme 6	Scheme 7
5	4.14×10^{-6}	-6.00×10^{-6}	1.13×10^{-5}	1.10×10^{-5}	1.19×10^{-5}
4	4.86×10^{-6}	-8.00×10^{-6}	1.56×10^{-5}	1.50×10^{-5}	1.56×10^{-5}
3	5.00×10^{-6}	-1.20×10^{-5}	1.94×10^{-5}	1.90×10^{-5}	1.94×10^{-5}
2	3.86×10^{-6}	-1.90×10^{-5}	2.36×10^{-5}	2.30×10^{-5}	2.31×10^{-5}
1	4.29×10^{-7}	-2.10×10^{-5}	2.84×10^{-5}	2.60×10^{-5}	2.81×10^{-5}
0	3.05×10^{-4}	4.44×10^{-4}	3.07×10^{-3}	3.63×10^{-3}	6.62×10^{-3}

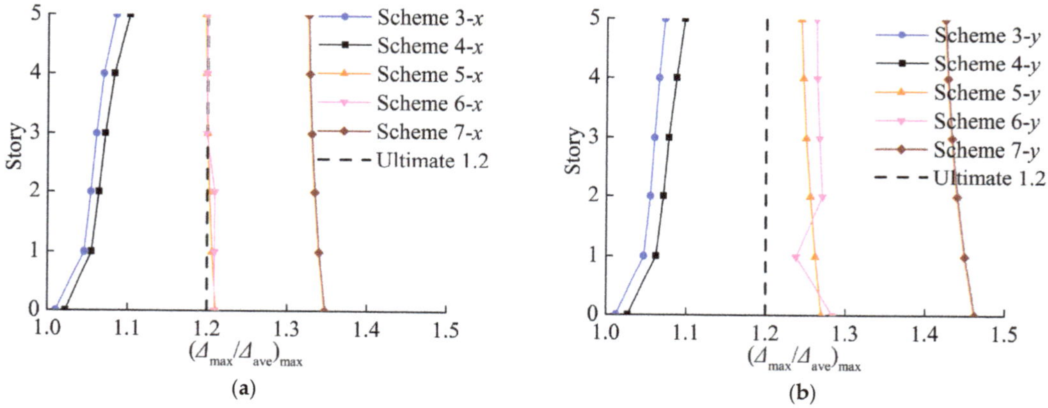

Figure 15. The maximum displacement ratio for schemes 3–7 in the (**a**) *x* direction and (**b**) *y* direction.

Furthermore, Table 8 reveals that θ_{max} for schemes 6 and 7 was significantly higher than for scheme 4. Particularly in the isolation layer, θ_{max} for scheme 4 was approximately 10% of that for schemes 6 and 7. As depicted in Figure 15, $(\Delta_{max}/\Delta_{ave})_{max}$ for schemes 6 and 7 was significantly greater than for scheme 4. Moreover, for schemes 6 and 7, $(\Delta_{max}/\Delta_{ave})_{max}$ exceeded the limit of 1.2, suggesting a negative impact on the structure due to torsion. These results indicate that the optimal design method of the viscous damper arrangement can better improve the torsional resistance efficiency than traditional conceptual design methods.

These findings suggest that compared to traditional design methods for asymmetric base-isolated structures, the two-stage optimal design method significantly reduces the torsional response under pulse-type earthquakes, while also ensuring a strong lateral resistance. Especially, the interstory drift ratio and the displacement ratio of the optimized asymmetric base-isolated structure were significantly reduced to acceptable limits. Additionally, during the second stage of optimization, the damper positions had a little impact on the structure's interstory drift. However, installing viscous dampers around the perimeter of the isolation layer negatively enlarged the rotation of the isolation layer, which increased the corner isolator displacements. This adverse effect was primarily due to the added stiffness of the dampers when they were distributed around the isolation layer's perimeter. After the first stage of optimization, the CSI and the CMS essentially coincided. The added damper stiffness increased the distance between the CSI and the CMS, thus exacerbating the torsional response. Consequently, it is recommended to install the viscous dampers directly under the CMS, along the x and y directions of the isolation layer, during the second stage of the optimization design.

7. Conclusions

In this study, we presented a two-stage optimal design method for asymmetric base-isolated structures subjected to pulse-type earthquakes. The advantages of this proposed method were demonstrated through an examination of a typical five-story L-plan asymmetric base-isolated structure. Furthermore, we analyzed the most appropriate positioning of viscous dampers during the second stage of the design. The key conclusions from our study are:

1. During the first stage of the design, the NSGA-II algorithm provides a more efficient and quicker means of finding an optimal isolator arrangement compared to other algorithms. This optimized arrangement minimizes the rotation of the isolation layer and the interstory rotation of the superstructure, thereby reducing the negative effects caused by the torsional response to negligible levels.
2. Our findings revealed that distributing viscous dampers across the perimeter of the isolation layer during the second stage of optimization can negatively impact the torsional response. To maintain the effectiveness of the torsional resistance achieved in the first stage and significantly decrease the overrun of the translational response of asymmetric base-isolated structures caused by pulse-type earthquakes, it is advisable to install the viscous dampers in the isolation layer along the x and y directions, directly below the CMS in the second stage.
3. Compared to traditional conceptual design methods for isolators and viscous dampers, the two-stage optimal design method proposed herein can more effectively enhance the lateral resistance and torsional resistance of asymmetric base-isolated structures. As per the case study presented, both the maximum interstory drift ratio and the maximum displacement of the isolators in the optimal asymmetric base-isolated structure under pulse-type earthquakes were reduced within their limits, as compared to those in the structure designed using traditional conceptual design methods. Furthermore, the maximum rotation of the isolation layer and the maximum interstory rotation of the superstructure of the optimal asymmetric base-isolated structure were merely 14.16% and 51.28% of those of the structure designed using traditional conceptual design methods, respectively. Moreover, the maximum displacement ratio of the optimal asymmetric base-isolated structure was reduced within its limit following the two-stage optimal design, indicating that the adverse effect caused by torsional responses can be effectively disregarded.
4. In addition to damper positioning, future research could also target the optimization of damper parameters to achieve better lateral and torsional resistance. When site condition limitations are eliminated from the analysis of the structural response, the number of pulse-type earthquakes available for investigation grows considerably. At this point, it is crucial to investigate the impact of the characteristics of pulse-type earthquakes on the structural response. For example, Sharma et al. [51] found that a high directivity ratio of the pulse-type earthquakes has a significant effect on the structural response. Thus, future research should explore a new optimal design method that not only provides better lateral and torsional resistance to asymmetric base-isolated structures under pulse-type earthquakes that are suitable for the site condition of the selected structure, but also allows these structures to perform well under pulse-type earthquakes with different characteristics.

Author Contributions: Conceptualization, J.Z. and A.Q.; methodology, J.Z. and A.Q.; software, J.Z.; validation, J.Z., A.Q. and M.Y.; formal analysis, J.Z.; investigation, J.Z.; resources, A.Q.; data curation, J.Z. and M.Y.; writing—original draft preparation, J.Z.; writing—review and editing, J.Z.;

visualization, J.Z.; supervision, A.Q.; project administration, A.Q.; funding acquisition, A.Q. All authors have read and agreed to the published version of the manuscript.

Funding: This research was funded by the National Natural Science Foundation of China (grant number 52278490).

Data Availability Statement: The data presented in the study are included in the article; further inquiries can be directed to the authors.

Conflicts of Interest: The authors declare no conflicts of interest.

References

1. Keri, L.R.; Gordon, P.W. A Review of Seismic Isolation for Buildings: Historical Development and Research Needs. *Buildings* **2012**, *2*, 300–325. [CrossRef]
2. Pınar, U. Investigation of a Base-Isolator System's Effects on the Seismic Behavior of a Historical Structure. *Buildings* **2021**, *11*, 217. [CrossRef]
3. Francisco, L.A.; Weng, D.G.; Li, T.; Bashar, A. Suitability of Seismic Isolation for Buildings Founded on Soft Soil. Case Study of a RC Building in Shanghai. *Buildings* **2020**, *10*, 241. [CrossRef]
4. Rong, Q. Optimum parameters of a five-story building supported by lead-rubber bearings under near-fault earthquakes. *J. Low Freq. Noise Vib. Act.* **2020**, *39*, 98–113. [CrossRef]
5. Mazza, F. Seismic demand of base-isolated irregular structures subjected to pulse-type earthquakes. *Soil Dyn. Earthq. Eng.* **2018**, *108*, 111–129. [CrossRef]
6. Mazza, F.; Labernarda, R. Effects of nonlinear modelling of the base-isolation system on the seismic analysis of r.c. buildings. *Procedia Struct. Integr.* **2018**, *11*, 226–233. [CrossRef]
7. Zhao, D.H.; Wang, H.W.; Qian, H.; Liu, J.M. Comparative vulnerability analysis of decomposed signal for the LRB base-isolated structure under pulse-like earthquakes. *J. Build. Eng.* **2022**, *59*, 105106. [CrossRef]
8. Mazza, F.; Vulcano, A. Effects of near-fault earthquakes on the nonlinear dynamic response of base-isolated r.c. framed buildings. *Earthq. Eng. Struct. Dyn.* **2012**, *41*, 211–232. [CrossRef]
9. Mazza, F.; Mazza, M.; Vulcano, A. Nonlinear response of r.c. framed buildings retro-fitted by different base-isolation systems under horizontal and vertical components of near-fault earthquakes. *Earthq. Struct.* **2017**, *12*, 135–144. [CrossRef]
10. Necmettin, G. Effects of near-fault pulse-like earthquakes on seismically isolated buildings. *J. Build. Eng.* **2022**, *52*, 104508. [CrossRef]
11. Nagarajaiah, S.; Reinhorn, A.M.; Constantinou, M.C. Torsion in base-isolated structures with elastomeric isolation systems. *J. Struct. Eng.* **1993**, *119*, 2932–2951. [CrossRef]
12. Cavdar, E.; Ozdemir, G. Amplifcation in maximum isolator displacement of an LRB isolated building due to mass eccentricity. *Bull. Earthq. Eng.* **2022**, *20*, 607–631. [CrossRef]
13. Tena, C.A.; Zambrana, R.C. Dynamic torsional amplifications of base-isolated structures with an eccentric isolation system. *Eng. Struct.* **2006**, *28*, 72–83. [CrossRef]
14. Ye, K.; Li, L. Research on seismic response of asymmetric base-isolated structures under near-fault pulse-like earthquakes. *Eng. Mech.* **2009**, *26*, 163–169. (In Chinese)
15. Carlos, E.S.; Jose, L.A.; Juan, C.D. Torsional balance of seismically isolated asymmetric structures. *Eng. Struct.* **2013**, *46*, 703–717. [CrossRef]
16. Hwang, J.S.; Hsu, T.Y. Experimental study of isolated building under triaxial ground excitations. *J. Struct. Eng.* **2000**, *126*, 879–886. [CrossRef]
17. Tena, C.A.; Gomez-Soberon, L.A. Torsional response of base-isolated structures due to asymmetries in the superstructure. *Eng. Struct.* **2002**, *24*, 1587–1599. [CrossRef]
18. Tena, C.A.; Esamilla-Cruz, J.L. Torsional amplifications in asymmetric base-isolated structures. *Eng. Struct.* **2007**, *29*, 237–247. [CrossRef]
19. Laguardia, R.; Morrone, C.; Faggella, M.; Gigliotti, R. A simplifed method to predict torsional effects on asymmetric seismic isolated buildings under bi-directional earthquake components. *Bull. Earthq. Eng.* **2019**, *17*, 6331–6356. [CrossRef]
20. Kilar, V.; Koren, D. Seismic behaviour of asymmetric base isolated structures with various distributions of isolators. *Eng. Struct.* **2009**, *31*, 910–921. [CrossRef]
21. Qi, A.; Shang, H.J.; Fan, H.W. Study on torsion of high-rise isolation structures. *World Earthq. Eng.* **2011**, *27*, 121–130. (In Chinese)
22. Kelly, J.M. The role of damping in seismic isolation. *Earthq. Eng. Struct. Dyn.* **1999**, *28*, 3–20. [CrossRef]
23. Politopolous, I. A review of adverse effects of damping in seismic isolation. *Earthq. Eng. Struct. Dyn.* **2008**, *37*, 447–465. [CrossRef]
24. He, W.L.; Agrawal, A.K. Analytical model of ground motion pulses for the design and assessment of seismic protective systems. *J. Struct. Eng.* **2008**, *134*, 1177–1188. [CrossRef]
25. Mazza, F.; Vulcano, A. Nonlinear response of r.c. framed buildings with isolation and supplemental damping at the base subjected to near-fault earthquakes. *J. Earthq. Eng.* **2009**, *13*, 690–715. [CrossRef]

26. Wang, C.N.; Yang, F.C.; Nguyen, V.T.T.; Vo, N.T.M. CFD Analysis and Optimum Design for a Centrifugal Pump Using an Effectively Artificial Intelligent Algorithm. *Micromachines* **2022**, *13*, 1208. [CrossRef]
27. Tran, V.T.; Le, M.H.; Vo, M.T.; Le, Q.T.; Hoang, V.H.; Tran, N.T.; Nguyen, V.T.; Nguyen, T.A.T.; Nguyen, H.N.; Nguyen, V.T.T.; et al. Optimization design for die-sinking EDM process parameters employing effective intelligent method. *Cogent Eng.* **2023**, *10*, 2264060. [CrossRef]
28. Ali, E.C.; Avşar, O.; Hasancebi, O. Optimum design of seismic isolation systems using metaheuristic search methods. *Soil Dyn. Earthq. Eng.* **2020**, *131*, 106012. [CrossRef]
29. Ali, S.; Hadi, E.; Joong, H.K. Water cycle algorithm for solving constrained multi-objective optimization problems. *Appl. Soft Comput.* **2015**, *27*, 279–298. [CrossRef]
30. Kaveh, A.; Zakian, P. Improved GWO algorithm for optimal design of truss structures. *Eng. Comput.* **2018**, *34*, 685–707. [CrossRef]
31. Shobhit, C.; Elangovan, R.; Sukumar, N. Multi-objective Building Design Optimization under Operational Uncertainties Using the NSGA II Algorithm. *Buildings* **2020**, *10*, 88. [CrossRef]
32. Deb, K.; Agrawal, S.; Pratap, A.; Meyarivan, T. A fast elitist non-dominated sorting genetic algorithm for multi-objective optimization: NSGA-II. *Lect. Notes Comput. Sci.* **2000**, *1917*, 849–858. [CrossRef]
33. Deb, K.; Goel, T. Controlled elitist non-dominated sorting genetic algorithms for better convergence. *Lect. Notes Comput. Sci.* **2001**, *1993*, 67–81. [CrossRef]
34. Deb, K.; Jain, H. An evolutionary many-objective optimization algorithm using reference-point-based nondominated sorting approach, Part I: Solving problems with box constraints. *IEEE Trans. Evol. Comput. A Publ. IEEE Neural Netw. Counc.* **2014**, *18*, 577–601. [CrossRef]
35. Deb, K.; Pratap, A.; Agarwal, S.; Meyarivan, T. A fast and elitist multi-objective genetic algorithm: NSGA-II. *IEEE Trans. Evol. Comput.* **2002**, *6*, 182–197. [CrossRef]
36. Song, Z.R.; Zhai, C.H.; Ma, Y.H.; Wang, Z.Q.; Pei, S.S. Multi-stage and multi-objective design optimization for improving resilience of base-isolated hospital buildings. *Eng. Struct.* **2024**, *304*, 117644. [CrossRef]
37. Xu, Z.D.; Huang, X.H.; Xu, F.H.; Yuan, J. Parameters optimization of vibration isolation and mitigation system for precision platforms using non-dominated sorting genetic algorithm. *Mech. Syst. Signal Process.* **2019**, *128*, 191–201. [CrossRef]
38. Li, H.; Yu, Y.; Li, J.C.; Askari, M. Multi-objective optimisation for improving the seismic protection performance of a multi-storey adaptive negative stiffness system based on modified NSGA-II with DCD. *J. Build. Eng.* **2021**, *43*, 103145. [CrossRef]
39. Zhong, W.J.; Liu, Y.H.; Mercan, O.; Zhou, F.L. Investigations of the effects of a passive bumper on the seismic response of base-isolated buildings: Experimental study and parameter optimization. *J. Vib. Control* **2023**, *29*, 2842–2853. [CrossRef]
40. *JGJ297-2013*; Technical Specification for Seismic Energy Dissipation of Buildings. Building Industry Press of China: Beijing, China, 2013.
41. *DBJ53/T-125-2021*; Technical Specification for Application of Building Seismic Energy Dissipation. Yunnan Science and Technology Press: Kunming, China, 2021.
42. *An Ultimate Integrated Software Package for the Structural Analysis and Design of Buildings*, ETABS v21.1.0; Computers and Structures, Inc.: Berkeley, CA, USA, 2021. Available online: https://www.csiamerica.com/products/etabs (accessed on 1 June 2022).
43. MathWorks Math. Graphics. Programming, URL. 2002. Available online: https://www.mathworks.com/products/matlab.html?s_tid=hp_products_matlab (accessed on 1 June 2022).
44. *GB/T51408-2021*; Standard for Seismic Isolation Design of Building. China Planning Press: Beijing, China, 2021.
45. *GB50011-2010*; Code for Seismic Design of Buildings. Building Industry Press of China: Beijing, China, 2010.
46. Zhang, J.Y.; Qi, A.; Yang, M.Y. Effects of rotational component of earthquakes on seismic responses of asymmetric base-isolated structures. *Structures* **2023**, *59*, 105676. [CrossRef]
47. *UBC97*; Uniform Building Code. International Conference of Building Officials: Whittier, CA, USA, 1997.
48. *Pacific Earthquake Engineering Research (PEER) Center (2014) Open System for Earthquake Engineering Simulation*, v.2.4.3. 2014; Pacific Earthquake Engineering Research Center: Berkeley, CA, USA, 2014. Available online: http://ngawest2.berkeley.edu (accessed on 1 June 2022).
49. Wang, Y. Research on Seismic Performance of Base-Isolated Structure under Pulse-Like Very Rare Ground Motion. Master's Thesis, Fuzhou University, Fuzhou, China, 2021. (In Chinese)
50. Zhou, Y. *Structural Design Theory and Application of Viscous Damping Structures*; Wuhan University of Technology Press: Wuhan, China, 2013. (In Chinese)
51. Sharma, V.; Shrimali, M.K.; Bharti, S.D.; Datta, T.K. Seismic fragility evaluation of semi-rigid frames subjected to near-field earthquakes. *J. Constr. Steel Res.* **2021**, *176*, 106384. [CrossRef]

buildings

MDPI

Article

Deformation Characteristics and Influence Factors of Shear Force Lateral Stiffness Matching Index for Non-Rigid Plate Bent Frame Structures

Yulong Li [1], Pengfei Zhao [2,*], Wen Xue [1,2], Qiang Zhang [1,2], Changjie Ye [1,2] and Ming Ma [1,2]

1 China Academy of Building Research, Beijing 100013, China; liyulong_cabr@163.com (Y.L.); xuewen@cabrtech.com (W.X.); zhangqiang@cabrtech.com (Q.Z.); yechangjie@cabrtech.com (C.Y.); maming@cabrtech.com (M.M.)
2 National Engineering Research Center of Building Technology, Beijing 100013, China
* Correspondence: zhaopengfei@cabrtech.com

Abstract: The period ratio and the drift ratio are commonly used as plane regularity control indices for multi-story buildings. However, they fail to reasonably reflect the regularity of lateral force-resisting component configuration and deformation characteristics in non-rigid plate bent frame structures. This study focuses on the analysis of non-rigid single-span bent frames, examining the variation patterns of a suitable regularity index for non-rigid plate bent frame structures, referred to as the shear force lateral stiffness matching index, under various parameters. Additionally, it introduces indices to quantify the deformation response of non-rigid plate bent frame structures, providing a detailed analysis of the impact of factors such as eccentricity, torsional stiffness, and roof slab stiffness on the deformation characteristics of non-rigid plate bent frame structures and the shear force lateral stiffness matching index. The results show that the shear force lateral stiffness matching index can reflect the inconsistency in the horizontal displacement response of lateral force-resisting components caused by deformations in the roof slab. The proposed indices for torsional and bending deformations accurately quantify the roof slab's deformation response, revealing the horizontal deformation characteristics of lateral force-resisting components in non-rigid frames. When eccentricity is present, the stiffness of the roof slab has a non-monotonic effect on the torsional component of the structural seismic response.

Keywords: shear force lateral stiffness matching index; non-rigid plate bent frame structures; lateral force-resisting components; deformation characteristics; torsional stiffness

1. Introduction

Despite the relatively late development of long-span buildings, their innovative designs and superior performance have facilitated rapid growth, establishing them as a significant structural form in urban infrastructure with substantial potential for future applications [1,2]. Long-span buildings encompass various forms, among which non-rigid plate bent frame structures are widely utilized due to their flexibility and adaptability to different architectural and engineering demands [3]. With the increasing requirements for earthquake resistance, anti-seismic design has become an essential component in the development of non-rigid plate bent frame structures [4].

The criteria for determining structural regularity, controlling structural torsion, and implementing construction measures are critical aspects of anti-seismic design. The definition of torsional irregularity encompasses discriminative indicators for controlling structural torsion, primarily including the drift ratio (η) (maximum displacement at the floor end divided by the average displacement of the floor), eccentricity. UBC97 [5], NEHRP [6], IBC2003 [7], EC [8], NZS4203 [9], and the "Code for anti-seismic design of Buildings" [10] specify that when $\eta > 1.2$, the structure is considered irregular. NZS4203 stipulates that

$\eta > 1.4$ defines an irregular structure, while Europe's EC8 ensures plan regularity by controlling eccentricity. Furthermore, China's "Technical Specification for Concrete Structures of Tall Buildings" [11] introduces requirements for maintaining the period ratio, defined as the ratio of the first mode period dominated by torsional motion (T_t) to the first mode period dominated by translational motion (T_1). The specification uses both the drift ratio and the period ratio as dual control indicators for structural regularity. This dual requirement for controlling structural irregularities is unique to China and is not found in the codes of other countries.

Recent studies have investigated the effectiveness of control measures for planar regularity indicators such as the drift ratio and the period ratio. Xu et al. [12], based on seismic measurements of three multi-story buildings, demonstrated the necessity of controlling the period ratio in seismic design for high-rise buildings and suggested that the control method for the period ratio should be adjusted for unconventional structures. Cai et al. [13], through the analysis of high-rise concrete structures, pointed out that in some cases, the dual control measures of coupled period ratio and displacement ratio might no longer be reasonable. Liu et al. [14] proposed an algorithm for the period ratio under different eccentricities. Their results indicated that in some situations, the period ratio values may not correlate with the structural regularity, and the period ratio may decrease as eccentricity increases. Wei et al. [15] conducted a theoretical analysis of the physical significance and control objectives of the period ratio for high-rise structures. They pointed out that many factors influence the period ratio in high-rise structures, but only by increasing the torsional stiffness while maintaining translational stiffness can a reduction in the period ratio significantly improve the seismic performance of high-rise structures. Liu et al. [16] observed that in uneven-height ground structures, there is a significant inherent eccentricity in the shorter direction, which may lead to an increase in the relative eccentricity to meet the period ratio limit, thereby distorting the control of torsional effects. Wang et al. [17] noted that when a structure is vertically uniform or has minimal changes, the drift ratio can reflect the overall torsional effects of the structure; however, when there are significant vertical irregularities, especially at the top, the drift ratio may fail to reflect the overall torsional vibration effects of the structure accurately. Han et al. [18], based on a complex eccentric steel frame structure, proposed influencing parameters for the dynamic characteristics of single-layer biaxial eccentric structures. The analysis showed that as eccentricity increases, the torsional effects of the structure also increase; when the primary torsional frequency ratio approaches 1, the coupled translational–torsional effects are maximized, while the influence of the secondary torsional frequency ratio is relatively small.

Current research on torsional regularity indicators in seismic design primarily focuses on rigid roof panels, such as concrete roof panels, steel–concrete composite roof panels, and prestressed concrete roof panels. Previous research has already considered the impact of roof panel rigidity or its in-plane deformation on the structure. Jahami et al. [19] pointed out that shear reinforcement can effectively improve the stiffness and load-bearing capacity of post-tensioned prestressed concrete slabs under impact loads, thereby enhancing their impact resistance. Paul et al. [20] proposed a nonlinear elastoplastic finite element model to analyze the behavior of side panel connections and explored the effects of side panel thickness (in-plane stiffness), size, and spacing on connection stiffness. Sánchez et al. [21] proposed a numerical calculation method for the optimal design of planar semi-rigid steel frames, incorporating second-order effects to account for the influence of panel stiffness. He et al. [22] defined a horizontal load transfer coefficient to quantify the impact of in-plane stiffness on horizontal load distribution, explored the relationship between in-plane stiffness and the lateral stiffness ratio of vertical components and load transfer coefficients, and defined the concept of rigid floors based on lateral stiffness ratio and load transfer coefficients.

In fact, the rigidity or flexibility of the roof panel directly determines the distribution method of horizontal loads. The current studies on the impact of torsional regularity indicators in anti-seismic design are mostly based on multi-story structures, with very few studies focusing on non-rigid plate bent frame structures. Currently, the relevant codes [5–11,23,24] do not specify methods for evaluating the plane configuration regularity of non-rigid plate bent frame structures. In practical anti-seismic design, the methods and limit indices for calculating the regularity of non-rigid plate bent frame structures often apply indiscriminately to those used for multi-story buildings, such as the drift ratio and the period ratio. However, existing research indicates that due to the sparse configuration of lateral force-resisting components and relatively low in-plane stiffness of the roof slab in non-rigid plate bent frame structures, the drift ratio cannot accurately reflect the torsional effects and internal force response characteristics of structures [25,26]. Moreover, the period ratio limit specified in the "Technical Specification for Concrete Structures of Tall Buildings" is based on studies of non-coupled period ratios. However, non-rigid plate bent frame structures often exhibit coupling between translational and torsional modes. Research shows that the coupled period ratios may not accurately reflect the torsional characteristics of the structure [13,25]. Indiscriminately applying the regularity indices for multi-story buildings to non-rigid plate bent frame structures often results in cases where the lateral stiffness requirements are met, but the regularity indices exceed the limits [27–29].

In addition, the seismic response of horizontal components in non-rigid plate bent frame structures is minimal, so the focus should be on the lateral force-resisting components. These lateral force-resisting components not only provide lateral stiffness but also contribute significantly to torsional stiffness. The regularity in the arrangement of the lower support components in non-rigid plate structures affects not only the overall torsional effects but also the distribution of seismic responses in the lateral force-resisting components due to in-plane bending deformations, as non-rigid plate bent frame structures are not suitable for the rigid diaphragm assumption. Therefore, in the anti-seismic design of non-rigid plate bent frame structures, it is essential to focus on the calculation methods and control indicators of the regularity of the lateral force-resisting member layout.

Recent research has introduced a reasonable index, known as the shear force lateral stiffness matching index, for determining the regularity of lateral force-resisting component configurations in non-rigid plate bent frame structures [30]. The index can reflect the regularity of lateral force-resisting component configuration by describing the mismatch between the internal forces and the lateral stiffness of the components under seismic loading. It also considers the effects of roof slab bending deformations. Given that the seismic response of non-rigid plate bent frame structures is not significant, it is essential to pay greater attention to their lateral force-resisting components. The endpoints at the connections between these components and the roof panels are utilized to calculate the shear force lateral stiffness matching index and to extract the deformation characteristics of the roof panels.

This paper primarily focuses on the single-span, single-story frame structures commonly found in public buildings. Building on the dynamic characteristics of these structures and referencing [30], it further explores the properties and influencing factors of the shear force lateral stiffness matching index. Additionally, it proposes indices to quantify the roof slab deformation in non-rigid frame structures, providing references for the seismic conceptual design of long-span structures.

2. Definition of Plane Regularity Index

2.1. Shear Force Lateral Stiffness Matching Index

Ideally, a structure should experience only lateral movement during an earthquake. Under such conditions, the ratio of the shear force in the lateral force-resisting members to the total shear force of the structure equals the ratio of their lateral stiffness to the total stiffness. However, when the structure undergoes torsion, the lateral force-resisting members must bear not only the translational shear force but also the additional shear force caused by torsional effects, as shown in Figure 1. When translational–torsional coupling occurs in a structure, the lateral displacements of the lateral force-resisting components are not uniform. In this situation, components located farther from the center of rigidity experience increased lateral displacement due to torsion, leading to a significant increase in the horizontal seismic shear force they must resist. This can easily result in local failure due to exceeding the allowable strength and deformation limits, potentially causing the entire structure to collapse due to the failure of one side's components. In contrast, components closer to the center of rigidity undergo smaller total deformation, and they may be destroyed as the entire structure collapses, even before reaching their load-bearing capacity limits.

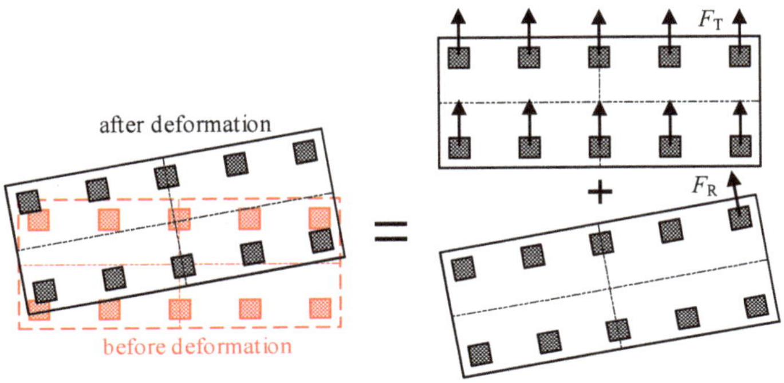

Figure 1. Additional shear force of lateral force resistant members due to torsion.

Members located further from the rigid center with less stiffness will undergo greater lateral displacement, resulting in a substantial increase in the shared horizontal seismic shear force. This increase can cause the members to exceed their allowable resistance and deformation limits, leading to severe local damage, the failure of individual members, and potentially the collapse of the entire structure.

Reference [30] proposed an index to characterize the match between the shear force and lateral stiffness of lateral force-resisting components in non-rigid plate bent frame structures, known as the shear force lateral stiffness matching index β. This index is derived from the results of bidirectional seismic response spectrum enveloping. The shear force lateral stiffness matching index is defined as:

$$\beta = \max((F_i/F_0)/(K_i/K_0)) \quad (1)$$

where F_i and K_i represent the shear force and lateral stiffness of the i-th lateral force-resisting column in a main direction under seismic loading, respectively. F_0 and K_0 represent the total shear force under seismic loading and total lateral stiffness of the structure in the same direction, respectively.

Under seismic loading, if the structure experiences only translational motion, the deformations of all lateral force-resisting components are consistent, and $\beta = 1$; if torsion is involved, $\beta > 1$. Therefore, β essentially represents the inconsistency in the deformation of lateral force-resisting components under seismic loading. A larger β indicates that the structure exhibits less translational motion. Factors influencing β include not only the torsional response of the structure but also the in-plane non-uniform deformation of the roof slab. Utilizing this index to represent the regularity of lateral force-resisting component configuration in non-rigid plate bent frame structures is considered reasonable [30].

2.2. Roof Slab Deformation Indices

As part of the horizontal load-bearing system in non-rigid plate bent frame structures, the roof slab significantly influences the cooperative behavior of vertical lateral force-resisting components. When the roof slab is non-rigid, the in-plane deformation of the roof slab cannot be neglected when considering the horizontal force distribution among lateral force-resisting components in non-rigid plate bent frame structures.

The deformation of a spatial element is generally a combination of rigid body translation, in-plane rigid body rotation, and in-plane bending deformation [31–33]. To further analyze the impact of these deformations (Figure 2) on β, we propose the following indices to characterize structural deformation: the point of maximum displacement lies between the displacements at both ends, with the corresponding interpolation representing the torsional deformation Δ_{int}. The difference between the maximum displacement and the torsional deformation is the bending deformation Δ_{be}. The ratio of the bending deformation to the maximum displacement is referred to as the bending deformation ratio P_{b}.

$$\Delta_{\text{int}} = \frac{(u_{B,n} - u_{B,1}) \times x}{L} + u_{B,1} \tag{2}$$

$$\Delta_{\text{be}} = u_{\max} - \Delta_{\text{int}} \tag{3}$$

$$p_{\text{b}} = \frac{\Delta_{\text{be}}}{u_{\max}} \tag{4}$$

where $u_{\max}$ represents the maximum displacement at the top of the column under unidirectional seismic loading. $u_{\text{B},1}$ represents the displacement at the left end post, and $u_{\text{B},n}$ represents the displacement at the right end post, with the subscript n indicating the number of longitudinal spans. L is the dimension of the structure in the main direction, and x is the distance from the column with the maximum displacement to the leftmost column line.

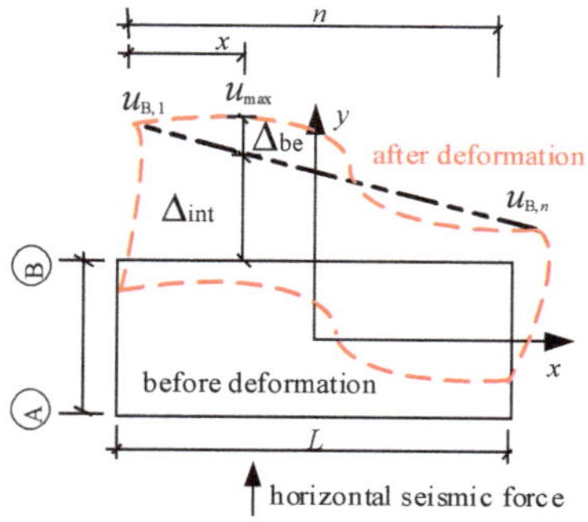

Figure 2. Structural torsion under horizontal seismic loading.

Equation (2) represents the rigid body linear displacement and rigid body rotational displacement of the structure, namely, torsional deformation and translational displacement. However, the translational displacement of the structure generally does not change with the eccentricity. Therefore, Equation (2) can be considered a measure of the torsional deformation of the structure. Equation (3) corresponds to the residual deformation after removing the torsional deformation at the point of maximum displacement. This residual deformation quantifies the extent of the planar bending deformation of the roof panel, thus reflecting the bending deformation of the roof panel.

3. Properties and Influencing Factors of β

3.1. Example Design

Taking the airport concourse and the exhibition hall of the convention center as reference objects, their main feature is the extensive use of single-story frame structures [3], with rectangular or quasi-rectangular roof slabs. Based on the "Code for Anti-seismic design of Buildings" [10] and the "Standard for Design of Steel Structures" [23], a rectangular single-story frame structure example was designed as shown in Figure 3.

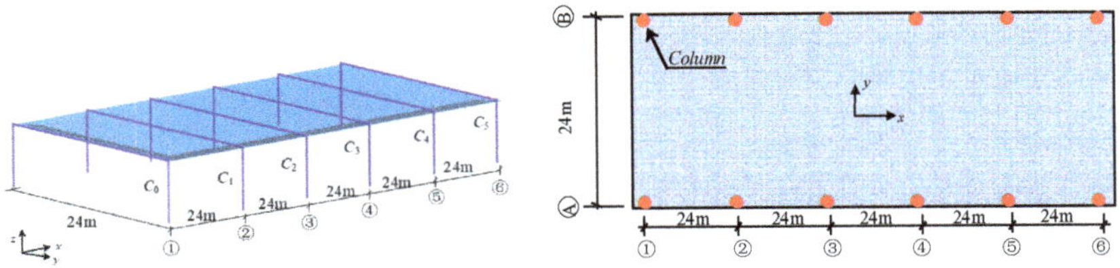

Figure 3. Schematic of example model.

The structure uses Q355B with a transverse column spacing of 24 m and one span and a longitudinal column spacing of 24 m and five spans. The height of the roof-supporting columns is 5 m. The roof load is simplified to a permanent load of 1.5 kN/m^2 and a live load of 0.5 kN/m^2. The seismic precautionary intensity is 8 degrees (0.20 g), the site category is type III, the design seismic grouping is Group I, the seismic influence coefficient is 0.16, and the site period is 0.45 s.

To investigate the impact of eccentricity on β, eccentric loads were applied, as shown in Figure 4. By applying an additional Z-direction concentrated load at the top of the lateral force-resisting components along axes ① to axis ③, the centroid of the structure deviates from the geometric center while the position of the rigid center remains unchanged. The magnitude of the additional concentrated load is a multiple of the structural mass m_0.

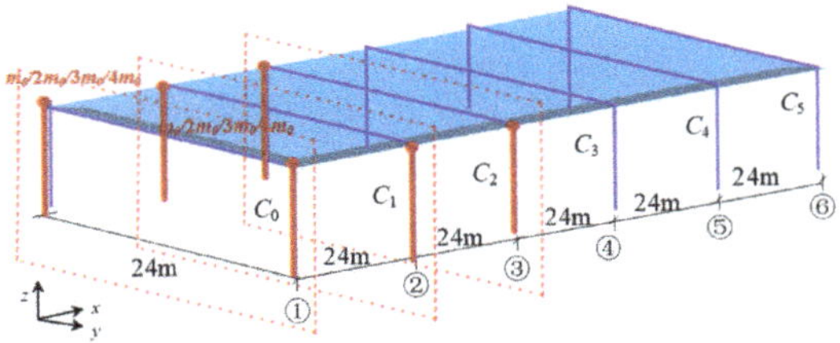

Figure 4. Schematic of eccentric load application.

The model's support columns have cross-sectional specifications of 290 × 14, arranged uniformly to support the roof slab. To investigate the impact of torsional stiffness on β, the cross-sectional size of the corner columns is increased. Specifically, the cross-sectional size of the corner columns on axes ① and axes ⑥ is increased to 450 × 22, while the other support columns remain at 290 × 14.

Additionally, it is assumed that the 120 mm thick concrete slab has infinite in-plane stiffness, with its in-plane compressive stiffness set to 1.00*EA*. Other conditions are based on a 120 mm thick concrete slab, with adjustments made by varying the panel thickness to explore the impact of roof stiffness. The varying calculation parameters are shown in Table 1.

Table 1. Parameters of the calculation model.

Calculation Parameters	Parameter
Cross-sectional size of the corner columns	290 × 14 450 × 22
Panel stiffness	1.00*EA* (Rigid plate, 120 mm thick concrete plate) 0.05*EA* (6 mm thick concrete plate) 0.01*EA* (1.2 mm thick concrete plate) 0.005*EA* (0.6 mm thick concrete plate)
Eccentricity	e = 2.77 m, 5.08 m, 7.04 m, 8.71 m, 10.1 7 m, 15.25 m, 18.31 m, 20.35 m. From axes ① to axes ③, the additional loads at the top of the lateral force-resisting components are 0.2m_0, 0.4m_0, 0.6m_0, 0.8m_0, 1.0m_0, 2.0m_0, 3.0m_0, 4.0m_0, respectively.

3.2. Properties of β

To analyze the impact of lateral force-resisting component configuration on β, a panel stiffness of 0.01EA is used as an example. As shown in Figure 5, when the section size of corner columns increases from 290 × 14 to 450 × 22, β significantly increases.

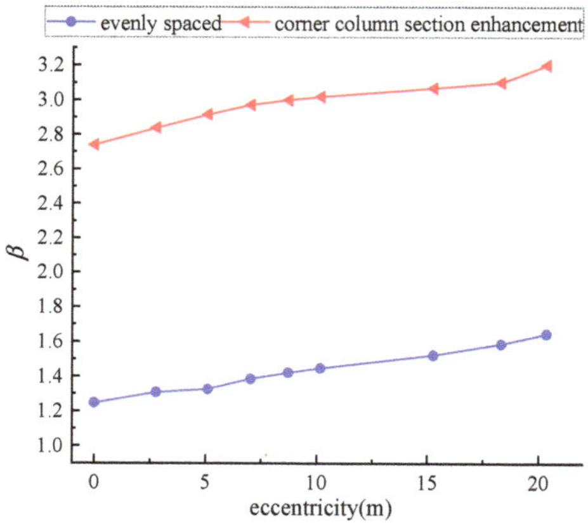

Figure 5. Comparison of β for non-rigid plate bent frame structures (0.01EA) with different lateral force-resisting component configurations.

Taking support columns with a section size of 290 × 14 as an example, the influence of panel stiffness and eccentricity on β is investigated. Figure 6 demonstrates that, irrespective of the roof panel's rigidity or elasticity, β increases with increasing eccentricity. For small eccentricities, β is negatively correlated with panel stiffness; at the same eccentricity, increased panel stiffness results in a smaller β. However, as eccentricity continues to increase, this trend reverses, and β begins to increase with panel stiffness at large eccentricities.

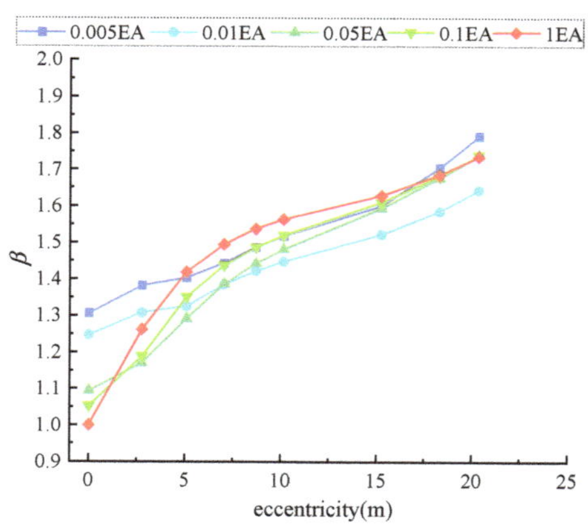

Figure 6. β for different panel stiffness with uniformly arranged lateral force-resisting components.

These results indicate that β can simultaneously account for the influence of lateral force-resisting component configuration and panel stiffness on structural regularity.

Figures 5 and 6 raise two issues that need to be discussed: (1) As seen in Figure 5, when the roof panel is non-rigid, β increases when the section size of the corner columns is enhanced, indicating an increase in the overall torsional stiffness of the structure; (2) In Figure 6, it is observed that with increasing eccentricity, the relationship between β and panel stiffness undergoes a shift. Initially, β decreases with increasing panel stiffness, but as eccentricity grows larger, this trend reverses and β begins to increase with panel stiffness, showing no consistent negative correlation.

This indicates that the changes in β involve the complex interplay of factors such as eccentricity, torsional stiffness, and roof panel deformation. The following parts will explore the impact of different factors on β.

3.3. Influence of Torsional Stiffness

When torsional effects are present, the contribution of vertical lateral force-resisting components to maintaining the overall stability of the building cannot be ignored [34]. With the structural panel stiffness set at 0.01EA, the impact of different torsional stiffness on various deformations of the roof panel is explored using two configurations: one with a uniform arrangement where all column sections are identical, and the other with enhanced section sizes at the corner columns.

As shown in Figures 7 and 8, with increasing eccentricity, both torsional deformation Δ_{int} and bending deformation Δ_{be} increase for structures with uniformly arranged lateral force-resisting components and with corner columns having enhanced section sizes. This indicates that when the roof panel is a non-rigid plate, the roof panel of structures with uniformly arranged lateral force-resisting components and strengthened corner columns will experience significant bending deformation in addition to torsional deformation.

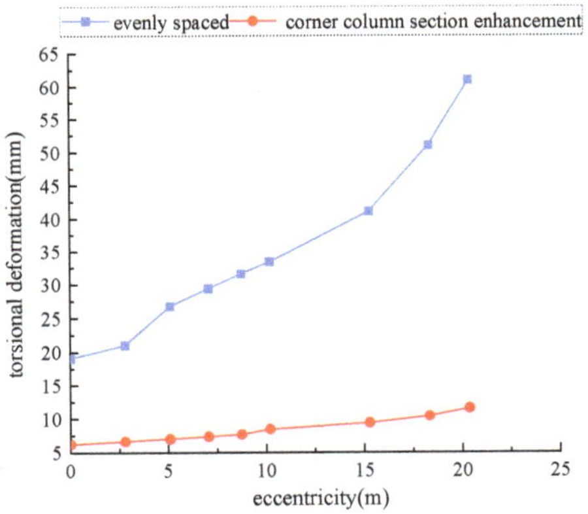

Figure 7. Torsional deformation Δ_{int} of structures with uniform distribution and strengthened corner columns (0.01EA).

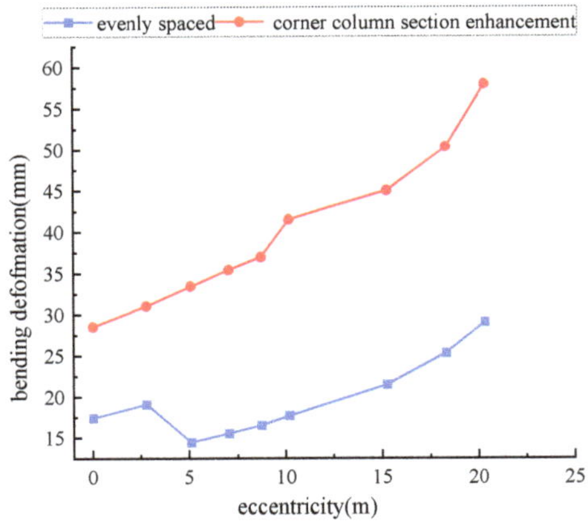

Figure 8. Bending deformation Δ_{be} of structures with uniform distribution and strengthened corner columns (0.01EA).

The result in Figure 9 shows that the primary deformation mode of the roof panel is bending deformation when corner columns are strengthened; however, it is torsional deformation when lateral force-resisting components are uniformly arranged. Strengthening the corner columns significantly increases the proportion of bending deformation P_b in the total deformation of the roof panel, indicating that additional non-uniform deformation arises due to the strengthening of the corner columns.

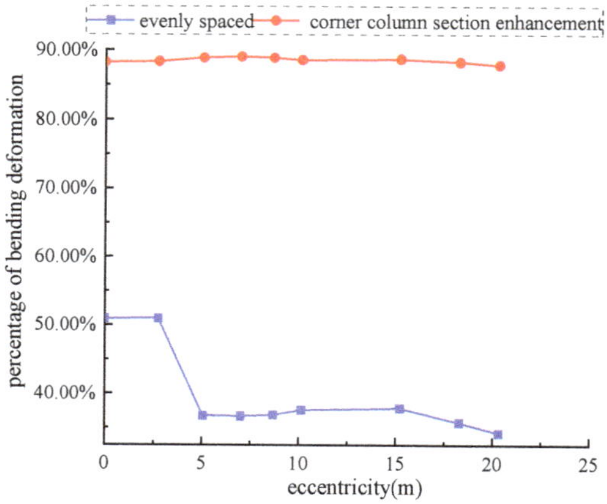

Figure 9. Proportion of bending deformation P_b in structures with uniform distribution and strengthened corner columns (0.01*EA*).

It is evident that the configuration of lateral force-resisting components significantly affects the deformation of the structure under seismic loading, thereby impacting β. For structures with strengthened corner columns, although the torsional stiffness of the structure increases, the distribution of lateral stiffness becomes more uneven, leading to greater bending deformation of the roof panel and a consequent significant increase in β.

3.4. Influence of Eccentricity

The eccentricity ratio affects the vibration characteristics of building structures, increasing the torsional effects [35]. Taking the configuration of uniformly arranged lateral force-resisting components as an example, Figures 10 and 11 illustrate the effects of different panel stiffness and eccentricities on the bending deformation Δ_{be} of the structure and the proportion of bending deformation P_b. Table 2 shows the bending deformation Δ_{be} of different axial columns for a structure with a roof slab stiffness of 0.01*EA* panel stiffness. From Figure 10 to Figure 11, it can be observed that the bending deformation of the structure under different roof panel stiffness and eccentricities exhibits a non-monotonic trend. This is related to the shift in the maximum displacement measurement point: as eccentricity increases, the maximum displacement point moves from the center of the roof panel toward the end column, i.e., from the point of concentrated uneven deformation toward the direction of eccentricity, as shown in Figure 12. Table 2 indicates that the bending deformation of a non-rigid plate bent frame structure's roof panel decreases from the center toward the ends. Therefore, as the maximum displacement point moves towards the post ends with increasing eccentricity, the bending deformation Δ_{be} of the structure and the proportion of bending deformation P_b exhibit a non-monotonic trend.

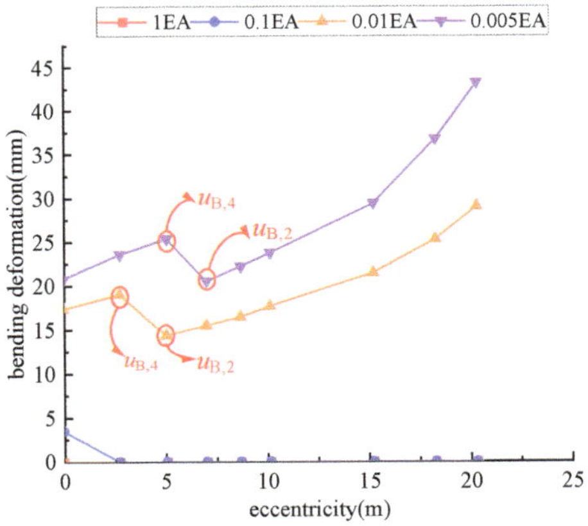

Figure 10. Bending deformation Δ_{be} of structures with different panel stiffness under uniform distribution of lateral force-resisting components.

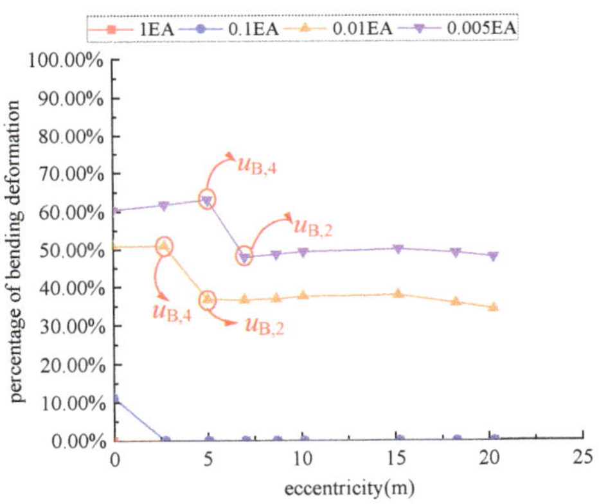

Figure 11. Proportion of bending deformation P_b in structures with uniform distribution of lateral force-resisting components.

Table 2. Bending deformation Δ_{be} for 290 − 0.01*EA* (mm).

e/m	Axes ①	Axes ②	Axes ③	Axes ④	Axes ⑤	Axes ⑥
0.00	0	10.95	17.42	17.42	10.95	0.00
2.77	0.14	13.03	19.07	16.95	9.74	0.07
5.09	0.25	14.43	20.11	16.47	8.87	0.09
7.04	0.27	15.52	21.05	16.32	8.45	0.09
8.72	0.28	16.52	22.07	16.45	8.26	0.08
10.17	0.00	17.69	23.34	16.78	8.16	0.00
15.26	0.00	21.46	26.13	16.16	6.93	0.00
18.31	0.00	25.32	27.86	14.78	5.93	0.00
20.35	0.00	29.07	29.53	14.04	5.71	0.00

Figure 13 shows the effects of different panel stiffnesses and eccentricities on the torsional deformation Δ_{int} of the structure. As seen in Figure 13, the torsional deformation of structures with different panel stiffness monotonically increases with increasing eccentricity.

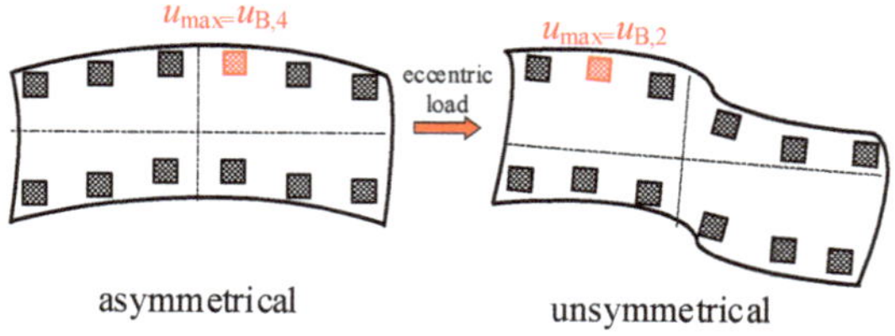

Figure 12. Shift in maximum displacement column measurement point with increasing eccentricity.

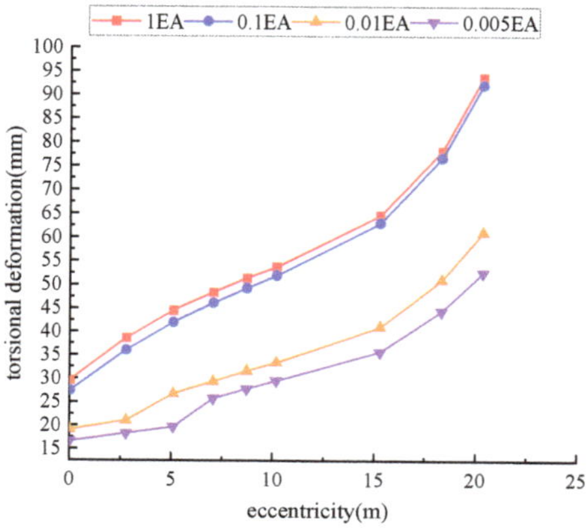

Figure 13. Torsional deformation Δ_{int} of structures with different panel stiffness under uniform distribution of lateral force-resisting components.

3.5. Influence of Panel Stiffness

The roof slab primarily bears the vertical loads while determining the distribution of horizontal loads among the vertical lateral force-resisting components [36]. Taking the configuration of uniformly arranged lateral force-resisting components and an eccentricity upper limit of 8.720 m (eccentricity ratio of 7.3%) as examples, Figures 14–17 illustrate the effects of different panel stiffness and eccentricities on the torsional deformation Δ_{int}, base shear force, bending deformation Δ_{be}, and the proportion of bending deformation P_b of the structure, respectively.

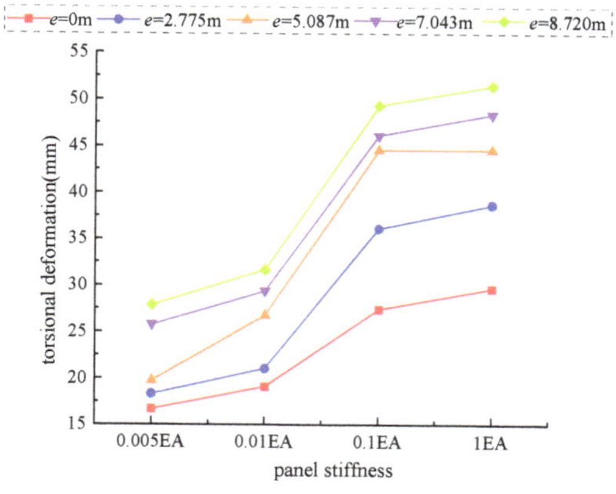

Figure 14. Influence of different panel stiffness on torsional deformation Δ_{int}.

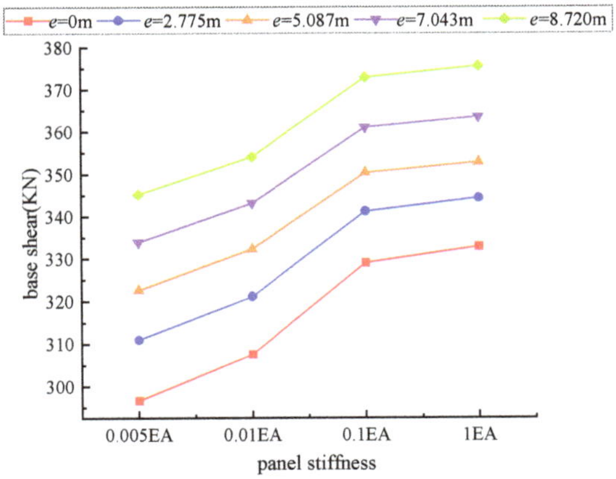

Figure 15. Influence of different panel stiffness on base shear.

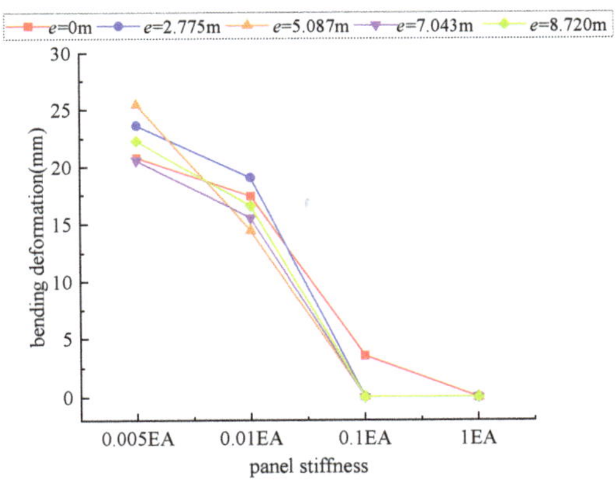

Figure 16. Influence of different panel stiffness on bending deformation Δ_{be}.

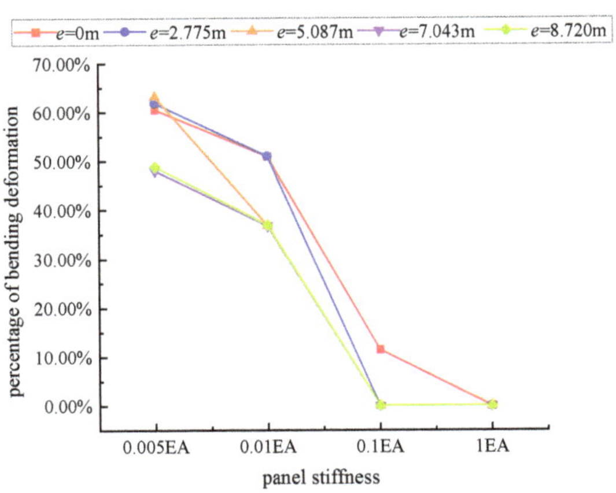

Figure 17. Influence of different panel stiffness on the proportion of bending deformation P_b.

As shown in Figure 17, with increasing eccentricity, the deformation at the maximum displacement point of a non-rigid plate gradually transitions from bending to torsional deformation. Figures 14–16 demonstrate that at a constant eccentricity, increasing panel stiffness leads to higher base shear forces, greater torsional deformation, and reduced bending deformation. This indicates that the trends of torsional and bending deformations are inversely related to panel stiffness.

Combining Sections 3.3 and 3.4, the trend of β exhibiting a shift in behavior with increasing eccentricity can be elucidated as follows: Both overall structural torsion and roof bending deformation induce a discrepancy between the shear force sustained by lateral force-resisting components and their lateral stiffness. At a constant eccentricity, with increasing panel stiffness, structural torsional deformation increases while bending deformation decreases. Compared to rigid plate bent frame structures, non-rigid plate bent frame structures exhibit additional bending deformation at small eccentricities, resulting in consistently higher β values for non-rigid plate bent frame structures than for rigid plate bent frame structures. Hence, at small eccentricities, β is negatively correlated with panel stiffness. However, as eccentricity increases, the proportion of structural torsional deformation rises while the proportion of bending deformation declines. When the primary deformation mode of the roof panel transitions to torsional deformation, the β for rigid plate bent frame structures exceeds that for non-rigid plate bent frame structures. Therefore, at large eccentricities, panel stiffness and β are not necessarily negatively correlated.

4. Conclusions

This study evaluated the applicability of the shear force lateral stiffness matching index in non-rigid frame structures and analyzed its variation patterns under different parameter conditions. Through numerical simulation, the deformation indices of the roof slab in frame structures were obtained under various parameters, revealing the influence mechanisms of factors such as structural eccentricity, torsional stiffness, and roof slab stiffness on the shear force lateral stiffness matching index.

(1) The deformation response of the roof slab under seismic loading can be divided into torsional deformation Δ_{int} and bending deformation Δ_{be}. The proposed torsional and bending deformation indices can accurately quantify these two deformation responses, describing the horizontal deformation characteristics of lateral force-resisting components in non-rigid plate bent frame structures.

(2) Although strengthening the corner column sections can increase the structural torsional stiffness, it also increases the in-plane bending deformation. Strengthening the corner column sections increases the unevenness of the lateral stiffness distribution, resulting in a significant increase in the shear force lateral stiffness matching index.

(3) For non-rigid plate bent frame structures, when the eccentricity is small, the maximum displacement under seismic loading is primarily controlled by bending deformation. As eccentricity increases, the main component of maximum displacement gradually transitions from bending deformation to torsional deformation.

(4) The shear force lateral stiffness matching index considers the inconsistency in the horizontal displacement response of lateral force-resisting components caused by various deformations of the roof panel. At a constant eccentricity, with increasing panel stiffness, torsional deformation increases while bending deformation decreases. The combined effect results in panel stiffness not necessarily being negatively correlated with β.

(5) When eccentricity is present in the structure, selecting an appropriate panel stiffness is essential to ensuring the consistency of lateral force-resisting component deformation. A larger panel stiffness does not necessarily result in a larger translational component of the structural seismic response.

(6) This study explored the influence mechanisms of various parameters on the shear force lateral stiffness matching index. The analysis was limited to single-span, single-story frames, and future research could extend to more complex cases, particularly those involving complex plans and multi-span structures. Additionally, the current study mainly focused on elastic seismic responses; future research could explore the

nonlinear behavior of the shear force lateral stiffness matching index under elasto-plastic seismic conditions to optimize seismic design further.

Author Contributions: Conceptualization, P.Z.; methodology, P.Z.; formal analysis, Y.L.; investigation, Y.L.; writing—original draft, Y.L.; writing—review and editing, Y.L., W.X., Q.Z., M.M. and C.Y.; project administration, Y.L.; funding acquisition, P.Z. All authors have read and agreed to the published version of the manuscript.

Funding: The authors express their sincere gratitude for the financial support provided by the National Key Research and Development Program of China [Grant No. 2022YFC3002300].

Data Availability Statement: Data are contained within the article.

Conflicts of Interest: Authors Yulong Li, Pengfei Zhao, Wen Xue, Qiang Zhang, Ming Ma and Changjie Ye were employed by the company China Building Technique. The remaining authors declare that the research was conducted in the absence of any commercial or financial relationships that could be construed as a potential conflict of interest.

References

1. Krishna, P. A Review of Developments in Steel: Implications for Long-Span Structures. *Trans. Indian Inst. Met.* **2021**, *74*, 1055–1064. [CrossRef]
2. Dong, S.; Xing, D.; Zhao, Y. Application and Development of Modern long-span Spatial Structures in China. *Front. Struct. Civ. Eng.* **2012**, *6*, 224–239. [CrossRef]
3. Cheng, H. Modern Light Steel Structure Buildings and Their Application in China. *China Build. Met. Struct.* **2022**, *7*, 39–41.
4. Zhao, P. Design of Special-Shaped Components in long-span Structures. *Build. Struct.* **2013**, *43*, 36–40.
5. International Code Council. *1997 Uniform Building Code, Vol. 2: Structural Engineering Design Provisions*; International Conference of Building Officials: Whittier, CA, USA, 1997.
6. FEMA. *NEHRP Recommended Provisions for Seismic Regulations for New Buildings and Other Structures (FEMA 450)*; Building Seismic Safety Council (BSSC): New York, NY, USA, 2003.
7. International Code Council. *International Building Code 2003*; International Code Council: Washington, DC, USA, 2003.
8. Cheng, S. *Structural Eurocodes (EC8)*; Institute of Engineering Earthquake Research, China Academy of Building Research: Beijing, China, 1997.
9. *NZS4203: 1992*; Code of Practice for General Structural Design and Design Loadings for Buildings. Standards Association of New Zealand: Wellington, New Zealand, 1992.
10. *GB 50011-2010*; Code for Anti-Seismic Design of Buildings. China Architecture & Building Press: Beijing, China, 2016.
11. *JGJ3-2010*; Technical Specification for Concrete Structures of Tall Buildings. China Architecture & Building Press: Beijing, China, 2010.
12. Xu, P.; Huang, J.; Wei, C. Torsional vibration effects of high-rise building structures under earthquake action. *Build. Sci.* **2000**, *1*, 4–9.
13. Cai, J.; Pan, D.; Huang, Y. Study on the control of torsional vibration effects in high-rise building structures. *Eng. Mech.* **2007**, *7*, 116–121.
14. Liu, J.; Li, Y. Influence of Eccentricity Variation on Period Ratio of High-Rise Building Structures. *Wuhan Univ. Technol. J.* **2010**, *32*, 55–60.
15. Wei, L.; Wang, S.; Wei, C. Control issues of translational period and torsional period ratio in high-rise buildings. *Build. Struct.* **2014**, *1–3+13*, 44.
16. Liu, L.P.; Li, A.L.; Li, Y.M.; Fan, S.; Han, J. Applicability Analysis of Torsional Control Indicators in Uneven-Height Grounded Frame Structures. *Earthq. Eng. Eng. Vib.* **2014**, *34*, 680–685.
17. Wang, J.; Zhao, D.; Wang, D.; Liu, G. Study of Torsional Response Control Index of the Spreading Layer upon Layer. *Build. Sci.* **2016**, *32*, 25–29.
18. Han, Y. Study on Seismic Torsional Effects of Complex Eccentric Structures. Master's Thesis, Tianjin University, Tianjin, China, 2017.
19. Jahami, A.; Temsah, Y.; Khatib, J.; Baalbaki, O.; Darwiche, M.; Chaaban, S. Impact behavior of rehabilitated post-tensioned slabs previously damaged by impact loading. *Mag. Civ. Eng.* **2020**, *93*, 134–146.
20. Paul, B.; Roy, K.; Ghosh, K.; Fang, Z.; Maity, S.; Lim, J.B.P. Numerical Modelling of Side-Plate Connection for Cold-Formed Steel Nested Tapered Box Beam Portal Frame. In *Recent Developments in Structural Engineering*; Goel, M.D., Kumar, R., Gadve, S.S., Eds.; Volume 1 SEC 2023 Lecture Notes in Civil Engineering; Springer: Singapore, 2024; Volume 52.
21. Sánchez-Olivares, G.; Espín, A.T. Design of planar semi-rigid steel frames using genetic algorithms and Component Method. *J. Constr. Steel Res.* **2013**, *88*, 267–278. [CrossRef]
22. He, M.; Ma, Z.; Ma, R.; Li, Z. Light Wood-steel Hybrid Performance on Transferring Horizontal Loads. *Tongji Univ. Nat. Sci. Ed.* **2014**, *42*, 1038–1043.

23. *GB 50017-2017*; Standard for Design of Steel Structures. China Architecture & Building Press: Beijing, China, 2018.
24. *JGJ 257-2012*; Technical Specification for Cable Structures. China Architecture & Building Press: Beijing, China, 2012.
25. Shao, F.; Chen, Z.; Ge, H. Parametric analysis of the dynamic characteristics of a long-span three-tower self-anchored suspension bridge with a composite girder. *ABEN J.* **2020**, *1*, 10. [CrossRef]
26. Zhao, P.; Chen, X.; Liu, X. Preliminary Study on Calculation Method and Influence of Regularity indices for long-span Structures. *Archit. Sci.* **2022**, *38*, 17–26.
27. Liu, Y.; Xiong, J.; Mo, L. Structural Design of a Commercial Complex in Changsha. *Build. Struct.* **2021**, *51*, 36–41+6.
28. Yu, Z. Design and Analysis of the Main Structure of the Expo Cultural Center. *J. Build. Struct.* **2010**, *31*, 95–102.
29. Liang, Q.; Wu, J.; Lu, G.; Hu, J. Structural Design and Analysis of a Super-High Building in Nanjing, China. *Sustainability* **2023**, *15*, 6521. [CrossRef]
30. Zhao, P.; Xue, W.; Li, Z. Study on the Regularity Indicator for Single-Span Frame Structures Considering Planar Bending Deformation. *J. Build. Struct.* **2024**, *45*, 50–61.
31. Sun, P. Deformation Decomposition and Basic Application of Elastic Structures. Master's Thesis, Zhengzhou University, Zhengzhou, China, 2016.
32. Tseng, W.; Tarn, J. Exact Elasticity Solution for Axisymmetric Deformation of Circular Plates. *J. Mech.* **2015**, *31*, 617–629. [CrossRef]
33. Ye, T.; Jin, G.; Liu, Z. Analysis of the Influence of Transverse Shear and Stretching Deformation on the Vibration Characteristics of Laminated Plates. *J. Vib. Shock* **2019**, *38*, 118–125.
34. De-La-Colina, J.; Valdés-González, J.; González-Pérez, C.A. Dependency of the Accidental Torsion Building Response on Both Live-to-Dead Load Ratio and Material Stiffness Variation. *J. Civil Eng.* **2024**. [CrossRef]
35. Tabatabaei, R. Torsional Vibration of Eccentric Building Systems. In *Recent Advances in Vibrations Analysis*; InTech: London, UK, 2011.
36. Sucuoğlu, H.; Akkar, S. Analysis Procedures and Seismic Design Principles for Building Structures. In *Basic Earthquake Engineering*; Springer: Cham, Switzerland, 2014.

buildings

MDPI

Article

Shaking Table Tests and Numerical Study on the Seismic Performance of Arc-Shaped Shear Keys in Highway Continuous-Girder Bridges

Liang Huang, Wenze Wang, Shizhan Xu *, Bo Wang and Zisheng Li

College of Civil Engineering, Zhengzhou University, Zhengzhou 450001, China; huangliang@zzu.edu.cn (L.H.); wangwenze@gs.zzu.edu.cn (W.W.); wangbo@zzu.edu.cn (B.W.); lizisheng@gs.zzu.edu.cn (Z.L.)
* Correspondence: xushizhan@zzu.edu.cn

Abstract: Typical forms of seismic damage to laminated-rubber-bearing girder bridges in the transverse direction are falling beams, girder displacement, and bearing damage. However, the damage to piers and foundations is generally lighter. This is mainly due to slippage of the bearings. Therefore, we propose a new type of arc-shaped shear key to improve the lateral seismic performance. A 1/12-scale highway continuous-girder bridge isolated by different shear keys was tested utilizing a 4 m × 4 m shaking table with six DOFs. The seismic responses of the bridge were analyzed in terms of phenomenon, displacement, strain, and acceleration. The main girder and pier exhibited different seismic responses because the bridge had different stops. A numerical simulation based on FEM showed that the established finite element model can well reproduce the displacement time history of the main girder and the cap girder. By analyzing the finite element model, the relative displacement of the bearing under different seismic waves was obtained. A comparison between the measured and FEM responses showed that the arc-shaped shear key can well limit the displacement of the main girder and the bearing. In addition, it does not significantly amplify the seismic response of the substructure. The arc-shaped shear key dissipates more energy while limiting the displacement of the main girder, and the comprehensive seismic performance is better than that of the rubber pad shear key.

Keywords: highway continuous-girder bridge; arc-shaped shear key; shaking table test; seismic response; FEM

1. Introduction

In the infrastructure of China's transportation system, bridges play roles as nodes and hubs [1]. Among them, medium- and small-span simple-bearing girder and continuous-girder bridges are widely used in highway construction because of their practicality and convenience, which greatly promotes sustainable economic and social development. As an important part of transportation infrastructure, bridges consist of a superstructure with a large mass and piers with less lateral stiffness. Because of the wide distribution of earthquake zones in China, the seismic safety of bridges in areas with high seismic intensity has attracted the attention of researchers [2,3]. Restoring bridges to normal use after a strong earthquake is more challenging than maintaining bridges that are in use [4,5]. For long-distance transportation, if a section of the bridge is damaged in an earthquake, the normal operation of the entire lane will be interrupted. For example, in the Lushan and Wenchuan earthquakes, most bridges were destroyed and traffic lanes were interrupted. Therefore, seismic design is very important for the seismic safety of bridges in the event of a strong earthquake [6–8].

A series of studies has been carried out on the seismic problems of bridges, including numerical simulations of bridges during earthquakes [9,10], interactions between bearings and bridge structures [11–13], and interactions between shear keys and bridge structures [14]. Research shows that antiseismic isolation bearings can significantly improve the seismic resistance of a bridge. Commonly used antiseismic isolation bearings, such as friction pendulum bearings (FPBs), lead rubber bearings (LRBs), and plate rubber bearings, have satisfactory seismic isolation effects. However, the displacement and residual displacement of these bearings in strong earthquakes are much larger [15–17], resulting in more severe girder damage. Therefore, many scholars have proposed to limit the displacement of bridges by shear keys. Wu [18] introduced a novel base-supported dual damping system, demonstrating its capacity to mitigate the main beam's displacement by 20–47% when subjected to El Centro seismic waves. Qi [19] and Wei [20] conducted a comprehensive study on the seismic performance of friction pendulum bearings (FPBs) under real earthquake conditions, examining the bridge's seismic response in light of experimental observations, displacements, accelerations, and strains. Their experimental findings indicate that the FPBs, when compared to traditional planar sliding bearings, exhibited reduced peak and residual displacements along with enhanced energy dissipation capabilities. Zhuang [21] innovatively combined the friction pendulum bearing with shape memory alloy to devise a shock absorption device, effectively managing peak and residual displacements even under the impact of strong earthquakes. The study's outcomes suggest that this new seismic isolation bearing significantly attenuates the bridge's seismic response. However, these advanced bearings are not without drawbacks, notably their high manufacturing costs. Moreover, the challenge of retrofitting many existing bridges, which are already in service, with these new bearings presents a significant barrier to improving their seismic resilience.

The rational design of a concrete shear key has long been a controversial issue, as their role in the seismic performance of bridges is unclear and their seismic requirements are difficult to determine accurately. Bozorgzadeh et al. [22] and Silva et al. [23] proposed a design method in which reinforced concrete shear keys act as sacrificial elements in an earthquake. Li Jianzhong et al. [24] rationally designed the mechanical properties of the lateral limiter to allow the bearings to undergo large slips during strong earthquakes, and the lateral limiting device is damaged or sacrificed, protecting the substructures, such as the pier and foundation, from serious damage. Liu Xiaoxian et al. [25] established a finite element model of a simply supported girder bridge to study the influence of X-shaped elastoplastic steel shear keys on the lateral seismic response of bridge piers. Tang Zhenhui [26] proposed different calculation methods and conducted experimental studies for different forms of seismic steel shear keys. Huang [27] proposed a new type of shear key consisting of multiple X-shaped steel plates arranged in parallel. A quasi-static test showed that this shear key exhibited good deformation ductility and stable energy dissipation. Researchers have devised various configurations of innovative seismic stops, intended to bolster bridges' seismic performances by facilitating energy dissipation and employing sacrificial components. Nonetheless, these blocks are often characterized by a relatively low design strength. Consequently, once the block sustains damage, the plate rubber bearings are prone to slippage, leading to an overall deficiency in the bridge system's self-centering capacity. Li Zhongxiong [28] established a three-dimensional analysis model considering the slip phenomenon of plate rubber bearings and the nonlinear hysteresis characteristics of the stopper. He analyzed the influence of the shear key parameters on the seismic performance of the bridge, and took the pier height into consideration. In addition, Wu Gang [29] and Song Shuai [30] studied the effects of performance degradation and bearing sliding on bridge shear keys and piers, respectively. The notion that the support's sliding effect and the sacrificial nature of the concrete retaining block can significantly mitigate the seismic forces experienced by a bridge's substructure has been posited. However, the majority of these seismic design strategies remain confined to the conceptual phase of seismic design. They have not been subjected to rigorous numerical analysis or empirical validation through experimental outcomes. The degradation and damage of bearing performance have a great influence on the lateral seismic response of the bridge. The shear key is the last line of defense preventing the bridge from falling laterally [31–33]. However, when an actual earthquake occurs, the bearing and the shear key are interacting. Therefore, the seismic isolation system composed of suitable shear keys and plate rubber bearings [34,35] can not only limit the displacement of the superstructure but also reduce the damage of the substructure under a certain intensity earthquake, which is also the focus of this paper.

The focus of this study is the enhancement of the seismic performance of bridges by utilizing the limiting and deformation characteristics of shear keys to ensure that the bearings move within an allowable range, thereby significantly improving the bridges' capacity to withstand seismic events. Considering the compressive strength of the arch structure and the ductility of steel, a concrete shear key integrated with an arc-shaped steel plate was developed. Taking shear keys of various forms as the research objects, the limiting capacities and seismic performances of the arc-shaped shear keys were studied via shaking table tests [36]. For this purpose, a 1/12-scale, three-span continuous beam bridge and concrete blocks of different strengths were designed. The 4 m × 4 m shaking table at the South China University of Technology [37] was used to excite the bridge with real ground motion. In addition, the shaking table test was effectively simulated by ANSYS. Finally, the limiting capacity and seismic performances of bridges with different shear keys were compared.

2. Experimental Program

2.1. Engineering Background

In this paper, a typical highway girder bridge was taken as the research object. The superstructure of the bridge was a 40 m span continuous beam, and the general arrangement is shown in Figure 1. The main girder adopts a multi-box and multi-chamber girder, which was composed of 4 precast girders, with a calculated span of 40 m, a width of 11.75 m, and a height of 2.0 m. The main girder was made of C50 concrete and HRB335 steel bar, and the total weight of the single-span superstructure was about 894.6 t. The substructure was a double-column cylindrical pier, with a pile spacing of 7.0 m in the transverse direction of the bridge, a clear height of 22 m, and a diameter of 1.8 m. The tie girder between the piers was 1.4 m wide and 1.6 m high. The cap girder was 1.7 m high, 2 m wide, and 11.32 m long. The substructure adopted C30 concrete (Zhuoteda Energy, Guangzhou, China) and HRB335 steel bar (Yimin steel processing, Guangzhou, China). A typical cross-section of the main girder and pier is shown in Figure 1. The two abutments used GYZF4350 × 76 PTFE sliding-plate rubber bearings (Xinqin, China), and the other piers used GYZ450 × 114 plate rubber bearings (Chunshi, China). Table 1 shows the calculation parameters for the plate rubber bearing.

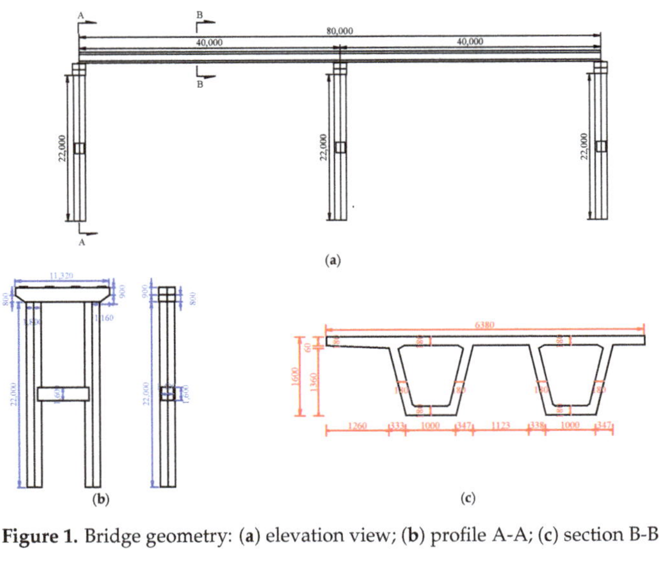

Figure 1. Bridge geometry: (**a**) elevation view; (**b**) profile A-A; (**c**) section B-B (half) (unit: mm).

Table 1. Calculation parameters for the plate rubber bearings.

Bearing Model	GYZ D450 × 114	GYZF4 D350 × 76
Dynamic shear modulus (KN/m^2)	1200	1200
Area (m^2)	0.152	0.0691
Total thickness of rubber layer (m)	0.082	0.053
Horizontal stiffness (KN/m)	2225	2056
Vertical stiffness (KN/m)	1,201,590	1,254,334

2.2. Design of the Bridge Model

Shaking table tests accurately reflect the dynamic responses of structures to earthquakes, providing an important method for assessing their seismic performances [38]. The laboratory uses a 4 m × 4 m three-dimensional, six-degrees-of-freedom earthquake simulation shaking table. The maximum load of the system is 20 t, the maximum acceleration in the X and Y directions is 10 m/s^2, the maximum acceleration in the Z direction is 20 m/s^2, and the maximum overturning moment can reach 45 ton·m. The operating frequency of the shaking table is 0.1~50 Hz, and it can input triangular waves, random waves, simple harmonic vibrations, and seismic waves, as shown in Figure 2.

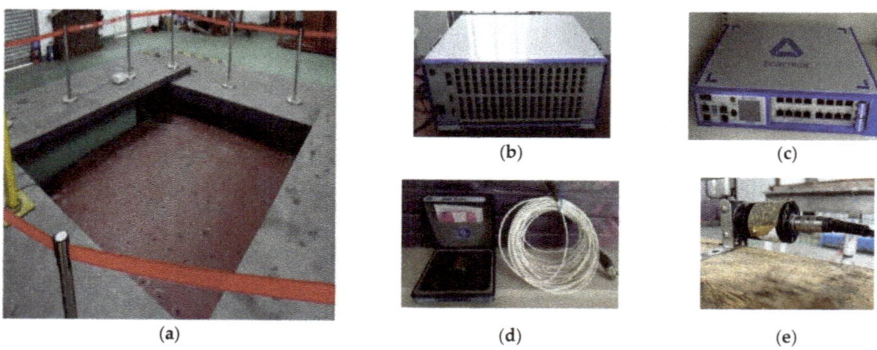

Figure 2. Schematic diagram of the shaking table system: (**a**) shaking table; (**b**) displacement data acquisition system; (**c**) accelerometer data acquisition system; (**d**) accelerometer; (**e**) displacement transducers.

According to the model similarity ratio principle [39], the complete model of the shaking table test should satisfy Equation (1).

$$S_E/(S_\rho S_a S_l) = 1 \quad (1)$$

S_E: Similarity ratio of the modulus of the elasticity;
S_ρ: Similarity ratio of the density of the material;
S_α: Similarity ratio of the acceleration;
S_l: Similarity ratio of the geometric dimensions.

Considering the size of the shaking table and the prototype size and structure of the bridge, the scale model was planned to be made according on a 1:12 scale, and the ratio S_l was 1/12. The model is made of the same material as the actual structure, and the elastic modulus (S_E), stress (S_σ), and strain (S_ε) were all taken as 1. Based on Equation (1), when the acceleration scale factor is 1, the density ratio (S_ρ) of the model needs to be 12. In actual situations, it is very slow in satisfying the mass–density relationship of the model's material. Considering that the experimental model can be added to the mass, the acceleration ratio (S_α) was 2, and the density ratio (S_ρ) was 6. By determining the basic physical quantities in Equation (1) (Table 2), a scaled model of the continuous-girder bridge was designed. The design details of the bridge model are shown in Figure 3.

Table 2. Structural similarity relationships of the bridge scale models.

Property	Physical Quantity	Similar Indicators	Ratio	Remark
Geometric	Length	S_l	1/12	Size control
	Displacement	$S_d = S_l$	1/12	
Material	Elastic modulus	S_E	1	Material control
	Mass	$S_m = S_E\, S_l^2/S_\alpha$	1/288	
	Density	$S_\rho = S_m/S_l^3$	6	
	Stress	$S_\sigma = S_E$	1	
	Strain	$S_\varepsilon = S_\sigma/S_E$	1	
Load	Point load	$S_F = S_\sigma S_l^2$	1/144	
	Line load	$S_p = S_E\, S_l$	1/12	
Dynamic	Acceleration	S_α	2	Test control
	Moment	$S_M = S_E\, S_l^3$	1/1728	
	Period	$S_T = (S_l/S_\alpha)^{1/2}$	1/4.90	
	Frequency	$S_f = (S_l/S_\alpha)^{-1/2}$	4.90	
	Power velocity	$S_v = (S_l\, S_\alpha)^{1/2}$	1/4.90	
	Friction	$S_c = S_l^{1.5}\, S_\alpha^{-1/2}$	1/58.79	

Since the main purpose of this paper was to study the limit and seismic resistance of the arc-shaped shear keys, a series of simplifications was made to the original bridge type. First, the four-box section of the main girder was simplified to a two-box section, and the pier bearing was simplified to two. In view of the fact that only the lateral action of the bridge and the continuity of the main girder were studied, the model selected a single span (3334 mm), only one pier was retained, and the superstructure span adjacent to it was cut off in the middle of the span and replaced with a sliding bearing.

The effective heights of all piers were 1833 mm, the diameters were 150 mm, and the cross-sectional sizes of the connecting girders were 117 × 133 mm. All cap beam cross-sectional dimensions were 142 × 167 mm, and the spans were 943 mm. The side piers P1 and P3 were set with 50 mm thick antifalling-girder shear keys. Twenty-four bolts with diameters of 10 mm were embedded in the middle pier, P2, to fix 4 stops, and 1 stop was installed for every 6 bolts. Two PTFE plates (300 × 100 × 48 mm) were set up in the side piers P1 and P3 to simulate sliding bearings [40–42], and two plate rubber bearings with heights of 48 mm and diameters of 100 mm were set up in the middle pier P2 for bearing purposes. The concrete used in the main girder and piers were C50 and C30 respectively. In order to ascertain the real performances of the test materials, axial compression tests were carried out on a cubic concrete test block of 150 × 150 × 150 mm. The average strength of the C30 concrete was 31.0 MPa, and the average strength of the C50 concrete was 61.8 MPa.

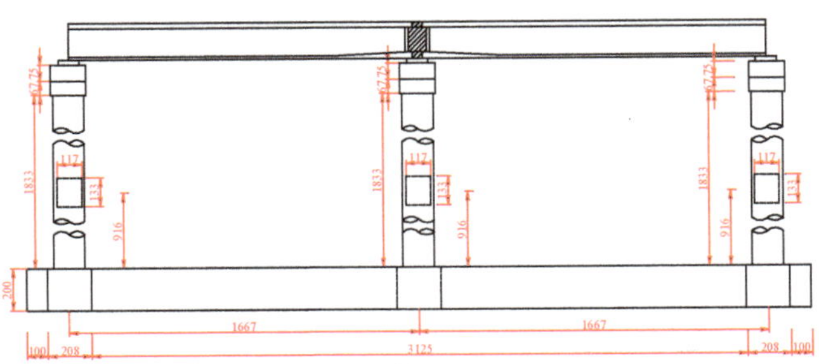

(a)

Figure 3. *Cont.*

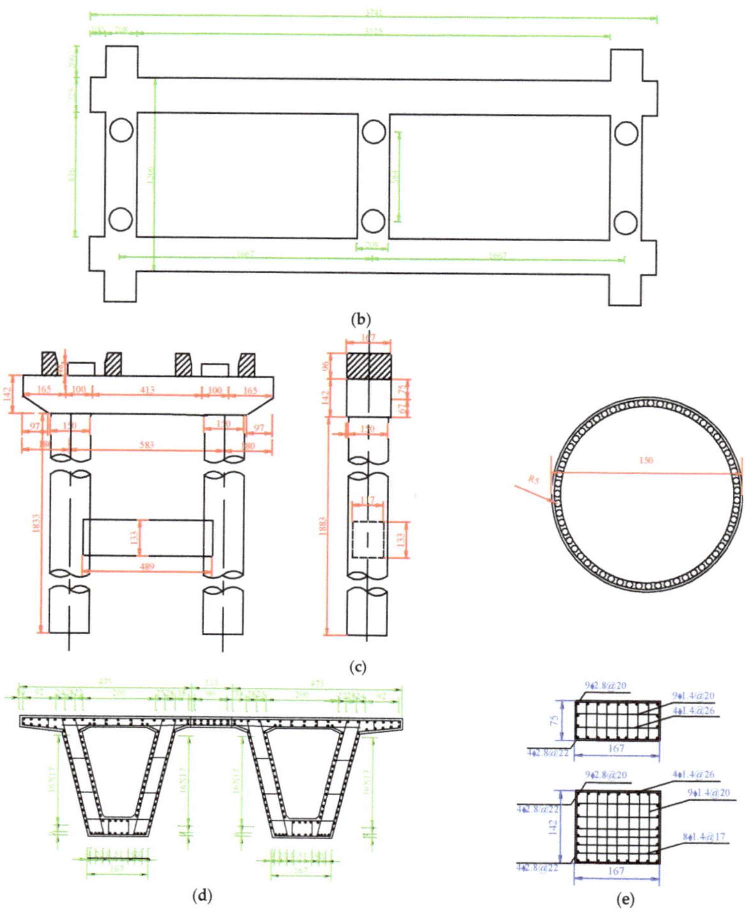

Figure 3. Design details of the bridge model: (**a**) bridge front view; (**b**) base top view; (**c**) bridge piers; (**d**) main girder; (**e**) tie girder and cap girder (unit: mm).

2.3. Design of the Shear Keys

Based on the purposes of "limitation" and "energy dissipation", a new type of arc-shaped seismic shear key is proposed. The arc-shaped shear key is composed of the following two components: an ordinary concrete shear key and an arc-shaped steel plate, whereby the latter is fabricated from structural steel, the compressive strength is 250 MPa, and the tensile strength is 460 MPa. The arc part is a hollow hemispherical structure, and the arc part and the steel plate are connected by welding. The steel plate is combined with the concrete shear key through the connecting steel bars reserved on its reverse side to form an arc-shaped seismic shear key. In addition, a concrete shear key with a rubber pad was selected as a comparison to study the performance of the arc-shaped shear key on the bridge limit and seismic resistance, and the settings of the shaking table test shear key is shown in Figure 4.

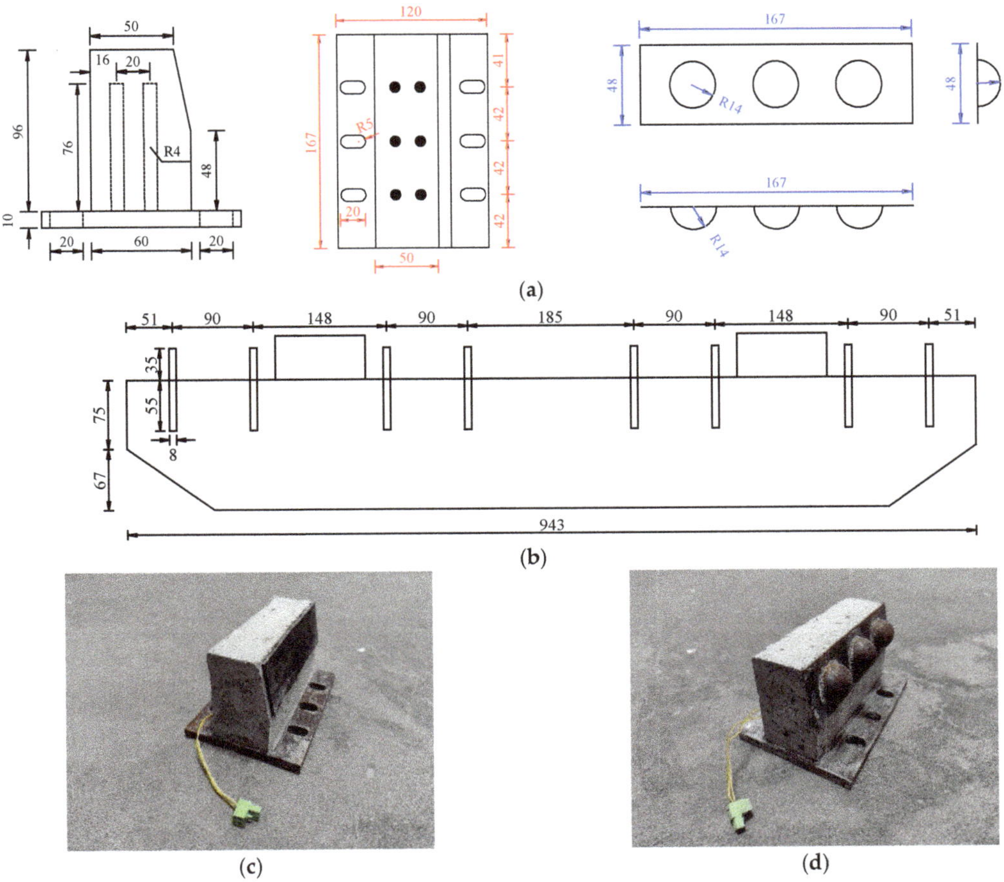

Figure 4. Design details of the shear key: (**a**) shear key size; (**b**) reserved holes for cap girder; (**c**) rubber pad shear key; (**d**) arc-shaped shear key (unit: mm).

2.4. Construction and Assembly Process

The scaled model of the bridge was cast according to the drawings. In the laboratory, the concrete surface of the strain measurement position was smoothed, and the bridge cap and pier were placed and fixed on the shaking table. After, plate rubber bearings and PTFE plates were placed on the middle pier and the side pier, and the main girder was placed on the pier for bearing leveling. When the main girder was adjusted in place, the shear key was fixed by the reserved bolts (Figure 4), and the bearing and PTFE plate were fixed with plaster and high-strength glue. The total weight of the single span of the prototype bridge's superstructure was 896.4 t. According to the mass similarity constant, S_m (Table 2), the total weight of the test model's superstructure was 3.1 t. The total weight of the superstructure after casting was 1.6 t, so 60 counterweights (25 kg each) totaling 1.5 t need to be placed on the main girder to better ensure the dynamic similarity of the bridge. The assembly of the bridge model is shown in Figure 5. In the process of installing the bridge model, the bridge substructure with the strain gauges pasted first needed to be safely hoisted onto the shaking table by means of a gantry crane. Bolts and preset holes were then used to securely connect the substructure to the shaker. After the substructure was stably positioned, PTFE slide plates and rubber bearings were installed in turn, and accurate leveling was carried out to ensure that the levelness of the structure met the experimental requirements. After the support system was stabilized, the stops were bolted to the piers to ensure the accuracy and stability of their position. Next, the hoisting work of the superstructure was carried out; in this process, it was necessary to ensure that the spacing between each stop and the main beam was consistent and, at the same time, ensure that the support system did not have any suspension phenomena so as to avoid affecting the experimental results. Finally, counterweights and sensors were installed on the bridge structure to complete the overall installation of the test model. All installation steps must be carried out in strict accordance with the requirements of the experimental design to ensure the accuracy and reliability of the experimental model.

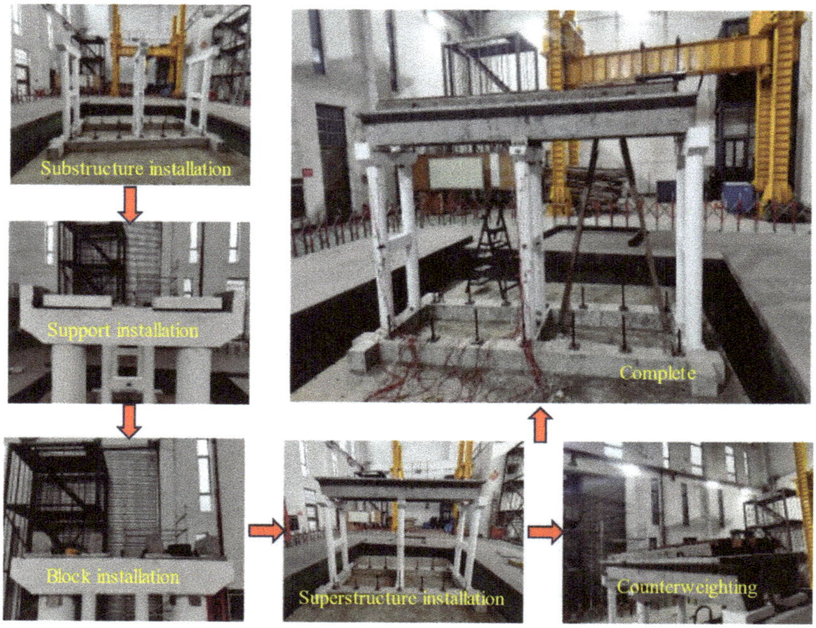

Figure 5. Installation of the bridge model on the shaking table.

2.5. Data Collection Measures

The dynamic response of the bridge model during an earthquake was measured using strain gauges, accelerometers and displacement meters [2,6,8], as shown in Figure 6. Concrete strain gauges were attached to the key parts of the piers [14,19,43] in a T-shape. The employed strain gauges were the 120-80AA model, engineered to operate effectively within a temperature range of −20 °C to 80 °C. The substrate of these gauges was composed of glass-fiber-reinforced epoxy, which offers a balance of strength and flexibility. The grid was fabricated from constantan, a high-precision alloy known for its thermal stability and low temperature coefficient of resistance. Additionally, they are characterized by a sensitivity coefficient of 2.11 ± 1%, which is critical for accurate measurement of strain under varying conditions. The accelerometers on the cap were used to measure the earthquake input of the shaking table, and the accelerometers on the main girder and piers were used to obtain the dynamic response of the bridge model during an earthquake. The employed accelerometers were precision instruments manufactured by PCB, a reputable United-States-based company. These accelerometers were designed with a robust measurement capability, spanning a range of ±50 g. They exhibit a sensitivity of 100 mV/g, indicative of their ability to provide accurate and reliable measurements even at low levels of acceleration. Furthermore, the dynamic data collection frequency of these accelerometers is versatile, accommodating frequencies from as low as 0.5 Hz up to a high-frequency range of 10,000 Hz. The displacement meters were installed on the main girder and cap girder to directly obtain the lateral displacement response of the bridge. The utilized displacement transducer is a type of rope-actuated sensor, designed with a measurement span of 0 to 2500 mm and boasting a linearity accuracy of ±0.15%, along with a repeatability specification of ±0.02%. Data acquisition and processing are facilitated by the Dewetron data logging system, an Austrian-engineered solution known for its precision and reliability.

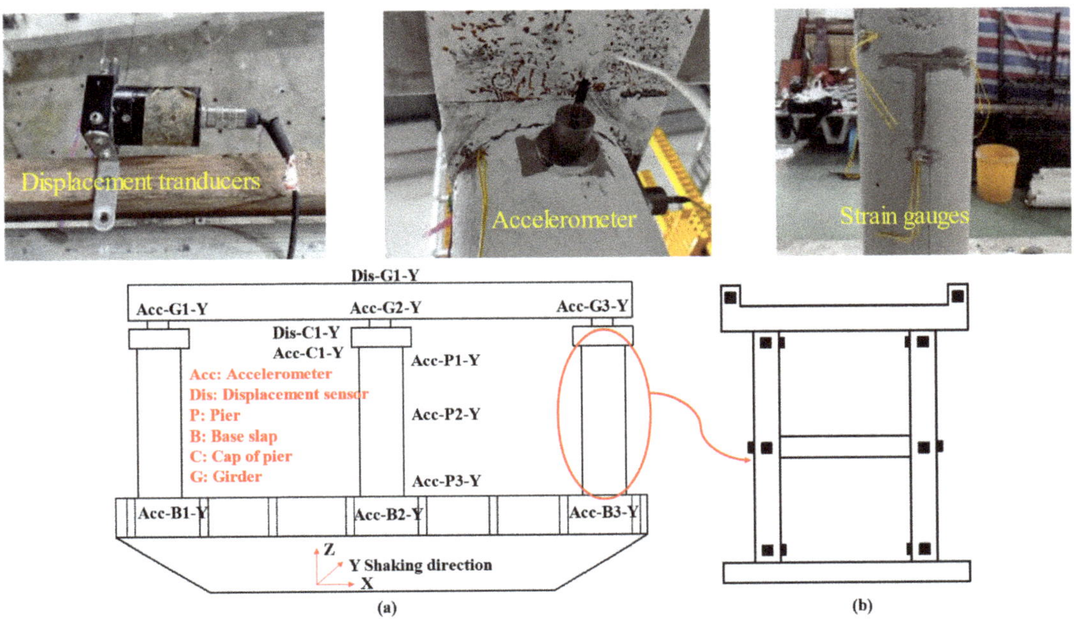

Figure 6. Arrangement of the sensors: (**a**) accelerometer and displacement transducers; (**b**) strain gauges.

For this experimental study, the prototype bridge was calibrated to a seismic fortification intensity of 8 on the intensity scale, classified under category B. It is tailored to a response spectrum with a characteristic period of 0.4 s and designed to withstand a peak ground acceleration of 0.2 g. In consultation with the Pacific Seismological Center, two appropriate natural seismic waveforms were identified, the El Centro and Cerro Prieto waves, complemented by an artificial waveform, for use as input excitations for the shaking table. These selections were meticulously chosen to align with the experimental setup's response spectrum, as illustrated in Figure 7. The El Centro wave was identified by the seismic activity triggered by the 1952 earthquake in the vicinity of El Centro, California, which occurred on 21 July. Ranking among the most severe earthquakes of the 20th century in Southern California, this event inflicted considerable damage to the local built environment. In contrast, the Cerro Prieto waves represent seismic recordings from the geothermally active region of Cerro Prieto in Mexico. This area, renowned for its geothermal energy, remains an important site for ongoing seismic surveillance and scholarly inquiry. Furthermore, the three selected seismic waves exhibit variations in both duration and peak ground acceleration (PGA), necessitating scaling and peak adjustment to meet experimental requirements. Given an acceleration similarity ratio of 2, the initial acceleration was set to 0.4 g, with subsequent adjustments to amplitudes of 0.4 g, 0.8 g, and 1.0 g, focusing primarily on the 0.4 g seismic wave in this study. The loadings were applied in the Y-direction. To standardize the experimental conditions, the duration of the loadings for the El Centro, Cerro Prieto, and artificial waves were meticulously calibrated to 10 s, 12.76 s, and 10 s, respectively, utilizing the experimental loading apparatus. Under the loading of the seismic wave with a PGA of 0.4 g, the actual peak acceleration was 0.37 g, achieving 92.5% of the target peak value, as shown in Figure 8.

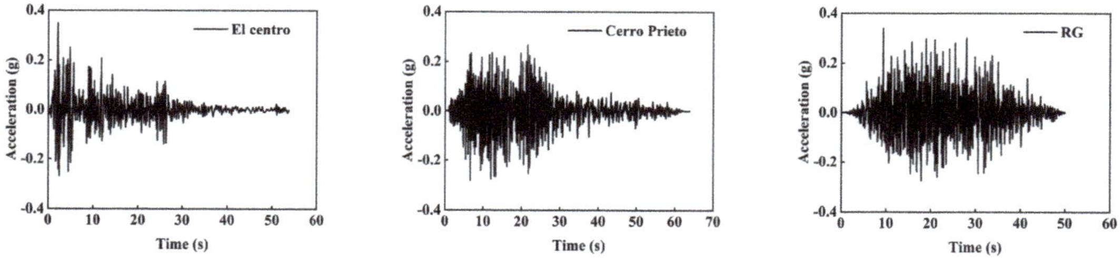

Figure 7. Original history curves of the seismic waves: (**a**) El Centro; (**b**) Cerro Prieto; (**c**) artificial wave.

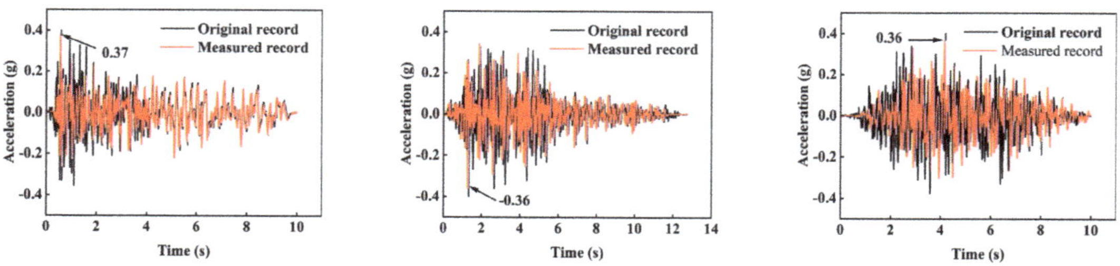

Figure 8. Shaking table seismic waves' inputs and outputs: (**a**) El Centro; (**b**) Cerro Prieto; (**c**) artificial wave.

3. Experimental Results

3.1. Experimental Phenomena

When the bridge lacked the shear key, the main girder had a large lateral displacement under the earthquake wave with a PGA of 0.4 g, but it did not collide with the antifalling-girder shear key. When the bridge was equipped with an arc-shaped shear key, the main girder collided with the stop. When the shear key was a concrete shear key with a rubber pad, the collision phenomenon of the main girder was similar to that of the arc-shaped shear key, but when the PGA was 1.0 g, obvious cracks appeared at the middle pier cap girder and tie girder, and no cracks appeared at the pier bottom, as shown in Figure 9. The cracks in the tie girder were caused by the violent seismic response of the pier in the Y direction and the excessive displacement of the pier. The cap girder was due to the fact that there were reserved bolts embedded in the cap beam during the model's development, and the collision force produced a bending and shear effect on the cap girder through the bolts, resulting in cracks. In addition, there were no cracks in either side pier. This is mainly due to the collision force of the shear key and the restoring force of the bearing in the middle pier, and the minimal friction force generated by the side pier due to the PTFE plate. There was no obvious residual deformation of the piers and bearings.

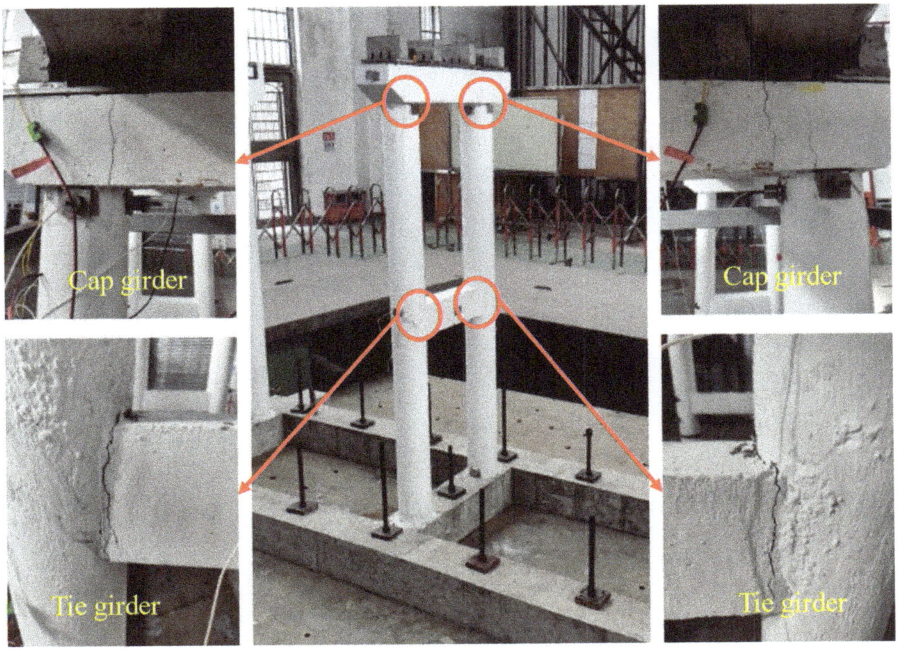

Figure 9. Crack in the bridge pier.

3.2. Structural Displacement Response

The displacement time history of the main girder is shown in Figure 10. The different settings for the forms of the shear keys are represented by A1, A2, and A3, where A1 is no shear key, A2 is the concrete shear key with a rubber pad, and A3 is the arc-shaped shear key. Furthermore, the seismic responses under the influence of the El Centro wave, Cerro Prieto wave, and artificial wave are represented by El, CP, and RG, respectively.

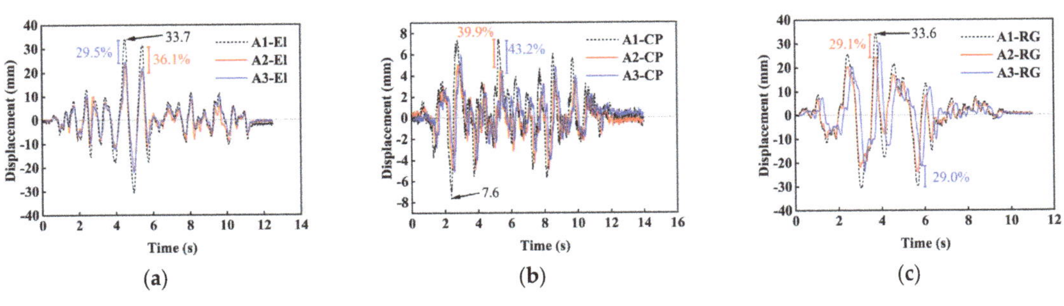

Figure 10. Displacement time history of the main girder in the Y direction: (**a**) El Centro; (**b**) Cerro Prieto; (**c**) artificial wave.

The displacement of the main girder shows the same trend under the action of the same seismic wave. For the rubber pad shear key, the reduction rate of the main girder displacement can reach 36.1%, 39.9%, and 29.1% in the case of El Centro, Cerro Prieto, and artificial wave. For the arc-shaped shear key, the reduction rate of the main girder displacement can reach 29.5%, 43.2%, and 29.0% in the case of El Centro, Cerro Prieto, and artificial wave. This shows that the shear key can significantly reduce the displacement of the main girder under earthquake action. In addition, the maximum displacement of the main girder of A3 was slightly larger than that of A2. This is mainly because of two aspects. First, the displacement of the main girder without a shear key was much larger than that with a shear key, which means that the settings for the shear key limits the lateral displacement of the main girder and reduces the displacement response of the main girder. In addition, in some cases, the maximum displacement of the main girder of A3 was greater than that of A2. This difference may be that the stiffness of the arc-shaped shear key was relatively small, and the deformation of the arc-shaped shear key during collision was more obvious, resulting in an increase in the maximum displacement of the main girder. The residual displacement of the main girder with different shear keys ranged from 0.2 mm to 2.1 mm, indicating that the main girder did not return to the initial position in the end.

The displacement time history of the cap girder is shown in Figure 11. The displacement time history curves of the cap girder and the main girder were similar, especially the time at which the displacement peak occurred, which was basically the same. The maximum displacement of the main girder was greater than the displacement of the cap girder, which also verifies the reliability of the collected data. However, the displacement of the cap girder without shear keys was much smaller than that of the cap girder with shear keys. Because of the limiting effect of the shear key, the main girder collides with the shear keys, and the collision force was transmitted to the cap girder through the shear key. Under the action of the lateral force, the seismic response of the pier intensifies, resulting in an increases in the displacement of the A2 and A3 cap girders. Furthermore, the El Centro wave induced peak displacements of 33.7 mm for the main girder and 25.7 mm for the cap girder. In contrast, the Cerro Prieto wave resulted in significantly lower peak displacements, at 7.6 mm for the main girder and 6.3 mm for the cap girder. Similarly, the artificial wave yielded peak displacements of 33.6 mm for the main girder and 29.9 mm for the cap girder. This shows that the displacement responses caused by different seismic waves are different, and the frequency of the scaled model is close to the frequency of the El Centro and the artificial wave, and is quite different from the frequency of the Cerro Prieto.

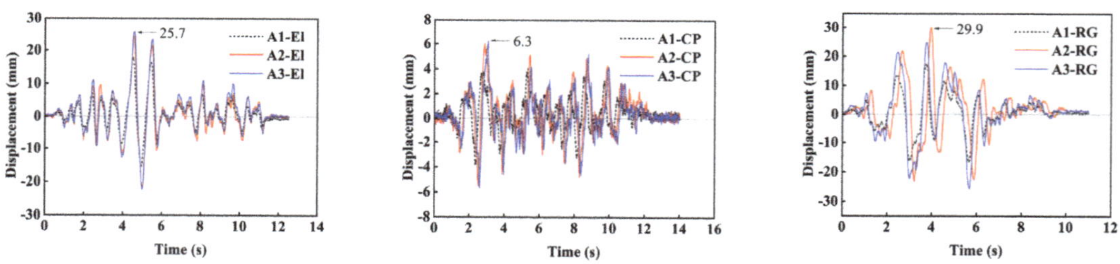

(a) (b) (c)

Figure 11. Displacement time history of the cap girder in the Y direction: (**a**) El Centro; (**b**) Cerro Prieto; (**c**) artificial wave.

3.3. Acceleration Response

Under the action of Cerro Prieto, the acceleration time history of the base, main girder, and cover girder in the Y direction of the different shear keys are shown in Figure 12. The trend in the cap time history curve was similar to that of the seismic waves (Figure 7b). Under the action of the bearing's restoring force and friction force, the acceleration time history curve trends in the main girder and the cover girder were similar for all time ranges. After adding the shear key, the acceleration time history trend did not change, but the peak value changed. For analytical convenience, an acceleration limit criterion of 0.1 g was provisionally adopted. The numbers of times the main girder accelerations of A1, A2, and A3 exceeds 0.1 g is 24, 54 and 26 respectively, and the number of times the cover girder acceleration exceeded 0.1 g were 0, 9, and 3, respectively. It can be found that due to the setting of the shear key, the main girder collided with the shear key, and the acceleration response of the main girder and the cover girder increased. The acceleration time course of the concrete shear key with a rubber pad significantly increased. This is due to the greater stiffness of the concrete shear key. Although the rubber pad can absorb shock, it does not produce large deformation in the impact direction. Therefore, the collision force is still transmitted to the shear key through the rubber pad, increasing the acceleration response of the substructure. Conversely, the arc-shaped stop girder exhibits a reduction in peak acceleration magnitude. This is due to the hollow design of the arc-shaped shear key, and the collision effect between the main girder, and the shear key is weakened through the deformation of the shear key, indicating that the arc-shaped shear key had a buffering effect.

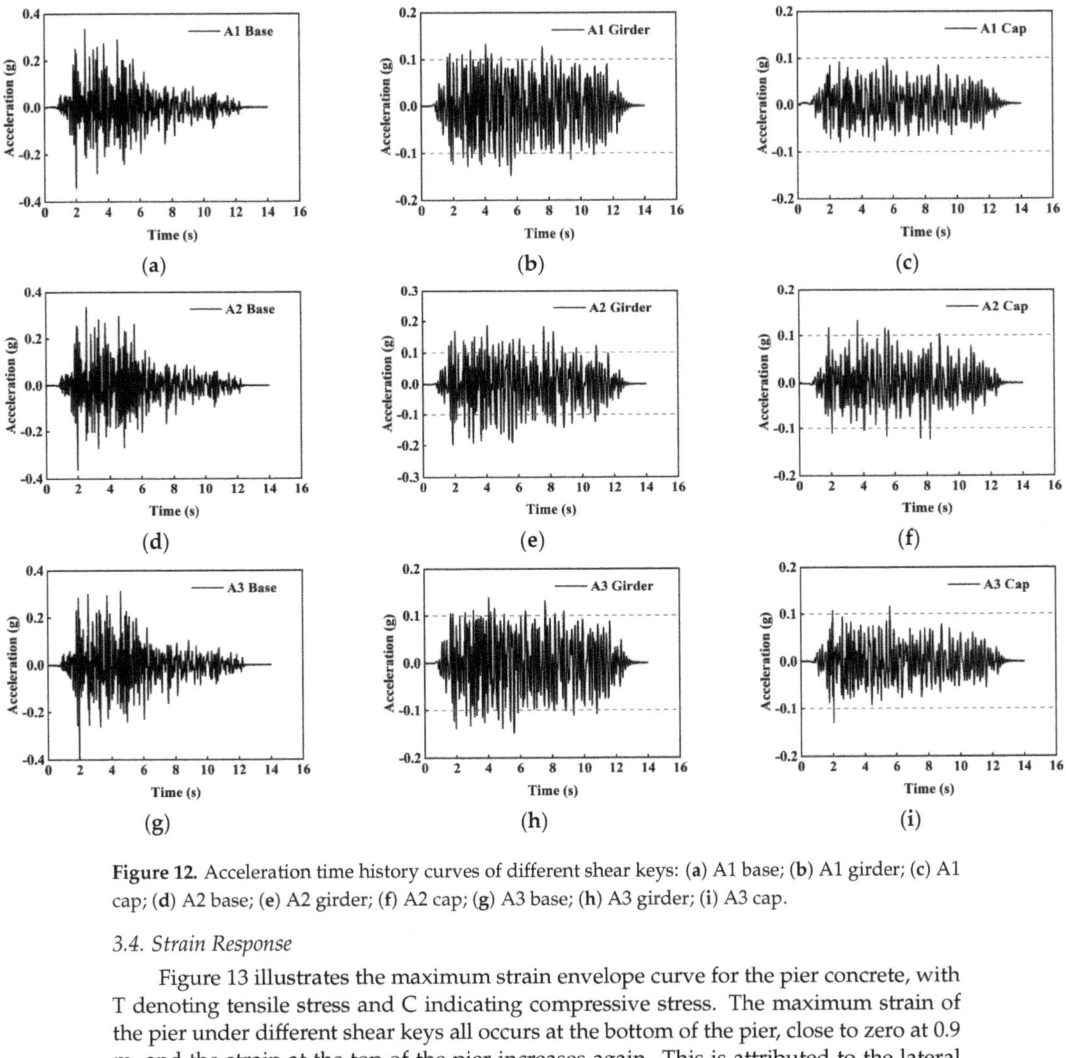

Figure 12. Acceleration time history curves of different shear keys: (**a**) A1 base; (**b**) A1 girder; (**c**) A1 cap; (**d**) A2 base; (**e**) A2 girder; (**f**) A2 cap; (**g**) A3 base; (**h**) A3 girder; (**i**) A3 cap.

3.4. Strain Response

Figure 13 illustrates the maximum strain envelope curve for the pier concrete, with T denoting tensile stress and C indicating compressive stress. The maximum strain of the pier under different shear keys all occurs at the bottom of the pier, close to zero at 0.9 m, and the strain at the top of the pier increases again. This is attributed to the lateral connection between the double-row piers, with the tie girder and pier forming a node, making it difficult for large lateral displacement differences to occur in the middle of the pier, so the strain is close to zero. The strain time history curve of the concrete at the bottom of the pier is shown in Figure 14. The maximum strains of the concrete for the cases of A1 and A3 are 313 με and 371 με, respectively. In the case of A2, the maximum strain of the concrete is 578 με, which is 1.8 times that of A1 and 1.5 times that of A3. This is because the rubber buffer capacity is limited, and the collision force is transmitted to the bottom of the pier to a great extent, causing the concrete strain to increase. The contours and extreme values of the concrete strain time history curves of A1 and A3 are similar, but the concrete strain of A3 is obviously greater than that of A1 in the nonextreme case. This is mainly due to the collision between the arc-shaped shear key and the main girder, which causes the arc-shaped shear key to deform. That is, the arc-shaped shear key weakens the collision force by deformation, and a small part of the force is transmitted to the bottom of the pier, so the overall force increases but the peak value does not increase too much. This shows that the arc-shaped shear key can protect the substructure of the bridge.

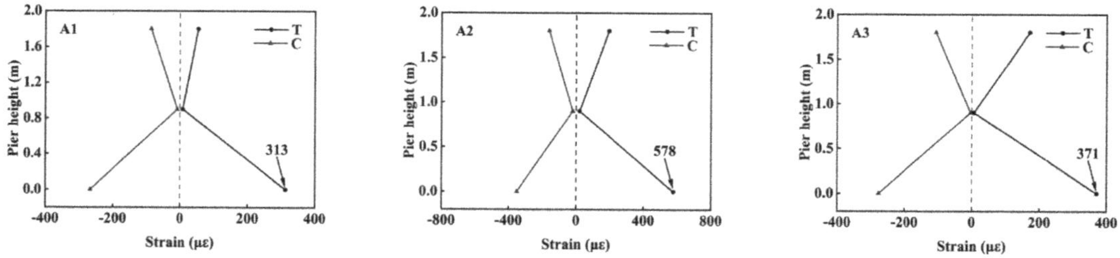

(a) (b) (c)

Figure 13. Strain envelope of the bridge pier: (**a**) A1; (**b**) A2; (**c**) A3.

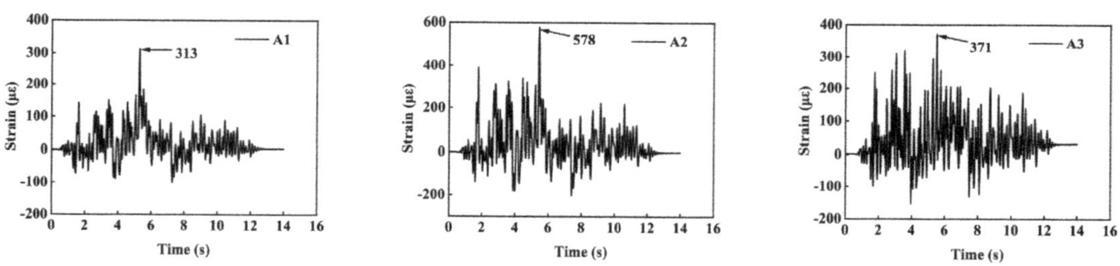

(a) (b) (c)

Figure 14. Strain time history of the concrete at the bottom of the pier: (**a**) A1; (**b**) A2; (**c**) A3.

4. Numerical Simulation

4.1. Finite Element Model of Bridge

According to the results of the shaking table test, the arc-shaped shear key has good limiting and shock-absorbing performances. In order to further understand the seismic performance of the arc-shaped shear key, a reasonable numerical simulation method is necessary. In this section, a three-dimensional model of a scaled bridge is established by finite element software [44,45], and the seismic performance of the shear key is verified by comparison with the test results, as shown in Figure 15. In this study, an isotropic elastic model was used to simulate the material properties, which determined the constitutive relationship based on the elastic modulus and Poisson's ratio of the material. During the simulation, we assumed that the bridge structure did not crack during loading and that all material behavior is in the elastic phase, i.e., the stress–strain response of the structure follows a linear relationship and is able to fully return to its original state after unloading. The main girder was made of C50 concrete with an elastic modulus of 35,000 MPa, a density of 2500 kg/m^3, and a Poisson's ratio of 0.18. The substructure was made of C30 concrete with an elastic modulus of 30,000 MPa, a density of 2300 kg/m^3, and a Poisson's ratio of 0.18. The sphere of the arc-shaped shear key was made of Q345 steel with a yield strength of 345 MPa, a density of 7.89 g/cm^3, and a Poisson's ratio of 0.3. The arc-shaped shear key is simulated by shell elements. In addition, other structures of the bridge are simulated by three-dimensional solid elements. The mesh division is mainly based on hexahedral elements, and is refined in important parts such as bearings and shear keys. The contact between the bearing and the cap girder and the main girder was binding contact. The shear key as in binding contact with the cap girder and in friction contact with the main girder. The dynamic coefficient was 0.1. The bottom of the shaking table moved in the Y direction. Considering the accuracy of the calculation, the bottom of the model was fixed by remote displacement, as follows: the displacement in the Y direction was free, the displacements in the X and Z directions were 0, and the rotation angles in the X, Y, and Z directions were 0. The acceleration was based on the seismic wave used under the experimental conditions, and the counterweight was applied by the load.

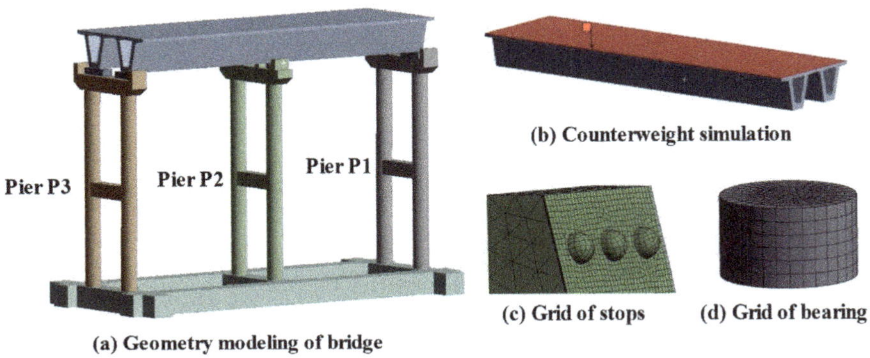

Figure 15. Scaled model finite element model.

4.2. Numerical Model Validation

A comparison of the displacements of the main girder between the test results and the numerical results is shown in Figure 16. It can be seen that under different seismic waves and shear keys, the finite element results of the main girder were consistent with the test results. However, when the duration exceeded 8 s, the fluctuation amplitude and frequency of the finite element results gradually weakened. This difference is due to the fact that there will be a small amount of energy loss in the numerical simulation, and the energy input by the seismic wave cannot be fully applied to the model, so the displacement time history of the main girder of the finite element will gradually weaken. In addition, there is a certain deviation in the peak displacement of the main girder, and the displacement of the main girder in the test was large. The reason is that, at the beginning of the test plan's formulation, the best efforts were made to follow the principle of similarity theory to accurately scale the prototype bridge. However, due to the limited size of the shaking table, the choice of available model materials was limited, resulting in a certain degree of quality defects in the actual bridge model. In the dimension of time, the experimental data and simulated data exhibited a slight temporal deviation. The output frequency of the simulation results had a significant impact on the progress of the model calculations. To ensure the accuracy of the results while achieving efficient computation, setting the output frequency of the results to 50 Hz. The collection frequency of the experimental data was 200 Hz, which is much higher than the simulation output frequency, leading to a temporal deviation between the experimental and simulated data.

In the comparative analysis of the experimental results with the finite element simulation outcomes, the following peculiar phenomenon was observed: at between 8 and 10 s in the artificial wave response, a divergence in the trend between the measured and simulated data was identified. Specifically, within this time frame, the experimental data indicated an increasing trend in the displacement, while the finite element results demonstrated a decreasing trend. This discrepancy suggests that there may be a discrepancy between the material response assumed in the simulation and the actual material response of the model. Furthermore, during the simulation process, the counterweights applied in the experiment were emulated in the finite element model by applying equivalent forces, resulting in an incomplete representation of the inertial effects of certain structural components. The combined effects of variations in material properties, incomplete consideration of inertial forces, and the substantial energy of the artificially input waves have led to a deviation in the trends between the experimental observations and the finite element simulation results. In addition, there are some factors that cannot be ignored in the test, such as the inaccurate installation of the shear key and the main girder, and the uneven friction performance of the rubber bearing. Nonetheless, the experimental data and the simulated data reveal a general agreement in the overall response trend, and the correlation between the two shows a high degree of agreement. This verifies the correctness and validity of the experimental data and numerical simulation results and, thus, confirms the reliability of the simulation model and the accuracy of the experimental method.

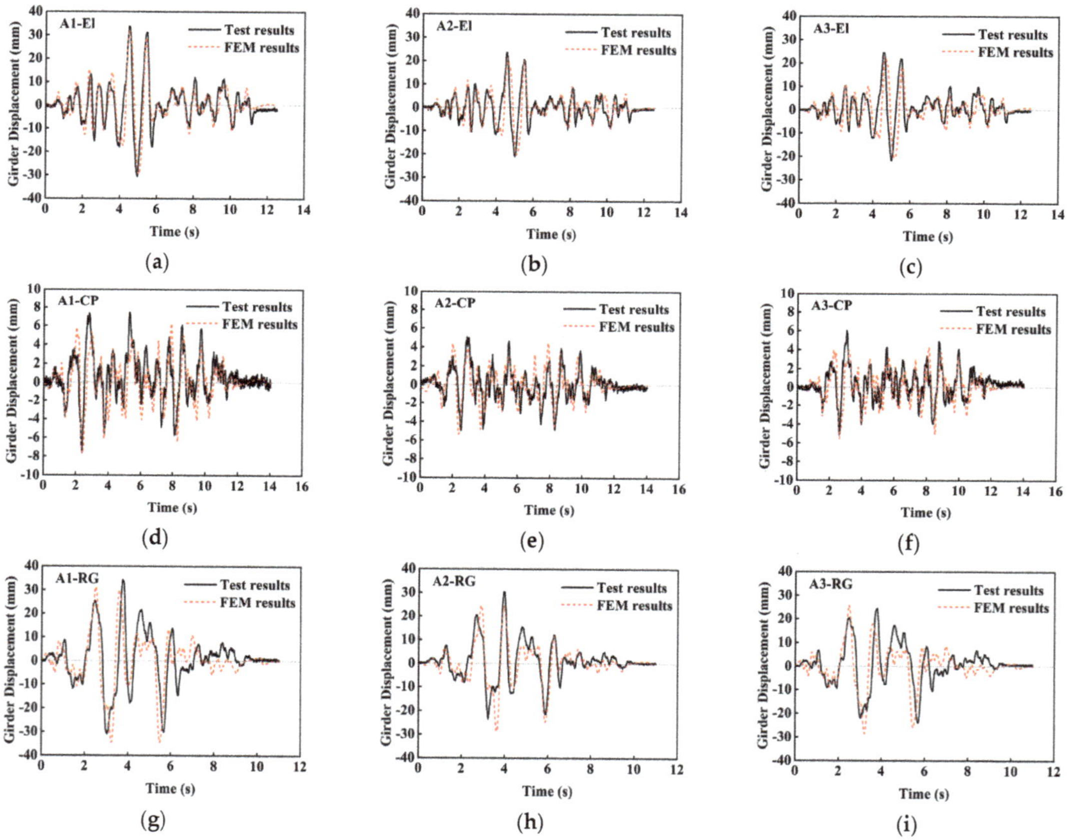

Figure 16. Test and FEM results in main girder displacements: (**a**) A1-El; (**b**) A2-El; (**c**) A3-El; (**d**) A1-CP; (**e**) A2-CP; (**f**) A3-CP; (**g**) A1-RG; (**h**) A2-RG; (**i**) A3-RG.

4.3. Comparison of Different Shear Keys

During the shaking table test, the data acquisition equipment was limited, and the relative displacement of the upper and lower surfaces of the bearing was difficult to measure. The scaled model was restored by finite element analysis, and the relative displacement time history of the bearing in the Y direction was obtained, as shown in Figure 17. According to the "Highway Bridge Seismic Design Code" (JTG/T 2231-01-2020) [43], the lateral allowable displacement of the actual bridge bearing (Table 1) is 60 mm. According to the scale ratio coefficient, the lateral allowable displacement of the scaled model bearing is

5 mm. As can be seen from Figure 17, under the action of El Centro, the peak displacement of the bearing without a shear key was 6.4 mm, and the peak displacements of the bearings with rubber pad shear key and arc-shaped shear key were 2.94 mm and 2.96 mm, which were reduced by 54% and 53.7%. Under the action of Cerro Prieto, the peak displacement of the bearing without a shear key was 6.31 mm, and the peak displacements of the bearings with rubber pad shear keys and arc-shaped shear keys were 1.28 mm and 1.55 mm, which were reduced by 79.7% and 75.4%. Under the action of artificial waves, the displacement peak value of the bearing without a shear key was 3.54 mm, which meets the requirements of the specification. The displacement's peak values for the bearing with a rubber pad shear key and the arc-shaped shear key were 1.27 mm and 1.55 mm, which were reduced by 64.1% and 56.2%. By comparison, it can be found that the shear key can significantly reduce the lateral displacement of the bearing under the action of earthquake, avoiding damage to the bearing due to excessive displacement.

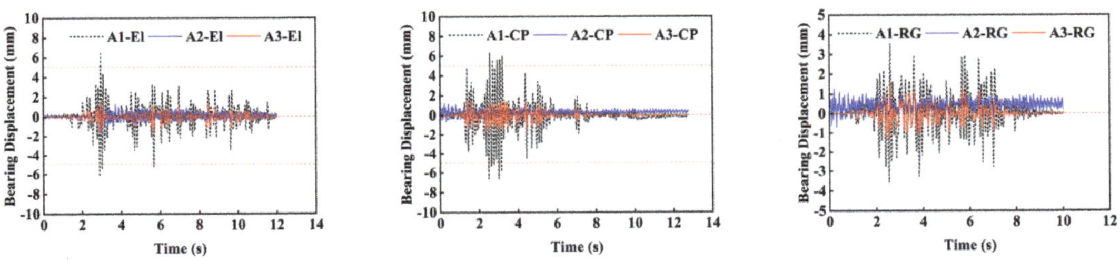

(a) (b) (c)

Figure 17. Relative displacement time history of the bearings in the Y direction: (**a**) El Centro; (**b**) Cerro Prieto; (**c**) artificial wave.

5. Conclusions

In this paper, shaking table tests were carried out on a typical highway continuous-beam bridge according on a scale of 1/12, and a seismic excitation of 0.4 g was carried out on the bridge equipped with different blocks under different seismic waves. Its main girder displacement, acceleration, cap girder displacement, and pier bottom strain were recorded and analyzed. In addition, the finite element model of the shaking table test was established, the excitation effect of the shaking table on the bridge was simulated and analyzed. The main conclusions are as follows:

(1) When PGA = 0.4 g, the main girder and the shear key produced obvious collision. Because of the interaction between the shear key and the main girder, the rubber pad shear key and the arc-shaped shear key both reduced the main girder displacement. In the cases of the El Centro wave, Cerro Prieto wave, and artificial wave, the main girder displacement reduction rates of the rubber pad shear key reached 36.1%, 39.9%, and 29.1%, and the main girder's displacement reduction rate for the arc-shaped shear key reached 29.5%, 43.2%, and 29.0%.

(2) Under the action of the collision force between the shear key and the main girder, the displacements of the cap girder with the rubber shear key and the arc-shaped shear key were greater than that of the cap girder without the shear key. Compared to without the stopper, the displacement of the cap girder with the rubber shear key was 174.8% of the former, and the arc-shaped shear key was 168.9% of the former. Therefore, when the shear key is used for bridge seismic resistance, the substructure's design needs to be carried out.

(3) Because of the deformation of the arc-shaped shear key, the acceleration responses of the main girder and the cap girder were similar to that without the stopper. However, compared with the rubber pad shear key, the acceleration response of the main girder was reduced by 27.8%, and the acceleration response of the cap girder was reduced by 15.4%. This shows that the arc-shaped shear key had better seismic performance than the traditional shear key.

(4) Under different shear keys, the maximum strain of the pier bottom concrete is different. The maximum strains of the pier bottom with the rubber pad shear key was 578 με, which is 1.8 times that without the stopper and 1.5 times that of the arc-shaped shear key. The maximum strains of the pier bottom of the arc-shaped shear key and without the stopper were close, being 1.18 times that of without the stopper. The increase in the concrete strain at the pier bottom was caused by the interaction between the shear key and the main girder. Because of the collision between the shear key and the main girder, the seismic response of the substructure was amplified, which manifested as an increase in the concrete strain at the pier bottom.

(5) The finite element method can correctly simulate the dynamic response of the highway continuous girder, and the displacement time history of the main girder, pier and bearing can be well reproduced through model calculation. Under the action of the 0.4 g seismic wave, due to the limiting capacity of the shear key, the relative displacement of the bearing was less than the displacement of the bearing without a stopper, and the reduction rate of the bearing's displacement reached more than 50%. The goal of protecting the bearings can, therefore, be achieved under earthquake action.

This paper observed interesting phenomena and obtained useful insights into the shear key limit and seismic performances of highway continuous-girder bridges. Nevertheless, the influence of different shear keys' design parameters on the bridge limit and seismic performance still needs further experiments for in-depth research. In addition, the impact of the collision between the main girder and the shear key also needs to be simulated and verified again at the energy level.

Author Contributions: Conceptualization, L.H.; methodology, W.W.; software, W.W.; investigation, W.W. and Z.L.; data curation, W.W.; writing—original draft preparation, W.W.; writing—review and editing, L.H.; supervision, L.H. and B.W.; funding acquisition, L.H. and S.X. All authors have read and agreed to the published version of the manuscript.

Funding: This research was funded by the National Natural Science Fund Committee of China (52479137).

Data Availability Statement: The original contributions presented in the study are included in the article, further inquiries can be directed to the corresponding author.

Conflicts of Interest: The authors declare no conflict of interest.

References

1. Ministry of Transport of the People's Republic of China. Statistical Bulletin on the Development of Transportation Industry in 2021. Available online: http://wap.china-railway.com.cn/crcwapEnglish/news_1293/202206/t20220607_121991.html (accessed on 25 May 2024).
2. Su, L.; Zhang, W.; Chen, Y.; Zhang, C. Seismic Performance and Parametric Study of a Winding Rope Fluid Viscous Damper (WRFVD) for Continuous Girder Bridges. *J. Earthq. Eng.* **2023**, *27*, 4390–4414. [CrossRef]
3. Xu, B.; Sun, C.; Song, S.; Zhang, X.; Zhao, B.; Zhang, W. Comparative Study on the Seismic Vulnerability of Continuous Bridges with Steel–Concrete Composite Girder and Reinforced Concrete Girder. *Buildings* **2024**, *14*, 1768. [CrossRef]
4. Song, S.; Tian, Z.; Zhao, Z.; Li, X.; Zhao, J.; Xu, B. Experimental Study on Seismic Performance of CFRP-Strengthened Recycled Concrete Columns with Different Levels of Seismic Damage. *Buildings* **2023**, *13*, 1470. [CrossRef]
5. Peng, S.; Xiong, Z. Experimental study on the seismic behavior of CFRP-strengthened seismic-damaged recycled aggregate concrete-filled rectangular steel tube frame columns. *J. Earthq. Eng.* **2022**, *45*, 103422. [CrossRef]
6. Chen, L. *Investigation of Highway Disasters in Wenchuan Earthquake-Bridge*; People's Communications Press: Beijing, China, 2012.
7. Li, J.; Peng, T.; Xu, Y. Damage investigation of girder bridges under the Wenchuan earthquake and corresponding seismic design recommendations. *Earthq. Eng. Eng. Vib.* **2008**, *7*, 337–344. [CrossRef]
8. Zhuang, W.; Liu, Z.; Jiang, J. Earthquake-induced damage analysis of highway bridges in wenchuan earthquake and countermeasures. *Chin. J. Rock Mech. Eng.* **2009**, *28*, 1377–1387.
9. Liu, Y.; Dang, K.; Dong, J. Finite element analysis of the aseismicity of a large aqueduct. *Soil Dyn. Earthq. Eng.* **2017**, *94*, 102–108. [CrossRef]
10. Meng, D.; Hu, S.; Yang, M.; Hu, R. Experimental and numerical study on the consequence of seismic-induced transversal poundings for a simply-bearinged bridge. *Structures* **2023**, *48*, 91–107. [CrossRef]

11. Peng, T.; Yu, X.; Wang, Z.; Han, L. Study of the seismic performance of expansion double spherical seismic isolation bearings for continuous girder bridges. *Earthq. Eng. Eng. Vib.* **2012**, *11*, 163–172. [CrossRef]
12. Barone, S.; Calvi, G.; Pavese, A. Experimental dynamic response of spherical friction-based isolation devices. *J. Earthq. Eng.* **2019**, *23*, 1465–1484. [CrossRef]
13. Xiang, N.; Li, J. Experimental and numerical study on seismic sliding mechanism of laminated-rubber bearings. *Eng. Struct.* **2017**, *141*, 159–174. [CrossRef]
14. Yue, K.; Xu, L.; Liu, J.; Fan, L.; Xu, L.H. Seismic performance of an energy dissipating shear key for highway bridges using butterfly-shaped steel plates. *Eng. Struct.* **2023**, *295*, 116885. [CrossRef]
15. Shahbazi, P.; Taghikhany, T. Sensitivity analysis of variable curvature friction pendulum isolator under near-fault ground motions. *Smart Struct. Syst.* **2017**, *20*, 23–33.
16. Liu, Y.H.; Tan, P.; Zhou, F.L.; Wenliuhan, H.; Mercan, O. Shaking table test for seismic simulation of continuous isolation girder bridge with high piers. *China J. Highw. Transp.* **2015**, *28*, 68–94.
17. Chen, Y.; Sun, H.; Feng, Z. Study on seismic isolation of long span double deck steel truss continuous girder bridge. *Appl. Sci.* **2022**, *12*, 2567. [CrossRef]
18. Wu, D. Research on seismic performance of new bridge pier seismic reduction isolation system based on shaking table test. *J. Vib. Eng.* **2022**, *24*, 1471–1485. [CrossRef]
19. Qi, Q.; Shao, C.; Cui, H.; Huang, H.; Wei, W.; Wang, C.; Zhuang, W. Shaking table tests and numerical studies on the seismic behaviors of FPB in railway continuous girder bridges. *Eng. Struct.* **2023**, *290*, 116318. [CrossRef]
20. Wei, B.; Tan, H.; Jiang, L.; Xiao, B.; Lu, A. Shaking table test of the seismic performance for railway simply-bearinged girder bridge isolated by self-centering bearing. *Eng. Struct.* **2024**, *300*, 117249. [CrossRef]
21. Zhuang, P.; Zhao, W.; Zhang, G.; Yang, T. Shaking table tests on seismic behaviors of a single-layer spherical lattice shell with shape memory alloy-controlled friction pendulum bearings. *J. Build. Eng.* **2024**, *82*, 108150. [CrossRef]
22. Bozorgzadeh, A.; Megally, S.; Restrepo, J.I.; Ashford, S.A. Capacity evaluation of exterior sacrificial shear keys of bridge abutments. *J. Bridg. Eng.* **2006**, *11*, 555–565. [CrossRef]
23. Silva, P.F.; Megally, S.; Seible, F. Seismic performance of sacrificial exterior shear keys in bridge abutments. *Earthq. Spectra* **2009**, *25*, 643–664. [CrossRef]
24. Li, J.; Tang, H. Study on transverse seismic design of small and mediumspan bridges with elastomeric bearing pads. *J. China Civ. Eng. J.* **2016**, *49*, 69–78.
25. Liu, X.; Li, J.; Chen, X. Effects of X-shaped elastic-plastic steel shear keys on transverse seismic responses of a simply-bearinged girder bridge. *J. Vib. Shock.* **2015**, *34*, 143–149.
26. Tang, Z. Experimental Study and Calculation Method on Bridge Seismic Shear Keys. Master's Thesis, Chongqing Jiaotong University, Chongqing, China, 2015.
27. Huang, X.G. Experimental and Theoretical Research on Anti-Falling Device of Continuous Girder Bridge. Doctoral Dissertation, Tongji University, Shanghai, China, 2009.
28. Li, Z. Study on the Effect of Side Retainers on the Seismic Isolation Effect of Continuous Girder Bridge with Rubber Bearings. Master's Thesis, Chongqing Jiaotong University, Chongqing, China, 2018.
29. Wu, G.; Wang, K.; Wang, Q.; Zhang, P. Transverse seismic response analysis for bridges considering performance degradation of bearings and shear keys. *J. Vib. Shock.* **2018**, *37*, 189–196.
30. Song, S.; Wang, S.; Wu, G. Study on seismic vulnerability of small and medium-span girder bridges considering bearing slip and shear key failure. *Technol. Earthq. Disaster Prev.* **2019**, *14*, 781–789.
31. Deng, Y.; Ge, S.; Ge, X. Mitigating Seismic Irregularity of Continuous Girder Bridge with Unequal Height Piers Through Differential Design of Shear Key Strength. *Arab. J. Sci. Eng.* **2023**, *48*, 13749–13769. [CrossRef]
32. Özşahin, E.; Pekcan, G. Inelastic seismic response of box-girder bridges due to torsional ground motions. *Eng. Struct.* **2020**, *218*, 110831. [CrossRef]
33. Zhang, Y.; Shi, Y.; Liu, D. Seismic Effectiveness of Multiple Seismic Measures on a Continuous Girder Bridge. *Appl. Sci.* **2020**, *10*, 624. [CrossRef]
34. Chen, W.; Bai, X.; Xu, T.; Ke, S.; Deng, K.; Xie, H. Damage control of a twin-column pier with a replaceable steel shear link in a cap beam under transverse seismic motion. *Adv. Bridge Eng.* **2021**, *2*, 11. [CrossRef]
35. Long, X.; Zhou, Q.; Ma, Y.; Gui, S.; Lu, C. Displacement-based seismic design of SMA cable-restrained sliding lead rubber bearing for isolated continuous girder bridges. *Eng. Struct.* **2024**, *300*, 117179. [CrossRef]
36. Li, H.; Xie, Y.; Gu, Y.; Tian, S.; Yuan, W.; DesRoches, R. Shake table tests of highway bridges installed with unbonded steel mesh reinforced rubber bearings. *Eng. Struct.* **2020**, *206*, 110124. [CrossRef]
37. He, J.; Chen, J.; Ren, X.; Li, J. A shake table test study of reinforced concrete shear wall model structures exhibiting strong non-linear behaviors. *Eng. Struct.* **2020**, *212*, 1104811. [CrossRef]
38. Shao, C.; Qi, Q.; Wei, W.; Hu, C.; Wang, M.; Xiao, L.; Xiao, Z. Shaking table test of seismic performance of a simply-bearinged girder bridge with single column. *J. Vib. Shock.* **2019**, *38*, 49–55.
39. Zhou, Y.; Lv, X. *Method and Technology for Shaking Table Model Test of Building Structures*; Science Press: Beijing, China, 2016.
40. Constantinou, M.; Mokha, A.; Reinhorn, A. Teflon bearing in base isolation II, Modeling. *J. Struct. Eng.* **1990**, *116*, 455–474. [CrossRef]

41. Mokha, A.; Constantinou, M.; Reinhorn, A.; Zayas, V.A. Experimental study of friction-pendulum isolation system. *J. Struct. Eng.* **1991**, *117*, 1201–1217. [CrossRef]
42. Tsopelas, P.; Constantinou, M.; Kim, Y. Experimental study of FPS system in bridge seismic isolation. *Earthq. Eng. Struct. Dyn.* **1996**, *25*, 65–78. [CrossRef]
43. *JTG/T 2231-01-2020*; Specifications for Seismic Design of Highway Bridges. Ministry of Transport of the People's Republic of China: Beijing, China, 2020.
44. Yan, X.-Y.; Zhao, Z.; Cao, S.-S.; Zhang, Y.-F.; Liu, C.-H. Seismic Response of the Continuous Rigid-Framed Bridge with Super-High Piers Based on Shaking Table Tests. *Buildings* **2024**, *14*, 1527. [CrossRef]
45. Guan, Z.; Li, J.; Qu, H. Shake table test and numerical study on a capable and resilient lateral seismic isolation system for long-span cable-stayed bridges. *Soil. Dyn. Earthq. Eng.* **2023**, *164*, 107629. [CrossRef]

buildings

MDPI

Article

Contribution of Torsional Vibration Modes and the Influence on Period Ratios in the Seismic Response of Elastic Plate Bent Frame Structures

Yulong Li [1], Pengfei Zhao [2,*], Wen Xue [1,2], Qiang Zhang [1,2], Changjie Ye [1,2] and Ming Ma [1,2]

1 China Academy of Building Research, Beijing 100013, China; liyulong_cabr@163.com (Y.L.); xuewen@cabrtech.com (W.X.); zhangqiang@cabrtech.com (Q.Z.); yechangjie@cabrtech.com (C.Y.); maming@cabrtech.com (M.M.)
2 National Engineering Research Center of Building Technology, Beijing 100013, China
* Correspondence: zhaopengfei@cabrtech.com

Abstract: The structural characteristics of large-span structures inherently differ from those of conventional multistorey structures, making it challenging to accurately describe the contribution of various vibration modes to the overall response using traditional dynamic response analysis methods. Based on the response spectrum method, this paper investigates the influence of the first torsional mode on the overall effects of large-span structures. It proposes a new metric, called the torsional mode contribution factor, to characterize the contribution of torsional modes. Focusing primarily on single-span frames, the study explores the impact of factors such as eccentricity ratio, aspect ratio, and roof stiffness on the torsional mode contribution factor. Additionally, the relationship between the period ratio and the torsional mode contribution factor is examined to assess the necessity of controlling the period ratio. The findings reveal that the contribution of torsional modes to the overall seismic response varies significantly under different conditions, such as eccentricity ratio, aspect ratio, roof stiffness, and torsional stiffness. The torsional mode's contribution is minimal for small eccentricity ratios, with the response primarily driven by translational modes. As eccentricity increases, translational-torsional coupling becomes more pronounced, amplifying the influence of torsional modes on the overall dynamic response. The study also highlights that increasing roof stiffness and aspect ratios can mitigate torsional effects to a certain extent. Still, excessive eccentricity ratios and stiffness may result in higher torsional contributions. Additionally, it is found that increasing torsional stiffness reduces the influence of torsional modes but does not eliminate the overall torsional deformation. The proposed torsional mode contribution factor offers an effective way to quantify these effects, demonstrating that traditional control methods, such as period ratio control, may not fully capture the torsional contributions.

Keywords: elastic plate bent frame structures; torsional mode; torsional mode contribution factor; mode participation factor; period ratio

1. Introduction

Seismic damage experiences reveal that buildings experience not only horizontal and vertical vibrations during earthquakes but also torsional vibrations [1]. Investigations have shown that torsional response during an earthquake significantly degrades seismic performance, leading to component failure [2] and structural damage or even collapse [3].

Key aspects of seismic design include criteria for determining structural regularity, controlling torsion, and implementing construction measures. Seismic codes generally restrict structural regularity through indicators such as the drift ratio [4–7] and eccentricity [8]. Additionally, the Chinese code JGJ3-2010 "Technical Specification for Concrete Structures of Tall Buildings" [9] specifies requirements for the period ratio (the ratio of the fundamental torsional period to the fundamental translational period) and uses both the period ratio and the drift ratio as dual control indicators for the planar regularity of structures.

Noncoupled modes of a building structure are either translational or torsional, and the corresponding noncoupled period ratio reflects the relative magnitude of lateral stiffness and torsional stiffness [10]. However, long-span buildings are complex, with dense periods and significant coupling between modes, resulting in coupled period ratios. In such cases, the period ratio does not have a one-to-one correspondence with torsional stiffness [11].

Currently, the research on torsional regularity indices in seismic design primarily focuses on rigid roofing systems such as concrete roof slabs, steel–concrete composite roof slabs, and prestressed concrete roof slabs. The "Code for Seismic Design of Buildings" [7] stipulates that for rigid roof slabs, horizontal loads are distributed according to the proportion of equivalent stiffness of the lateral force resisting elements; for flexible roof slabs, they are distributed according to the proportion of the representative value of gravity load on the tributary area of the lateral force resisting elements. For semi-rigid roof slabs, the horizontal force distribution can be taken as the average of the two distributions above the results. Previous research has considered the impact of the rigidity of the roof slab or its in-plane deformation on the structure. Ruggieri et al. [12] noted that, compared to rigid roofing, flexible roofing structures have longer resonant periods, and the flexibility of the roof significantly affects the acceleration at the top of the building and the floor response spectrum. Sánchez et al. [13] proposed a numerical calculation method for the optimal design of semi-rigid steel frames, considering the influence of panel stiffness by introducing second-order effects. Ruggieri et al. [14] pointed out that the in-plane deformation stiffness of the floor slab affects the overall behavior of the building, such as changing the torsional coupling effect of the building, thereby affecting the force distribution on columns and shear walls. He et al. [15] quantified the impact of in-plane stiffness on the distribution of horizontal loads by defining the horizontal load transfer coefficient. The study explored the relationship between in-plane stiffness, the lateral stiffness ratio of vertical elements, and the load transfer coefficient. It defined the concept of a rigid floor slab based on the values of the lateral stiffness ratio and load transfer coefficient.

For long-span buildings, controlling the period ratio depends on the contribution of torsional modes to the overall structural response. If torsional modes minimally contribute to the total effect under seismic action, controlling the period ratio will have little impact on reducing the dynamic response. Conversely, if torsional modes significantly contribute, examining the relationship between torsional mode contribution and the period ratio becomes necessary to determine how to adjust the period ratio to mitigate the overall structural response.

Several methods have been proposed to reflect the contribution of vibration modes. Wang et al. [16] conducted modal analysis on frame structures using Ritz vectors to describe the dynamic characteristics of the structure, pointing out that the closer the spatial distribution of dynamic loads is to the mode shape, the more easily the mode is excited. Oscar et al. [17], through nonlinear static and dynamic analysis of two 21-storey steel buildings located in soft soil regions of Mexico City, noted that higher order modes have a significant impact on the overall nonlinear dynamic response of the steel structures. Feng et al. [18] proposed the mode contribution ratio as a new criterion for identifying the dominant modes. The method is validated using 18 typical cases of single-layer reticulated domes, demonstrating the accuracy of the mode contribution ratio in predicting seismic responses compared to the traditional method of combining the first thirty modes. Sun et al. [19] studied the spectral characteristics of single-layer reticulated domes, finding that higher order modes contribute more to the structural response. Liao et al. [20], through numerical analysis of single-layer reticulated domes, pointed out that the size of a mode's participation factor does not necessarily correlate with its contribution to the dynamic response of the structure. In some cases, modes with large participation factors contribute little to the response, while modes with smaller participation factors contribute more.

This study focuses on the contribution of the first torsional mode to the seismic response of single-span elastic plate bent frame structures commonly found in public buildings, introducing the torsional mode contribution factor. By examining factors such as eccentricity, aspect ratio, and panel stiffness, the study reveals the mechanisms by which the first torsional mode affects elastic plate bent frame structures and identifies the lateral force resisting components significantly impacted by the torsional response. Additionally, the study explores the relationship between the torsional mode contribution factor and the period ratio to assess the necessity of controlling the period ratio. This research provides valuable reference points for identifying critical structural components in the seismic conceptual design of long-span buildings, thereby shortening the design cycle for structural optimization or reinforcement.

2. Torsional Mode Contribution Factor

2.1. Determination of Torsional Vibration Mode

According to the Chinese code JGJ3-2010 Technical Specification for Concrete Structures of Tall Buildings [9], "The principal vibration mode of torsional coupling can be identified by calculating the modal direction factor. When the torsional direction factor exceeds 0.5, the mode can be considered predominantly torsional". At present, the modal direction factor for each mode of a structure [21] can be computed as follows:

For a linear structural system discretized into n degrees of freedom, the equation of motion under seismic acceleration time history is $\ddot{u}_g(t)$ [22]:

$$[M]\ddot{u} + [C]\dot{u} + [K]u = -[M]r_e\ddot{u}_g(t) \tag{1}$$

where M, C, and K are the mass, damping, and stiffness matrices, respectively; u is the displacement vector of order n, and re represents the seismic direction vector.

The damping matrix C satisfies the conditions for classical damping. Therefore, Equation (1) can be decoupled in the space spanned by the undamped modal shapes, resulting in n independent single-degree-of-freedom equations of motion:

$$\ddot{q}_i(t) + 2\xi_i\omega_i\dot{q}_i(t) + \omega_i^2 q_i(t) = -\gamma_i\ddot{u}_s(t) \tag{2}$$

where $q_i(t)$ presents the generalized coordinate corresponding to the i-th undamped mode shape ϕ_i (normalized with respect to the mass matrix); ξ_i is the damping ratio of the i-th mode; ω_i denotes the natural circular frequency of the i-th undamped mode; γ_i is the mode participation factor, defined as $\gamma_i = \phi_i^T Mr$, which reflects the distribution of dynamic loads across the various modes.

The corresponding seismic force vector for the i-th order mode is:

$$\begin{gathered}\{F\}_i = [K]\{u\}_i = -\frac{\gamma_i}{\omega_i^2}[K]\{\phi\}_i = \\ -\frac{\gamma_i}{\omega_i^2}\omega_i^2[M]\{\phi\}_i = -\gamma_i[M]\{\phi\}_i\end{gathered} \tag{3}$$

The seismic force applied to j-th node by vibration i-th mode is:

$$F_{ji} = -\gamma_i m_j \phi_{ji} \tag{4}$$

The base shear for the i-th order vibration pattern is:

$$V_i = -\sum_{j=1}^{n} F_{ji} = \gamma_i \sum_{j=1}^{n} m_j\phi_{ji} = \gamma_i\{\phi\}_i[M]\{I\} = \gamma_i^2 \tag{5}$$

For unit seismic acceleration loads, the base shear of the entire structure is numerically equal to the total mass of the structure in a given direction and can be expressed by the following equation:

$$F_j = -m_j \tag{6}$$

$$V = -\sum_{j=1}^{n} F_j = \sum_{j=1}^{n} m_j = \{I\}^T[M]\{I\} \tag{7}$$

The mass participation coefficient of the *i*-th order vibration mode under seismic action in the *k*-th direction is:

$$\gamma_{mi} = \frac{\gamma_i^2}{m_k} \tag{8}$$

where γ_{mi} presents the modal mass participation factor used in GB 50011-2010 [7]. The torsional direction factor for the i-th mode can be expressed as:

$$\alpha_T = \frac{\gamma_{mT}}{\gamma_{mX} + \gamma_{mY} + \gamma_{mZ} + \gamma_{mT}} \tag{9}$$

where γ_{mX}, γ_{mY}, γ_{mZ} are the modal mass participation factors in the X, Y, and Z directions of the structure; γ_{mT} is the modal mass participation factor for torsional mode around the Z axis; α_T is the torsional direction factor. If $\alpha_T > 0.5$, the mode is determined to be a torsional mode.

2.2. Definition of Torsional Mode Contribution Factor

The response spectrum method was used to calculate the torsional response of the structure under coupled translational and torsional actions. The response spectrum curve defined in the "Code for Seismic Design of Buildings" GB 50011-2010 (2016 edition) [7] is given by:

$$\alpha_j = \left(\frac{T_g}{T_j}\right)^{0.9} \alpha_{\max} g \tag{10}$$

where T_g is the characteristic period of the site; $\alpha_{\max}$ is the maximum seismic influence coefficient; g is the gravitational constant. The subscript j-th indicates the mode number. The response spectrum curve is shown in Figure 1.

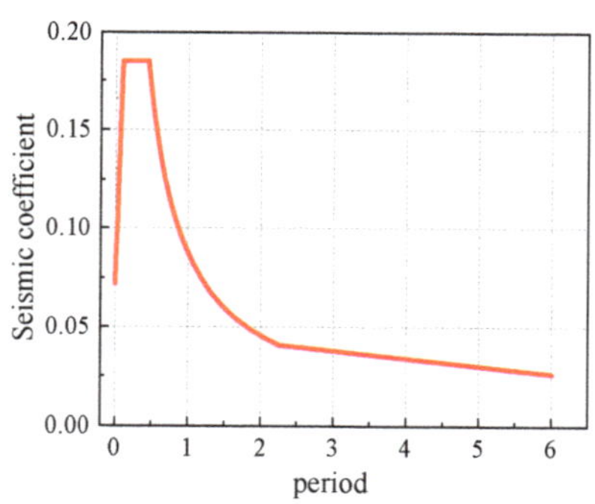

Figure 1. Acceleration response spectrum curve.

The modal displacement due to an earthquake is given by:

$$U_j = \frac{\gamma_{tj}}{\omega_j^2} \alpha_j \varphi_j \tag{11}$$

where ω_j is the coupling frequency; φ_j is the modal shape vector; γ_{tj} is the participation factor for the j-th mode, defined as $\gamma_{tj} = \varphi_j \{M\} \varphi_j^T$, where $\{M\}$ is the mass matrix of the structure.

Applying the fully quadratic combination to the modal combination of U_j, the torsional coupling displacements under horizontal seismic action are:

$$U_{EK} = \sqrt{\sum_{j=1}^{m}\sum_{k=1}^{m} \rho_{jk} U_j U_k} \tag{12}$$

where U_{EK} is the displacement of the standard value of seismic action; ρ_{jk} is the coupling coefficient of the j-th mode and the k-th mode, which is calculated by:

$$\rho_{jk} = \frac{8\sqrt{\zeta_j \zeta_k}(\zeta_j + \lambda_T \zeta_k)\lambda_T^{1.5}}{(1-\lambda_T^2)^2 + 4\zeta_j\zeta_k(1+\lambda_T^2)\lambda_T + 4(\zeta_j^2+\zeta_k^2)\lambda_T^2} \tag{13}$$

where ζ_j is the damping ratio of the mode; λ_T is the ratio of the self-oscillating period of the k-th mode and the vibration j-th mode, and the subscripts denote the ordinal numbers of the k-th mode.

The coupling coefficient ρ_{jk} of the first-order torsional vibration mode is set to zero, and then a fully quadratic combination of U_j is performed. Under horizontal seismic action, the displacement, excluding the torsional effect, is:

$$U_{REM} = \sqrt{\sum_{\substack{j=1 \\ j \neq j_t}}^{m} \sum_{\substack{k=1 \\ k \neq k_t}}^{m} \rho_{jk} U_j U_k} \tag{14}$$

where U_{REM} is the displacement excluding the torsional modes, and j_t, k_t are the modes primarily characterized by torsion. The torsional mode contribution factor can be defined as:

$$C_{tU} = \max\left\{\frac{U_{EK} - U_{REM}}{U_{EK}}\right\} \tag{15}$$

When calculating the torsional mode contribution factor, the principal directional displacement is extracted. For rectangular single-span frames, which typically have a symmetric planar layout, the mode participation factors in the principal direction are mainly contributed by the first-order translational mode and the first-order torsional mode. The principal directional displacement, calculated using the response spectrum method, can be considered by performing a complete quadratic combination (CQC) of the first-order translational mode and the first-order torsional mode. According to Equations (11) and (12), the CQC combination can be expanded as follows:

$$\begin{aligned} U_{1k} &= \tfrac{\gamma_{t1}}{\omega_1^2}\alpha_1\varphi_1 \\ U_{3k} &= \tfrac{\gamma_{t3}}{\omega_3^2}\alpha_3\varphi_3 \end{aligned} \tag{16}$$

$$U_{Yk} = \sqrt{\rho_{11}U_{1k}{}^2 + \rho_{33}U_{3k}{}^2 + 2\rho_{13}U_{1k}U_{3k}} \tag{17}$$

where U_{Yk} represents the principal directional displacement at node k, and U_{1k} and U_{3k} are the principal directional displacements at node k contributed by the first-order translational mode and the first-order torsional mode, respectively. ρ_{11}, ρ_{33} and ρ_{13} represent the autocorrelation coupling coefficients of the principal direction mode, the torsional mode, and the coupling between the principal direction mode and the torsional mode, respectively.

$$
\begin{aligned}
C_{tU} &= \max\left\{\frac{U_{EK}-U_{REM}}{U_{EK}}\right\} \\
C_{tUy} &= \max\left\{\frac{\sqrt{U_{1k}{}^{2}+U_{3k}{}^{2}+2\rho_{13}U_{1k}U_{3k}}-U_{1k}}{\sqrt{U_{1k}{}^{2}+U_{3k}{}^{2}+2\rho_{13}U_{1k}U_{3k}}}\right\} \\
C_{tUy} &= \max\left\{\frac{\sqrt{1+\left(\frac{U_{3k}}{U_{1k}}\right)^{2}+2\rho_{13}\frac{U_{3k}}{U_{1k}}}-1}{\sqrt{1+\left(\frac{U_{3k}}{U_{1k}}\right)^{2}+2\rho_{13}\frac{U_{3k}}{U_{1k}}}}\right\}
\end{aligned}
\tag{18}
$$

Let $\alpha = \frac{U_{3k}}{U_{1k}}$, then the following can be obtained:

$$
C_{tUy} = \max\left\{1 - \frac{1}{\sqrt{1+\alpha^{2}+2\rho_{13}\alpha}}\right\} \tag{19}
$$

As the seismic response of elastic plate bent frame structures is relatively minor, more focus should be placed on the lateral force resisting components. Thus, the connections between these components and the roof panels are used as reference points for calculating the torsional mode contribution factor.

2.3. Analytical Model

The study focuses on single-storey elastic plate bent frame structures with rectangular or quasi-rectangular building plans. These structures are widely used in public buildings, such as airport concourses and exhibition halls in convention centers. Additionally, as shown in Table 1, rectangular plans exhibit more pronounced torsional seismic damage under seismic action [23,24].

Table 1. Comparison of building plane failure rates under earthquake action.

Plan Shape	Rectangular	L-Shaped	II-Shaped	Cross-Shaped	H-Shaped	Others
Damage ratio (%)	85.0	4.0	8.5	0.05	0.15	2.4

Long-span buildings have complex forms, with connection methods including rigid connections and semi-rigid connections [25]. According to the "Code for Seismic Design of Buildings" (GB 50011-2010, 2016 edition) [7] and the "Code for Design of Steel Structures" (GB 50017-2017) [26], a rectangular-plan elastic plate bent frame structure was designed and analyzed, as illustrated in Figure 2. The structure is constructed using Q355B and has isotropic material properties. The spacing between transverse columns is 24 m, with a single span. The cross-section of all roof support columns measures 290 × 14, with a height of 5 m. The roof load is simplified to a permanent load of 1.5 kN/m^2 and a live load of 0.5 kN/m^2. The bottom of the support columns is rigidly connected to the ground, while the top is hinged to the roof panel.

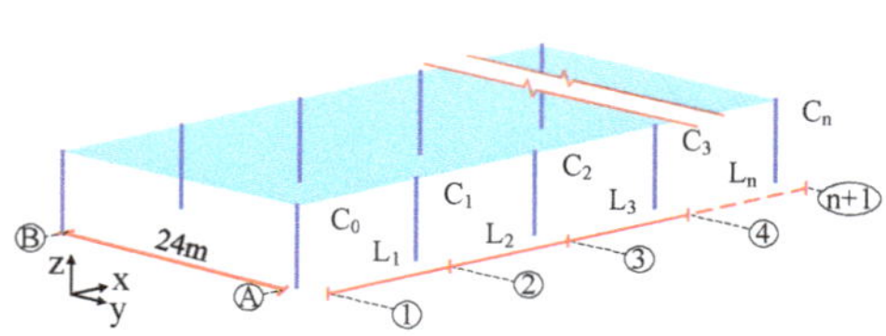

Figure 2. Schematic diagram of the example model.

The seismic precautionary intensity is 8 degrees (0.20 g), with a site category of type III, Group I seismic design, a seismic influence coefficient of 0.16, and a site period of 0.45 s. A bidirectional seismic analysis was conducted, and the results used the envelope of the bidirectional seismic response spectrum.

The structure's aspect ratio, ranging from 1.5 to 3, was adjusted by modifying the number of longitudinal spans or column spacing. Roof stiffness was evaluated by varying panel thickness, assuming a 120 mm thick concrete slab with infinite in-plane stiffness set at 1.00*EA*. In practical applications, eccentricity typically remains below 5%, but for this analysis, the upper limit was set at 20%. Detailed parameter variations are shown in Table 2.

Table 2. Parameters of the calculation model.

Calculation Parameters	Parameter
Longitudinal span × column spacing (Aspect ratio η)	3 × 12 m/4 × 12 m/5 × 12 m/6 × 12 m structures corresponding to aspect ratios of 1.5/2/2.5/3 respectively.
Panel stiffness	1.00*EA* (Rigid plate, 120 mm thick concrete plate) 0.05*EA* (6 mm thick concrete plate) 0.01*EA* (1.2 mm thick concrete plate) 0.005*EA* (0.6 mm thick concrete plate)
Eccentricity	2%, 4%, 6%, 8%, 10%, 12%, 14%, 16%, 18%, 20%.

The Rayleigh–Ritz method was employed to determine the vibration modes of the structure, ensuring the accuracy of the period ratios. The rigid center of the case study was located at the geometric center of the slab. By applying additional surface loads to specified areas, the distribution of the floor mass was adjusted, causing the centroid to deviate from its geometric center. This setup allowed the case study to simulate torsional effects under different eccentricities, as illustrated in Figure 3. The additional load was a multiple of the original structural mass m_0. The calculation formula for the eccentric load is as follows:

$$F_{Zi} = \frac{8\beta m_0}{(1-4\beta)S_h} \tag{20}$$

$$\beta = \frac{e}{L} \tag{21}$$

where F_{Zi} is the distributed eccentric force additionally applied to half of the side panels; e is the eccentric distance, which is the distance between the centroid and the rigid center within the plane of the structure; β is the eccentricity ratio, defined as the ratio of the static eccentric distance to the main dimension of the structure; L is the main dimension of the structure; m_0 is the total mass of the structure; and S_h is half the area of the roof slab.

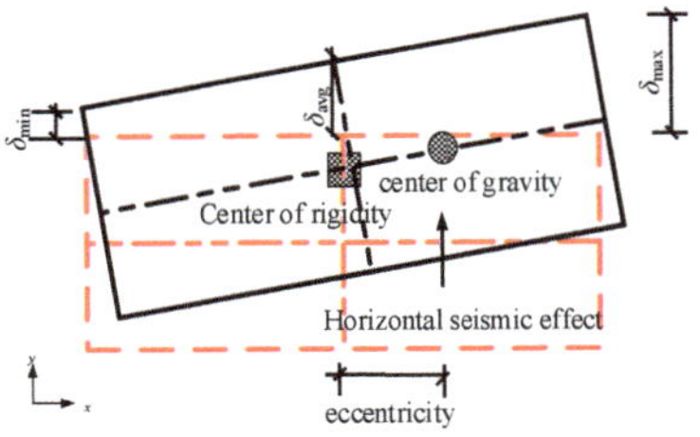

Figure 3. Schematic of structural torsion.

3. Properties of Torsional Mode Contribution Factor

3.1. Torsional Mode Contribution Factor Under Different Eccentricities

The eccentricity ratio affects the vibration characteristics of building structures and increases the torsional effect of the structure [27]. To examine the impact of the eccentricity ratio on the torsional mode contribution factor, the example of a panel stiffness of 0.01*EA* is used. Figure 4 shows the effect of the eccentricity ratio on the torsional mode contribution factor in elastic plate bent frame structures with different aspect ratios.

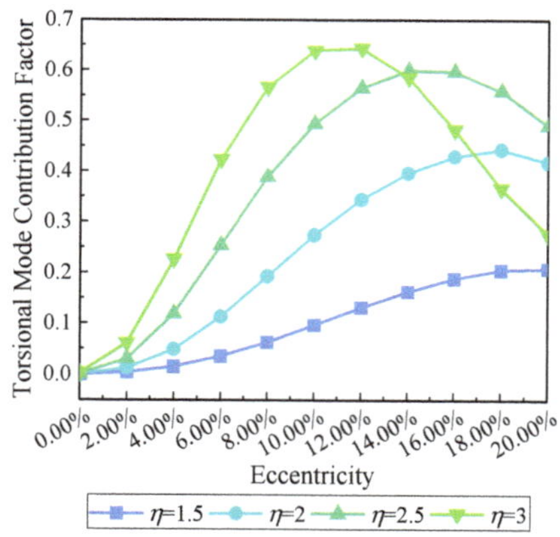

Figure 4. Impact of eccentricity on the torsional mode contribution factor.

As seen in Figure 4, for small eccentricities, the torsional mode contribution factor increases with an increasing eccentricity ratio. However, for large eccentricities, the torsional mode contribution factor decreases as the eccentricity ratio increases. This indicates that when the eccentricity ratio is small, the torsional effect of the structure is relatively weak, and the seismic response is dominated by translational motion. As the eccentricity ratio increases, the center of mass and center of stiffness gradually deviate, leading to a significant increase in the torsional effect of the structure under seismic action, and the torsional mode contribution factor increases accordingly. According to Figure 5, when the eccentricity increases beyond a certain point, the coupling effect between the translational and torsional modes becomes stronger, resulting in a reduction in the mode participation factor of the torsional mode. At the same time, the mode participation factors of the other modes, primarily governed by the translational mode, increase, and the torsional component of the torsional mode shifts to other modes.

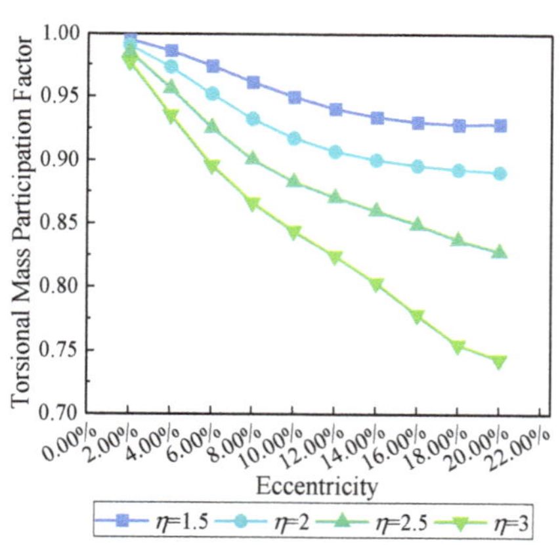

Figure 5. Torsional mass participation coefficient in torsional modes under different eccentricities.

3.2. *Torsional Mode Contribution Factor Under Different Aspect Ratios*

According to Clause 3 of the seismic design code GB50011-2010 Code for Seismic Design of Buildings [7], "the dynamic characteristics of a structure should be similar in the two principal axes". However, in practical engineering, when the building plan is rectangular or elliptical, the vibration periods of the two principal axes often differ due to one axis being longer than the other, resulting in significant differences in lateral stiffness in the two directions. To study the impact of the aspect ratio on the torsional mode contribution factor, the case with a panel stiffness of 0.01*EA* and a longitudinal column spacing of 12 m was used, with the longitudinal length varied by changing the number of longitudinal spans *n*. Figure 6 shows the influence of the aspect ratio on the torsional mode contribution factor of elastic plate bent frame structures under different eccentricity ratios.

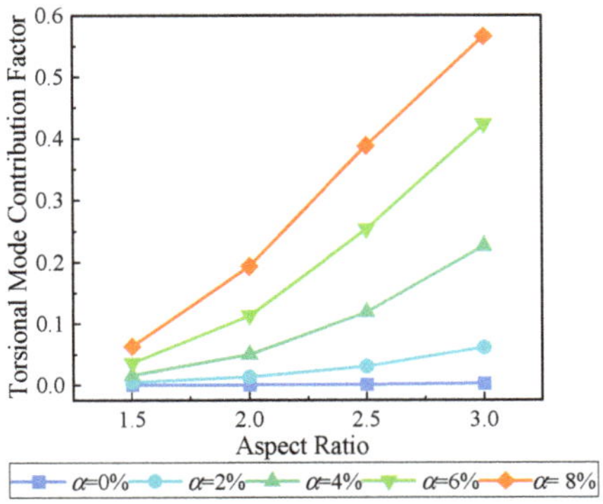

Figure 6. The impact of aspect ratio on the torsional mode contribution factor (small eccentricity).

From Figure 6, it can be seen that for small eccentricity ratios, the torsional mode contribution factor increases with an increasing aspect ratio, and the torsional mode contribution factor for structures with larger aspect ratios reaches its peak earlier. This indicates that the larger the aspect ratio, the smaller the lateral stiffness in the shorter direction, making the structure more susceptible to torsion.

From Figure 7, it can be observed that at large eccentricities, the torsional mode contribution factor initially increases and then decreases with the increase in eccentricity, with structures having larger aspect ratios exhibiting this decrease earlier. The reason is that as the aspect ratio increases, the mass distribution of the structure becomes more uneven under the same eccentricity ratio, resulting in a larger eccentric distance. This, in turn, enhances the coupling effect between translational and torsional modes, causing the torsional component of the torsional mode to migrate to other modes earlier (as shown in Figure 5), leading to a reduction in the torsional mode contribution factor earlier than in other cases.

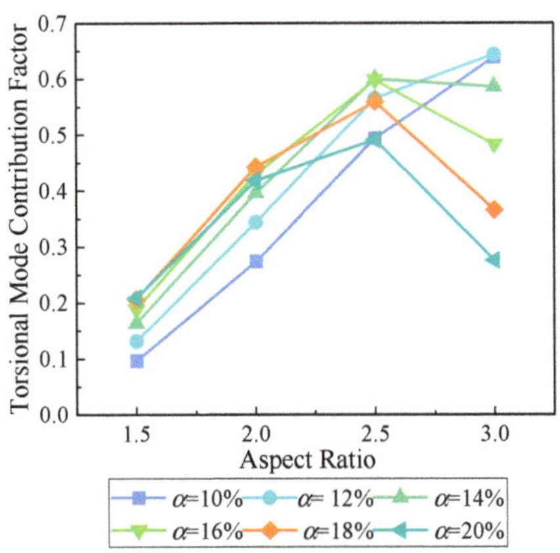

Figure 7. The impact of aspect ratio on the torsional mode contribution factor (large eccentricity).

3.3. Torsional Mode Contribution Factor Under Different Panel Stiffnesses

The roof panel, as a horizontal load transmitting system in spatial structures, coordinates the work of the vertical, lateral force resisting members [28]. For example, a structure with six longitudinal spans and a column spacing of 12 m (aspect ratio of 3) is used to illustrate this. Figures 8 and 9 show the influence of roof panel stiffness on the torsional mode contribution factor under different eccentricity ratios.

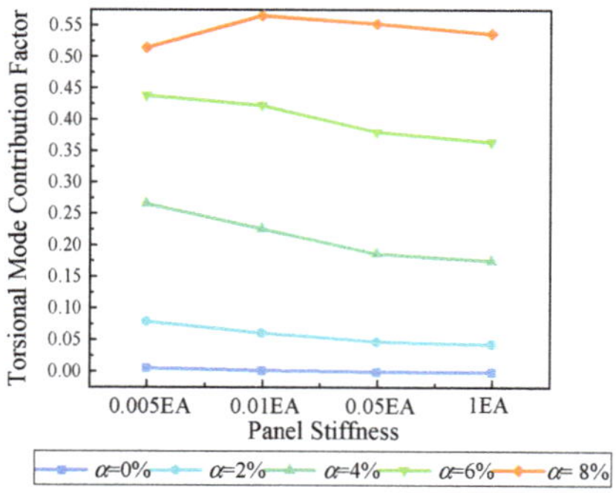

Figure 8. The impact of panel stiffness on the torsional mode contribution factor (small eccentricity).

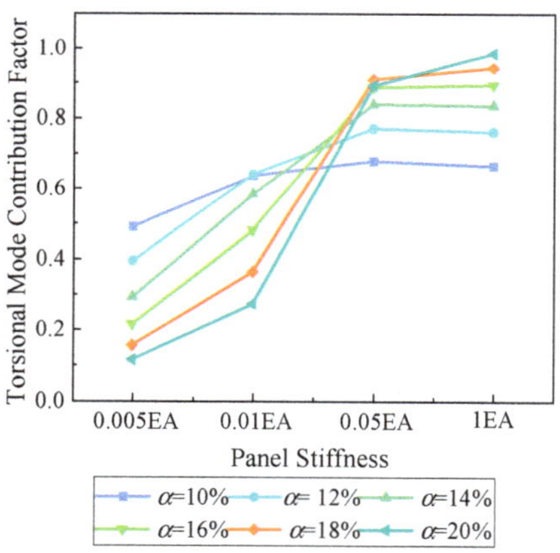

Figure 9. The impact of panel stiffness on the torsional mode contribution factor (large eccentricity).

As shown in Figure 8, for small eccentricities with a constant eccentricity ratio, the torsional mode contribution factor generally decreases as the panel stiffness increases. This indicates that at small eccentricity ratios, the flexible slab provides weaker constraints on the lower lateral force resisting members, making the structure more prone to torsion.

From Figure 9, it can be observed that for large eccentricity ratios, the torsional mode contribution factor generally increases with increasing panel stiffness, showing an overall inverse relationship between the torsional mode contribution factor and panel stiffness as the eccentricity ratio increases. This suggests that at large eccentricity ratios, the mass of the floors tends to concentrate on one side of the plane, making the translational-torsional coupling effect more significant.

As shown in Figures 10 and 11, even though flexible slabs provide weaker constraints on the lower lateral force resisting members at large eccentricity ratios, the translational-torsional coupling in flexible slab structures becomes more severe. The mode participation factors for both translational and torsional components in the torsional mode are significantly smaller than those in rigid slab structures, resulting in a smaller torsional mode contribution factor for flexible slabs compared to rigid slabs at large eccentricity ratios.

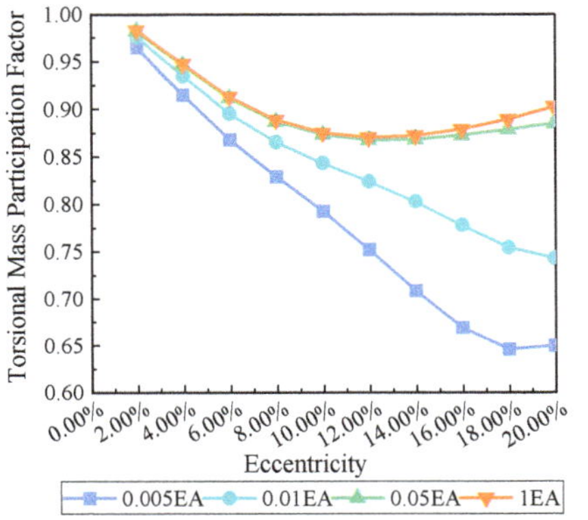

Figure 10. Torsional mass participation coefficient in torsional modes under different eccentricities.

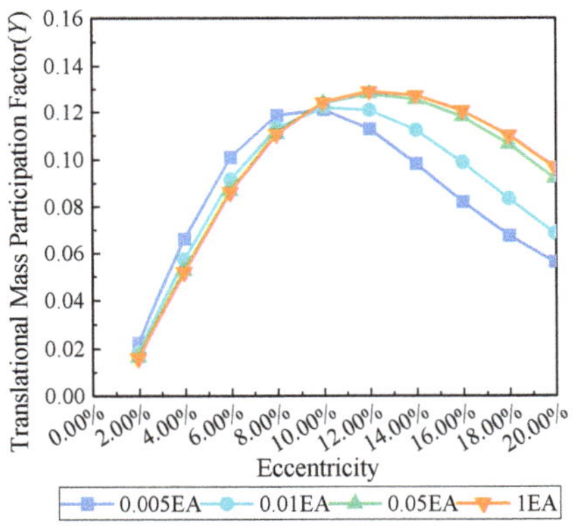

Figure 11. Translational (*Y*) mass participation coefficient in torsional modes under different eccentricities.

3.4. Torsional Mode Contribution Factor Under Different Torsional Stiffnesses

When the structure undergoes overall torsion, its members inevitably participate in resisting torsion [29]. Taking the example of a frame structure with six longitudinal spans and a column spacing of 12 m (aspect ratio of 3), the overall torsional stiffness of the structure is adjusted by changing the corner column cross-section specifications to investigate the effect of different torsional stiffnesses on the torsional mode contribution factor. Considering the practical engineering scenario, where accidental eccentricity is often included, the eccentricity ratio in this example is set to 6%. Figure 12 shows the influence of torsional stiffness on the Torsional Mode Contribution Factor of elastic plate bent frame structures with different panel stiffnesses under an eccentricity ratio of 6%.

As can be seen from Figure 12, regardless of the panel stiffness, the torsional mode contribution factor decreases as the torsional stiffness of the structure increases (i.e., as the corner column cross-sections increase). Furthermore, the larger the panel stiffness (as seen in Section 3.3), the smaller the torsional mode contribution factor.

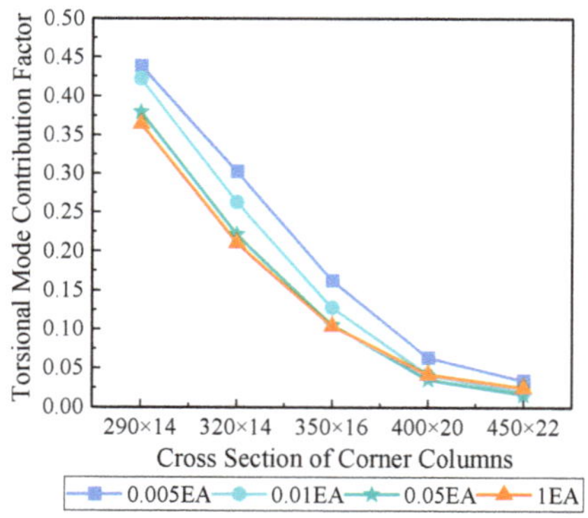

Figure 12. Influence of torsional stiffness on the torsional mode contribution factor.

Extracting the first three translational modes of the frame structure, with corner column cross-section specifications of 290 × 14 and an eccentricity ratio of 6% due to accidental eccentricity, reveals that the first-order translational mode in the *Y* direction exhibits a noticeable translational-torsional coupling, as shown in Figure 13.

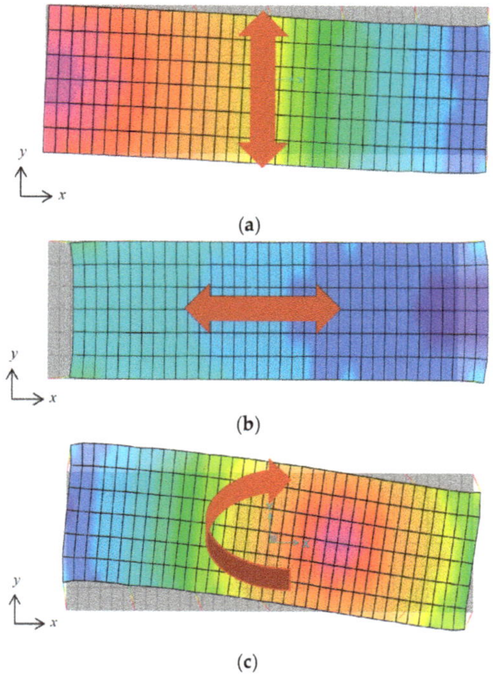

Figure 13. Diagrams of the first three translational modes in the example. (**a**) First-order translational mode (*Y* direction); (**b**) second-order translational mode (*X* direction); (**c**) first-order torsional mode.

Figures 14 and 15 show, respectively, the calculation of node $\alpha(U_{3k}/U_{1k})$ for the torsional mode contribution factor under different torsional stiffnesses in elastic plate bent frame structures with accidental eccentricity, and the coupling coefficient between the first-order translational mode (*Y* direction) and the first-order torsional mode. From Figure 14, it can be seen that α decreases as the torsional stiffness increases, indicating that when the torsional stiffness is low, the displacement in the principal direction (*Y* direction) is mainly contributed by the first-order torsional mode. However, as the torsional stiffness increases, the displacement contribution from the first-order translational mode (translational-torsional coupling) increases. Figure 15 shows that the coupling coefficient between the torsional mode and the first-order translational mode is relatively small, and the coefficient decreases as the torsional stiffness increases.

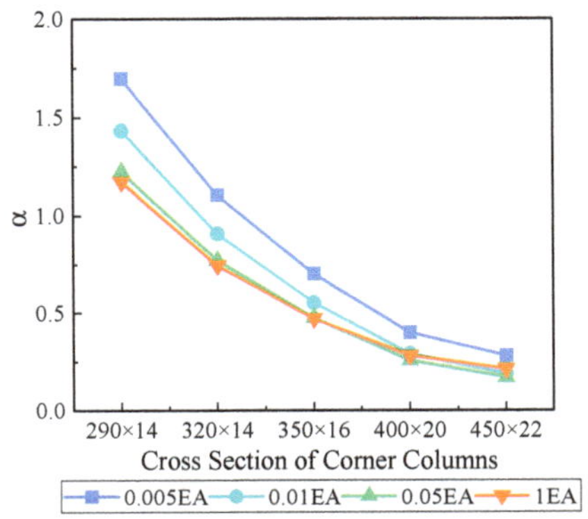

Figure 14. α of elastic plate bent frame structures with different torsional stiffnesses under accidental eccentricity.

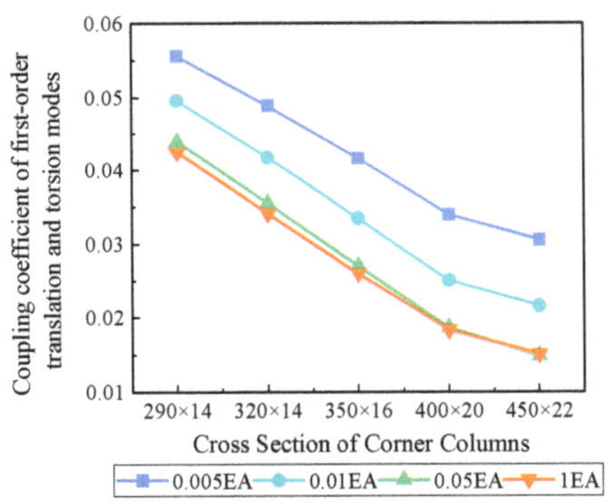

Figure 15. Coupling coefficient between the first-order translational mode (Y direction) and the first-order torsional mode in elastic plate bent frame structures with different torsional stiffnesses under accidental eccentricity.

In summary, as the torsional stiffness increases, both α and the coupling coefficient decrease, which, according to Equation (18), leads to a reduction in the torsional contribution. This indicates that when there is eccentricity in the structure, enhancing the structure's torsional stiffness can mitigate the impact of torsional vibration modes, but it cannot eliminate the structural torsional deformation. The reason is that when the frame structure has high torsional stiffness, the displacement response in the principal direction is mainly contributed by the first-order translational vibration mode (coupling of translation and torsion), and the contribution of other modes to the displacement response is limited.

4. The Relationship Between Torsional Mode Contribution Factor and Period Ratio

When discussing the relationship between the period ratio and the torsional mode contribution factor, cases where the eccentricity ratio is zero need to be excluded. This is because, in practical engineering, it is common for structural mass and stiffness to be distributed unevenly due to construction or usage, leading to a misalignment between the center of mass and the center of stiffness, making cases with an eccentricity ratio of zero rare. Additionally, when the eccentricity ratio is zero, the structure is in a fully symmetrical state, and the torsional effect is relatively weak or negligible. Using the example of a panel stiffness of 0.01EA, the aspect ratio is varied by changing the number of longitudinal spans n. Figures 16 and 17 show the relationship between the period ratio and the torsional mode contribution factor for elastic plate bent frame structures with different aspect ratios under varying eccentricity ratios.

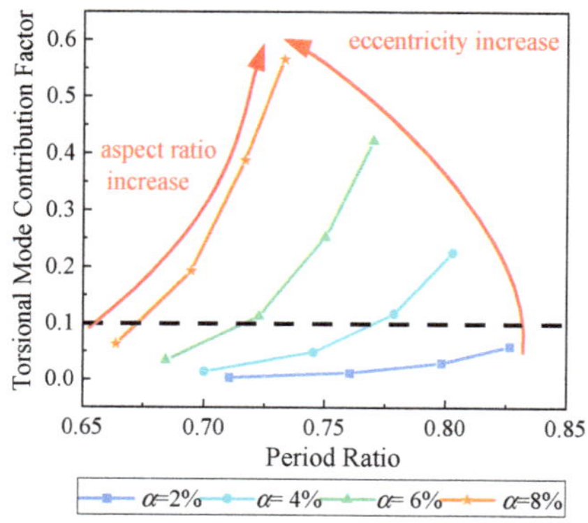

Figure 16. The relationship between the torsional mode contribution factor and period ratio under different eccentricities (small eccentricity).

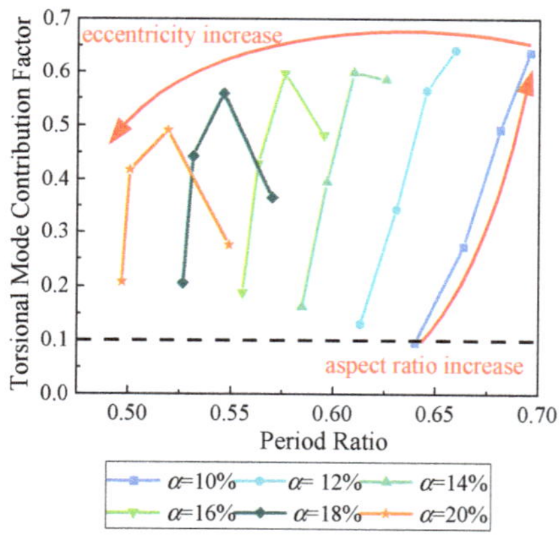

Figure 17. The relationship between the torsional mode contribution factor and period ratio under different eccentricities (large eccentricity).

As shown in Figure 16, for small eccentricity ratios with consistent values, the torsional mode contribution factor is positively correlated with the period ratio. Additionally, when the eccentricity ratio is less than 6%, the torsional mode contribution factor is generally below 10%, indicating that the contribution of the torsional mode to the overall effect is limited. From Figure 17, it can be observed that for large eccentricity ratios with consistent values, the torsional mode contribution factor is not necessarily positively correlated with the period ratio. Moreover, although the torsional mode contribution factor decreases as the eccentricity ratio increases, it remains above 10%, indicating that the torsional mode's impact on the overall effect cannot be ignored.

In summary, when the eccentricity ratio is small, the contribution of the torsional mode to the overall effect is not significant, and the dynamic response of the structure is mainly contributed by the first-order translational mode (translational-torsional coupling), with limited contributions from other modes to the displacement response. Even if the period ratio meets the code limits, it may not effectively limit the structure's torsional effect. On the other hand, when the eccentricity ratio is large, even if the period ratio meets the requirements, the contribution of the torsional mode to the overall dynamic response of the structure remains significant, which may not necessarily indicate that the structure has sufficient torsional stiffness to resist seismic loads.

5. Application of the Torsional Mode Contribution Factor

Unlike the period ratio and the drift ratio, current calculation software cannot compute the torsional mode contribution factor directly. The energy ratio of the structure can be calculated through the following steps:

(1) Create a three-dimensional model simulating the dynamic behavior of the structure, and then obtain the mass matrix and stiffness matrix of the structure using calculation software.
(2) Use the stiffness matrix and mass matrix to perform a response spectrum analysis of the mode shapes. Based on Equations (4)–(9), calculate the mode direction factors to determine the principal direction of each mode and identify the torsional modes.
(3) Set all coupling coefficients related to the torsional modes to zero according to Equation (13).
(4) Use the complete quadratic combination (CQC) method to combine the mode displacements after excluding the influence of torsional modes, thus determining the Torsional Mode Contribution Factor.

In seismic design, most structural calculation software provides mass matrices, mode vectors, and torsional coefficients. Therefore, designers only need to complete steps (2), (3), and (4).

6. Conclusions

The torsional mode contribution factor proposed in this paper effectively quantifies the contribution of torsional modes to the dynamic (displacement) response of the structure, revealing the effects of various parameters such as eccentricity ratio, aspect ratio, and roof panel stiffness on the contribution of torsional modes. The relationship between the period ratio and the torsional mode contribution factor is also explored, indicating that the period ratio may not fully reflect the torsional effect.

(1) The contribution of torsional modes to the dynamic response of the structure varies significantly under different parameters. The proposed torsional mode contribution factor can accurately quantify the impact of torsional modes on the overall effect of the structure.
(2) The eccentricity ratio has a significant impact on the translational-torsional coupling effect of the structure. For small eccentricities, the torsional mode contribution factor increases as the eccentricity ratio increases, but once the eccentricity ratio reaches a certain value, the torsional components of the torsional mode begin to migrate to other modes due to the translational-torsional coupling effect. At this point, the torsional mode contribution factor decreases with further increases in the eccentricity ratio but remains above 20%.
(3) The torsional mode contribution factor reflects the nonuniformity of stiffness in the two principal directions of the structure. The larger the aspect ratio, the weaker the lateral stiffness in the shorter direction and the more prone the structure is to torsion, resulting in a higher torsional mode contribution factor.
(4) When the eccentricity ratios are consistent and relatively small, the torsional mode contribution factor of flexible slab structures is greater than that of rigid slab structures, as flexible slabs provide weaker support constraints to the lower members, making the structure more prone to torsion. For larger eccentricity ratios, although flexible slabs provide weaker support constraints, the torsional-translational coupling effect

in flexible slab structures becomes more pronounced, with both the torsional and translational components of the torsional mode being smaller than in rigid slab structures, leading to a reduction in the torsional mode contribution factor, although it remains above 10%.

(5) When eccentricity exists in a structure, selecting an appropriate panel stiffness and aspect ratio ensures that the contribution of torsional modes remains at a lower level. It is not always the case that increasing panel stiffness or reducing the aspect ratio will result in a lower torsional mode contribution during seismic response.

(6) When considering accidental eccentricity, increasing the torsional stiffness of elastic plate bent frame structures can reduce the influence of the torsional mode. However, at this point, the displacement response of the frame structure is mainly contributed by translational-torsional coupling modes, so it is not possible to eliminate the torsional response of the structure.

(7) When the eccentricity ratio is small, although the torsional mode contribution factor increases as the eccentricity ratio increases, it remains below 10%, indicating that the torsional mode has a limited impact on the overall effect of the structure. In this case, the dynamic response of the structure is mainly contributed by the first-order translational mode (translational–torsional coupling), and controlling the period ratio has little significance. In cases with large eccentricity ratios, where the translational-torsional coupling is severe, the torsional mode contribution factor exceeds 10%, and the torsional mode's impact on the overall effect cannot be ignored. However, as the eccentricity ratio increases, the period ratio may decrease, meaning that to meet code requirements, the period ratio must be adjusted by increasing the eccentricity ratio, which may, in turn, increase the dynamic response.

(8) This study explored the influence mechanisms of various parameters on the torsional mode contribution factor. The analysis was limited to single-span, single-storey frames, and future research could extend to more complex cases, particularly those involving complex plans and multispan structures. Additionally, the current study mainly focused on elastic seismic responses; future research could explore the non-linear behavior of the torsional mode contribution factor under elastoplastic seismic conditions to optimize seismic design further.

Author Contributions: Conceptualization, P.Z.; Methodology, P.Z.; Formal analysis, Y.L.; Investigation, Y.L.; Writing-original draft, Y.L.; Writing-review & editing, Y.L., W.X., Q.Z., M.M. and C.Y.; Project administration, Y.L.; Funding acquisition, P.Z. All authors have read and agreed to the published version of the manuscript.

Funding: This research was funded by the National Key Research and Development Program of China [Grant No. 2022YFC3002300].

Data Availability Statement: The original contributions presented in the study are included in the article, further inquiries can be directed to the corresponding author.

Conflicts of Interest: Authors Yulong Li, Pengfei Zhao, Wen Xue, Qiang Zhang, Ming Ma and Changjie Ye were employed by the company China Building Technique. The remaining authors declare that the research was conducted in the absence of any commercial or financial relationships that could be construed as potential conflicts of interest.

References

1. Guéguen, P.; Astorga, A. The Torsional Response of Civil Engineering Structures during Earthquake from an Observational Point of View. *Sensors* **2021**, *21*, 342. [CrossRef] [PubMed]
2. Yao, Y.; Huang, H.; Zhang, W.; Ye, Y.; Xin, L.; Liu, Y. Seismic performance of steel-PEC spliced frame beam. *J. Constr. Steel Res.* **2022**, *197*, 107456. [CrossRef]
3. Paulay, T.; Priestley, M.J.N. *Seismic Design of Reinforced Concrete and Masonry Buildings*; Wiley: New York, NY, USA, 1992.
4. FEMA. *NEHRP Recommended Provisions for Seismic Regulations for New Buildings and Other Structures (FEMA 450)*; Building Seismic Safety Council (BSSC): New York, NY, USA, 2003.
5. International Code Council. *International Building Code 2003*; International Code Council: Washington, DC, USA, 2003.

6. *NZS4203: 1992;* Code of Practice for General Structural Design and Design Loadings for Buildings. Standards Association of New Zealand: Wellington, New Zealand, 1992.
7. *GB 50011-2010;* Code for Anti-Seismic Design of Buildings. China Architecture & Building Press: Beijing, China, 2016.
8. Cheng, S. *Structural Eurocodes (EC8);* Institute of Engineering Earthquake Research, China Academy of Building Research: Beijing, China, 1997.
9. *JGJ3-2010;* Technical Specification for Concrete Structures of Tall Buildings. China Architecture & Building Press: Beijing, China, 2010.
10. Li, Y.; Han, J.; Liu, J. Study on Control Indicators of Torsional Period Ratio in Seismic Design of Building Structures. *J. Build. Struct.* **2009**, *30*, 77–85.
11. Alaa, K.M.; El-Kashif, K.F.; Salem, H.M. New Definition for Torsional Irregularity Based on Floors' Rotations of Reinforced Concrete Buildings. *J. Eng. Appl. Sci.* **2022**, *69*, 12. [CrossRef]
12. Ruggieri, S.; Vukobratović, V. Acceleration demands in single-storey RC buildings with flexible diaphragms. *Eng. Struct.* **2023**, *275*, 115276. [CrossRef]
13. Sánchez-Olivares, G.; Espín, A.T. Design of planar semi-rigid steel frames using genetic algorithms and Component Method. *J. Constr. Steel Res.* **2013**, *88*, 267–278. [CrossRef]
14. Ruggieri, S.; Porco, F.; Uva, G. A numerical procedure for modeling the floor deformability in seismic analysis of existing RC buildings. *J. Build. Eng.* **2018**, *19*, 273–284. [CrossRef]
15. He, M.; Ma, Z.; Ma, R.; Li, Z. Horizontal load transfer performance of lightweight steel wood hybrid floor. *J. Tongji Univ. (Nat. Sci. Ed.)* **2014**, *42*, 1038–1043.
16. Wang, Z.; Der Kiureghian, A. Multiple-Support Response Spectrum Analysis Using Load-Dependent Ritz Vectors. *Earthq. Eng. Struct. Dyn.* **2014**, *43*, 2283–2297. [CrossRef]
17. Oscar, A.L.; Cruz, M. Number of Modes for the Seismic Design of Buildings. *Earthq. Eng. Struct. Dyn.* **1996**, *25*, 837–855.
18. Feng, R.; Zhu, B.; Wang, X. A Mode Contribution Ratio Method for Seismic Analysis of Large-Span Spatial Structures. *Int. J. Steel Struct.* **2015**, *15*, 835–852. [CrossRef]
19. Sun, J.M.; Zhang, Q.L. Seismic Performance and Analysis of Large-Span Spatial Reticulated Shell Structures. *J. Nat. Disasters* **2011**, *20*, 193–198. [CrossRef]
20. Liao, B.; Luo, Y.; Wang, L.; Guo, X. Theoretical Study on Mass Participation Coefficient and Vibration Characteristics of Large-Span Spatial Structures. In Proceedings of the 14th Space Structure Academic Conference, Department of Architectural Engineering, Fuzhou, China, 2 November 2012.
21. Rama Rao, G.V.; Bharathi Priya, C.; Sreekala, R. Methodology for Uncoupling Translational and Torsional Modes in Framed Structures with Experimental Validation. *Sādhanā* **2023**, *48*, 111. [CrossRef]
22. Clough, R.W. *Dynamics of Structures*, 2nd ed.; McGraw Hill: New York, NY, USA, 1993.
23. Wang, Y.; Bai, X. Seismic Damage Characteristics of Reinforced Concrete Structures in the Taiwan 921 Earthquake. *Eng. Seismol.* **2001**, *1*, 3–7. [CrossRef]
24. Wang, Y. Seismic Damage of Buildings in the Taiwan 921 Earthquake. *Archit. Knowl.* **2000**, *1*, 14–15.
25. Lu, L.; Yuan, G.; Huang, Z.; Shu, Q.; Li, Q. Performance-based analysis of large steel truss roof structure in fire. *Fire Saf. J.* **2017**, *93*, 21–38. [CrossRef]
26. *GB 50017-2017;* Steel Structure Design Standard. China Architecture & Building Press: Beijing, China, 2018.
27. Tabatabaei, R. Torsional Vibration of Eccentric Building Systems. In *Recent Advances in Vibrations Analysis;* InTech: London, UK, 2011.
28. Sucuoğlu, H.; Akkar, S. Analysis Procedures and Seismic Design Principles for Building Structures. In *Basic Earthquake Engineering;* Springer: Cham, Switzerland, 2014.
29. De-La-Colina, J.; Valdés-González, J.; González-Pérez, C.A. Dependency of the Accidental Torsion Building Response on Both Live-to-Dead Load Ratio and Material Stiffness Variation. *J. Civ. Eng.* **2024**, *22*, 1919–1932. [CrossRef]

buildings

MDPI

Article

Effects of Soil–Structure Interaction on the Seismic Response of RC Frame–Shear Wall Building Structures Under Far-Field Long-Period Ground Motions

Ke Yang [1], Pengrong Cai [1], Zixuan Zhang [1], Qinglin Hou [1], Ruige Zheng [1], Bin Hao [2] and Bo Wang [1,*]

[1] School of Civil Engineering, Chang'an University, Xi'an 710061, China; yangkechd@chd.edu.cn (K.Y.)
[2] Shaanxi Architecture Seismic Reduction and Isolation Science Research Institute Co., Ltd., Xi'an 712000, China
* Correspondence: chnwangbo@chd.edu.cn

Abstract: This paper focuses on the effect of soil–structure interaction (SSI) on the seismic response of high-rise RC frame–shear wall structures under far-field long-period ground motions. Elastic–plastic time–history analyses were performed using ABAQUS. The effects of the ground motion type, soil type, and structural frequency on the seismic response are analyzed and quantitatively evaluated. On this basis, the influence mechanism of SSI on the seismic response under far-field long-period ground motions is discussed and revealed through a ground motion spectrum analysis. The results show that the consideration of the SSI effect leads to an increase in the displacement response and a decrease in the shear response. The SSI coefficient of the base shear is all less than 1, ranging from 0.5 to 1. The SSI effect under far-field long-period ground motions is more pronounced than that under ordinary ground motions. The shear force reduction in the current code may not be applicable to the structural design considering the SSI effect under far-field long-period ground motions. The displacement response amplification of the SSI effect on loess soil (Site 2) is more remarkable than that on sand soil (Site 1). The SSI effect can reduce the structural frequency, especially for the structures with fewer floors on the softer soil site. The "bimodal characteristic" of the acceleration response spectrum for far-field long-period ground motions may lead to shear force amplification when SSI is considered.

Keywords: soil–structure interaction (SSI); far-field long-period ground motion; RC frame–shear wall structure; seismic response characteristics; elastic–plastic time–history analysis

1. Introduction

Far-field long-period ground motions (FLPGMs) refer to the far-field ground motions with predominant periods of 1–10 s or longer. The generation mechanism can be concluded as large earthquakes, effective long-distance propagation, and specific sites such as sedimentary basins or alluvial plains [1]. Far-field long-period ground motions have been recorded and observed in a number of devastating earthquakes in recent decades, including the 1985 Michoacan earthquake [2], the 1999 Chi-Chi earthquake [3], the 2003 Tokachi-oki earthquake [4], the 2011 Tohoku earthquake [5], and others. The examples of FLPGM-related destruction showed that high-rise buildings with longer periods are susceptible to resonance with far-field long-period ground motions, causing severe damage [6–11]. Figure 1 shows some photographs of the damage to high-rise buildings under far-field long-period ground motions.

(a) (b) (c)

Figure 1. High-rise buildings damage under far-field long-period ground motions: (**a**) high-rise building collapses in Mexico City earthquake, 1985; (**b**) Tokyo Tower before and after East Japan earthquake, 2011 [12]; and (**c**) high-rise building cracks in Wenchuan earthquake, 2008.

Soil–structure interaction (SSI) is a crucial concept in earthquake engineering that contemplates the interplay between structures and the soil supporting them. During an earthquake, the ground's motion transfers to the structure through the foundation. The structure responds according to its dynamic characteristics and the nature of the ground movement. Subsequently, the motion of the structure affects the ground motion, resulting in a complex interplay that is known as soil–structure interaction. High-rise buildings on soft soils are significantly affected by SSI [13]. The SSI effect reduces the frequency of the high-rise building structure in the soft soil site, and the period becomes longer [14]. In addition, the site soil filters out the high-frequency components in the seismic wave, leading to an increase in the low-frequency components [15]. Coupling long-period ground motions with the SSI effect, high-rise building structures are more prone to resonance, and cause the amplification of the seismic response and more serious damage.

The SSI effect on structures has been extensively studied in recent years. Forcellini [14] conducted the seismic fragility analysis on a 20-story building built on a pile foundation, taking SSI effects into consideration. It is pointed out that SSI helps to reduce the vulnerability of the system and the mechanisms of site amplification that are driven by SSI effects (kinematic and inertial interaction). Bolisetti et al. [16] investigated the influence of SSI on low- to medium-rise buildings through numerical simulations. The results show that SSSI has a negligible effect on the global spectral accelerations of the buildings in these arrangements. Askouni et al. [17] investigated the effect of SSI on the seismic response of ordinary asymmetric low-rise RC 3D buildings, while considering the elastoplastic behavior of RC elements and examining not only the global but also the detailed characteristics of the structure, such as the redistribution of the internal seismic forces due to SSI. Fatollahpour et al. [18] investigated the effects of structure–soil–structure interaction (SSSI) on the seismic response of two 20-story steel moment-resisting frames equipped with tuned mass dampers (TMDs). The results reveal that the consideration of SSSI can significantly increase the response of both structures.

However, existing studies on the effect of SSI have rarely involved the structures under far-field long-period ground motions. Based on the previous discussion, it is clear that there is a need for studies regarding the SSI effect on the structural response under far-field long-period ground motions. This paper focuses on high-rise RC frame–shear wall buildings subjected to far-field long-period ground motions. Through the elastic–plastic time–history analyses in the ABAQUS finite element program [19], the SSI effects on the seismic response of RC frame–shear wall structures under far-field long-period ground motions and ordinary ground motions are compared and analyzed, by varying the soil types (including the rigid base, sand site, and loess site) and structure periods. And the mechanism of the SSI effect on the structural response to far-field long-period ground motions is explored based on the ground motion spectrum analysis.

2. Structural Modeling and Ground Motion Selection

2.1. Structural Modeling and Validation

According to the current Chinese code [20–22], three RC frame–shear wall structures with 10, 20, and 30 stories, designated as S10, S20, and S30, were designed. The structural plane layout diagram is shown in Figure 2.

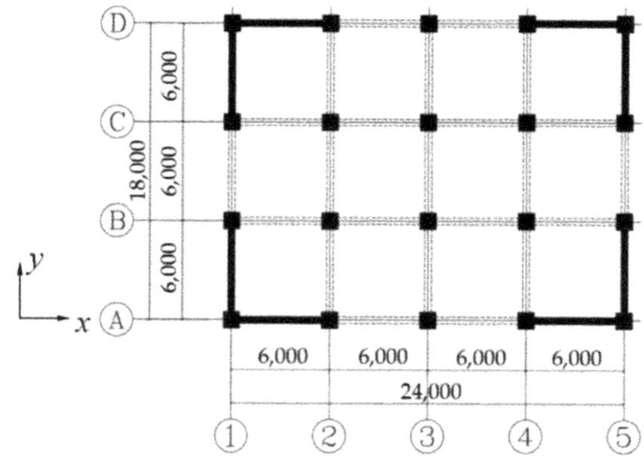

Figure 2. Structural plane layout diagram (unit: mm).

The first floor is 4.2 m high and the other floors are 3.3 m high. The total height of the three structures is 33.9 m, 66.9 m, and 99.6 m, respectively. The seismic precautionary intensity is 7 degrees, with a peak ground acceleration of 0.1 g. The structures are placed on the piled raft foundation, with a thickness of 800 mm, a width of 800 mm, and a pile length of 17 m. The piles were arranged in accordance with the frame columns on the structural plane layout. There are 20 piles in the plane with a width of 18 m and a length of 24 m. The structural load was 5 kN/m^2 for the dead load and 3.5 kN/m^2 for the live load. The design period reduction coefficient was set at 0.8, and the structural damping ratio was set at 0.05. The first and second floors were the shear wall strengthening areas.

The numerical simulation of the structural models was performed by ABAQUS software [19]. Figure 3 provides the analysis models with and without considering SSI.

The frame beams and columns were modeled by B31 beam elements [23,24], which is a Timoshenko beam that allows for transverse shear deformation. The floor slabs and shear walls were modeled by S4R shell elements in ABAQUS, which is a 4-node, quadrilateral, stress/displacement shell element with reduced integration and a large-strain formulation. The concrete material was simulated by the concrete damaged plasticity (CDP) model [20]. The steel reinforcement in the beam and shell elements was defined using the Rebar Layer command. The C3D8R solid element in ABAQUS, an 8-node linear brick reduced integration with hourglass control, was used to model the pile, raft, and soil. A tie constraint was applied to connect the structures above and below.

The Drucker–Prager (D-P) model [25] was used for the soil simulation. A viscoelastic artificial boundary was added at the soil boundary using spring damping. The soil region was set to be five times the size of the structure. For the contact interaction, the penalty method imposing frictional constraints was applied in the tangential direction, and hard contact was used for the normal behavior.

Sand soil and loess soil are the two different types of site soil considered in this study. Sand soil is characterized by its large particles and gritty texture. It is one of the most common soil types and is found in various regions around the world. Loess soil is a type of clay soil that is widespread in semi-arid and arid regions around the world. It is typically unsaturated and susceptible to collapse (i.e., a sudden decrease in soil volume) upon wetting [26]. Simply put, loess soil is softer than sand soil. Detailed parameters of soil sites are presented in Table 1. Referring to the existing studies [27,28], the soil size was set as 5 times the structure size to reduce the effect of infinite soil.

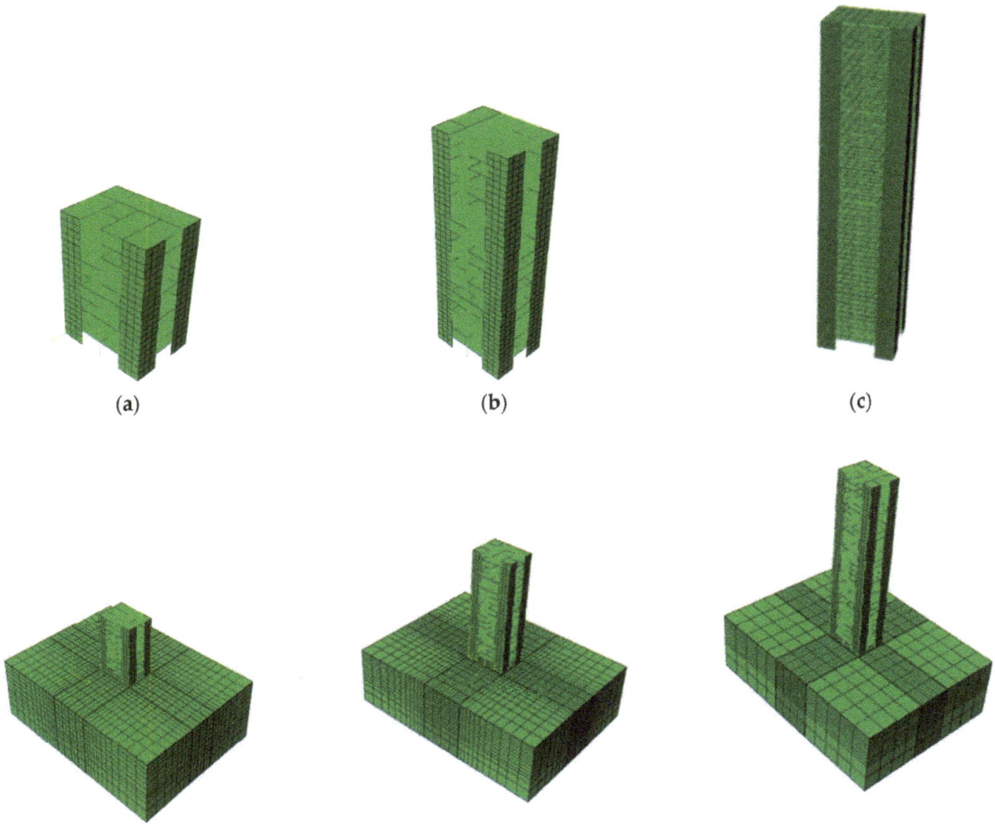

Figure 3. ABAQUS models: (**a–c**) S10, S20, and S30 without considering SSI; and (**d–f**) S10, S20, and S30 considering SSI.

Table 1. Soil site parameters.

Site Type	Soil Layer	Thickness (m)	Density (kg/m^3)	Elastic Modulus (MPa)	Poisson Ratio	Internal Friction Angle (°)	Shear Modulus (MPa)	Cohesion (MPa)
Site 1 (sand soil)	Sand 1	6	2000	94.08	0.20	15.0	39.20	25×10^{-3}
	Sand 2	6	2100	113.40	0.20	20.0	47.25	35×10^{-3}
	Sand 3	24	2200	135.17	0.20	25.0	56.32	45×10^{-3}
Site 2 (loess soil)	Loess 1	6	1830	75.50	0.31	27.1	28.80	26.5×10^{-3}
	Loess 2	6	1880	66.10	0.33	20.5	24.80	25.5×10^{-3}
	Loess 3	24	1980	83.70	0.30	19.5	32.20	55.5×10^{-3}

The existing experimental results were used to validate the numerical simulation method. The frame modeling approaches were validated by the one-third-scale RC frame structure shaking-table test by Bracci et al. [29,30]. Table 2 shows the modal analysis results' comparison of the simulation results and test results. It can be seen that the simulation error results are all less than 5%, within a satisfactory range for model validation [31]. Figure 4 shows the time–history results comparison of the top displacement and base shear. It can be observed in the figures that the peak values and their occurrence time are closely aligned.

Table 2. Modal analysis results comparison of the frame specimen.

Period (s)	The First Mode	The Second Mode	The Third Mode
Test results	0.568	0.188	0.127
ABAQUS results	0.575	0.179	0.123
Error (%)	1.2%	4.8%	3.1%

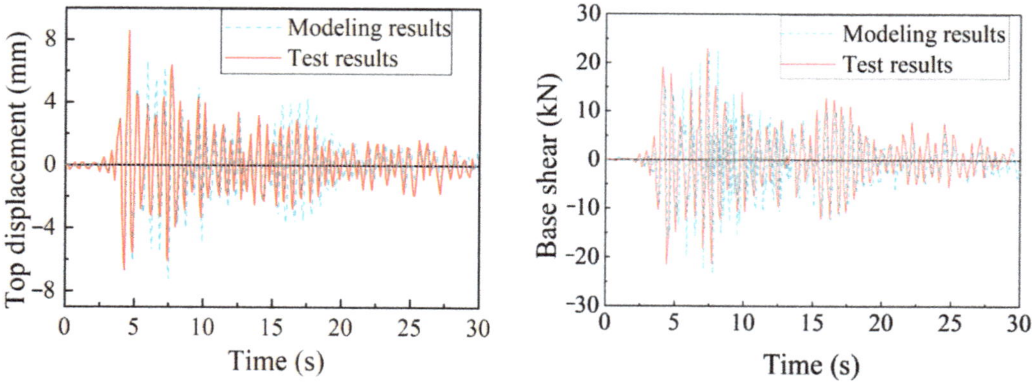

(a) (b)

Figure 4. Validation of the frame modeling approach: (**a**) top displacement comparison of the frame specimen; and (**b**) base shear force comparison of the frame specimen.

The shear wall modeling approaches were validated using the SW2-1 specimen from the low-cycle repeated shear wall loading tests by Zhang et al. [32]. Figure 5 shows the comparison of hysteresis curves. It can be observed that the simulation results show a good agreement with the test results in the shape of hysteresis curves and the maximum load-bearing capacity.

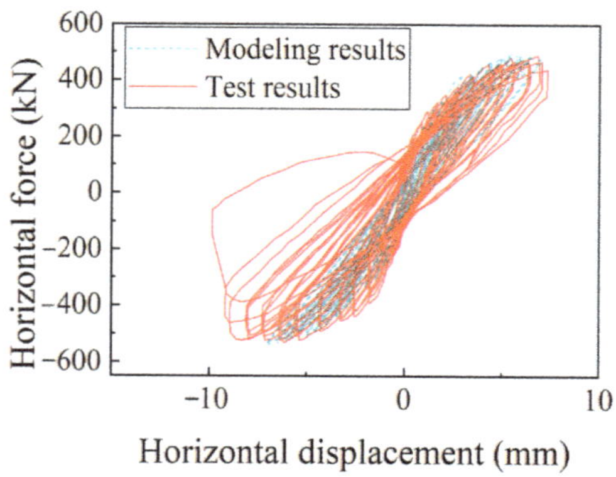

Figure 5. Validation of the shear wall modeling approach.

The SSI modeling approaches were validated by the shaking-table test of the regular steel moment-resisting frame considering SSI performed by Li et al. [33]. The comparison of the bottom acceleration results is shown in Figure 6. The simulation results basically agree with the test results. The peak acceleration during the test was 1.18 m/s^2, while the simulation result was 1.157 m/s^2, with a small error of 1.93%.

In conclusion, with an acceptable range of accuracy, the proposed finite element modeling approaches were reasonable for predicting the seismic behavior of RC frame–shear wall structures with and without considering the SSI effect.

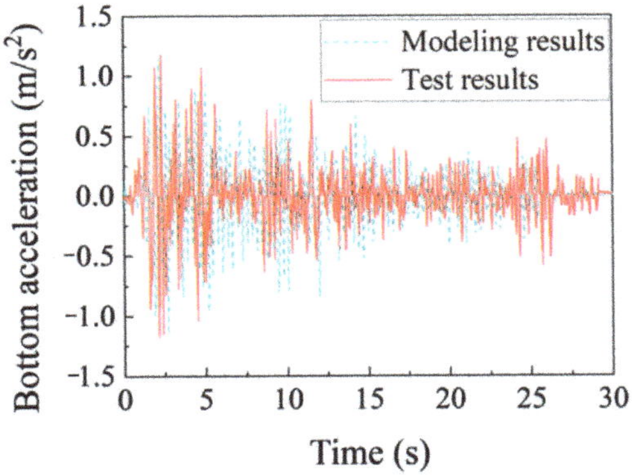

Figure 6. Validation of the modeling approach of structure considering SSI.

2.2. Ground Motion Selection and Amplitude Modulation

The seismic event parameter M_w-R [34–36] is used to choose the earthquake event. Combined with the generation mechanism, far-field long-period ground motions can be characterized by the site condition, long epicentral distance, long-period content on the response spectrum, "harmonic-like" cycles (amplified surface waves) on the acceleration time–history record, and so on. Using Li's categorization criteria (long-period: $\beta_l > 0.4$) [37] as the quantitative condition, the far-field long-period ground motion record selection is based on visual inspection and subjective evaluation. Twenty commonly used ordinary ground motion (OGM) and twenty far-field long-period ground motion records from the 1999 Chi-Chi earthquake (M_w 7.6) were selected. Chi-Chi earthquake data are from the PEER NGA-West 2 database (http://ngawest2.berkeley.edu/). Detailed information on the selected far-field long-period ground motion records and ordinary ground motions is given in Tables 3 and 4, respectively. The average acceleration response spectrum of two types of ground motions was compared with the code spectrum, as shown in Figure 7. It can be seen that far-field long-period ground motions have distinct low-frequency characteristics compared to ordinary ground motions.

Table 3. Basic information of selected far-field long-period ground motion records.

Number	Station-Component	Epicentral Distance (km)	Site Condition	PGA (cm/s^2)	PGV (cm/s)	PGD (cm)	Recording Duration (s)
1	CHY002-EW	43.97	E	108.18	42.81	38.29	150.00
2	CHY004-NS	71.63	E	93.46	14.73	15.84	150.00
3	CHY016-NS	97.59	E	104.73	16.31	15.35	150.00
4	CHY076-EW	63.53	E	61.07	23.08	20.66	90.00
5	CHY093-EW	71.92	E	52.73	19.94	15.10	150.00
6	CHY093-NS	71.92	E	64.52	14.37	17.48	150.00
7	ILA003-NS	143.65	E	63.17	19.11	11.27	134.00
8	ILA005-NS	137.44	D	78.81	15.63	11.60	171.00
9	ILA055-NS	140.81	E	68.04	23.34	20.89	203.00
10	ILA059-NS	136.25	E	57.33	11.48	12.52	153.00
11	TCU031-NS	80.09	D	104.32	46.95	28.11	90.00
12	CHY082-NS	55.03	E	61.01	20.94	24.75	90.00
13	ILA030-NS	136.15	E	69.92	18.67	10.10	90.00
14	ILA032-NS	133.67	D	39.65	8.64	2.40	89.01
15	ILA042-NS	135.89	E	57.79	12.88	10.71	90.00
16	TAP012-EW	151.24	E	60.14	19.11	14.59	127.00
17	TCU010-EW	132.38	C	74.86	33.20	41.06	90.00
18	TCU117-EW	48.33	E	97.98	43.52	33.80	90.00

Table 3. *Cont.*

Number	Station-Component	Epicentral Distance (km)	Site Condition	PGA (cm/s^2)	PGV (cm/s)	PGD (cm)	Recording Duration (s)
19	TCU118-EW	43.76	E	65.92	27.18	22.43	90.00
20	TCU118-NS	43.76	E	88.55	29.40	30.06	90.00

Table 4. Basic information of selected ordinary ground motion records.

Number	Station-Component	Epicentral Distance (km)	Site Condition	PGA (cm/s^2)	PGV (cm/s)	PGD (cm)	Recording Duration (s)
1	COALINGA	23.80	B	91.72	4.96	1.92	39.94
2	TIANJIN-NS	31.60	C	145.8	26.44	5.21	19.19
3	ELC270	12.98	D	233.49	31.30	18.45	40.00
4	EL-centro	12.98	D	293.42	31.39	21.17	31.02
5	TAFT	38.40	C	175.90	16.97	8.04	54.18
6	TABS	11.80	B	179.35	11.70	5.31	23.78
7	KOBE	22.50	C	241.98	18.74	6.36	40.94
8	ELC180	26.20	C	306.62	29.33	13.17	39.94
9	GRA074	31.20	C	64.88	7.08	13.14	39.94
10	CHALFANTB-BEN270	18.3	C	58.88	3.02	1.05	17.56
11	KERN-COUNTY	42.30	C	174.24	17.86	11.69	54.14
12	NORTHRIDGE	19.60	C	341.42	28.59	11.79	29.94
13	PETROLIA	25.80	B	578.14	39.79	38.01	59.82
14	SANTA-MONICA	17.30	B	362.62	25.33	31.70	59.82
15	SMART133O07EW	54.20	C	56.34	2.82	0.85	15.64
16	WHITTIER-NARROWS	23.10	B	214.47	8.93	1.66	39.96
17	CHY036-EW	16.04	C	266.36	40.66	14.99	89.00
18	CHY042-EW	27.47	D	97.54	14.79	12.53	89.00
19	HWA022-EW	58.76	C	118.68	14.22	12.02	76.00
20	HWA022-NS	63.21	C	80.26	10.89	14.42	120.00

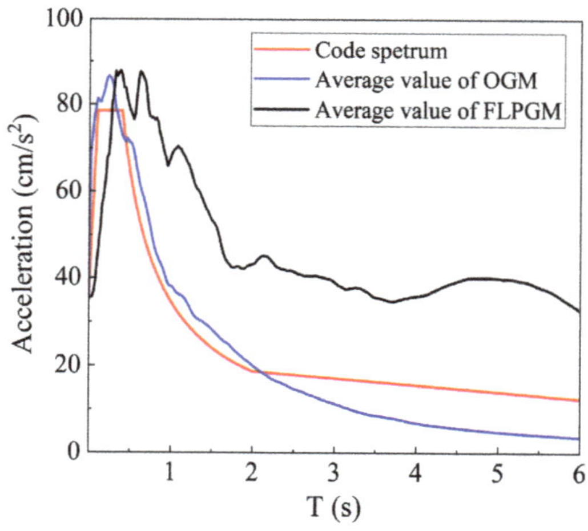

Figure 7. The acceleration spectrum comparison of the code spectrum, and the average spectrum of 20 records for both ordinary ground motion (OGM) and far-field long-period ground motion (FLPGM).

All ground motion records have been carefully checked for baseline correction and filtering. According to the designed seismic precautionary intensity of the structures, the peak accelerations of 20 far-field long-period and 20 ordinary ground motions were adjusted to 35 gal, 100 gal, and 220 gal, respectively. The structural seismic response under frequent, design-based, and rare earthquake levels was then compared and analyzed.

3. Seismic Response of RC Frame–Shear Wall Structures Considering SSI

This part conducts a comparative analysis of the seismic response of RC frame–shear wall structures (S10, S20, and S30) subjected to far-field long-period ground motions and ordinary ground motions, considering the SSI effect of sand soil (Site 1) and loess soil (Site 2). The influence of both ground motion types and soil types on the structural seismic response considering SSI is presented. The results shown below are all the average values of 20 ground motion records.

3.1. Floor Displacement

Figure 8 shows the floor displacement response considering SSI under two types of ground motions for the structures on sand soil (Site 1).

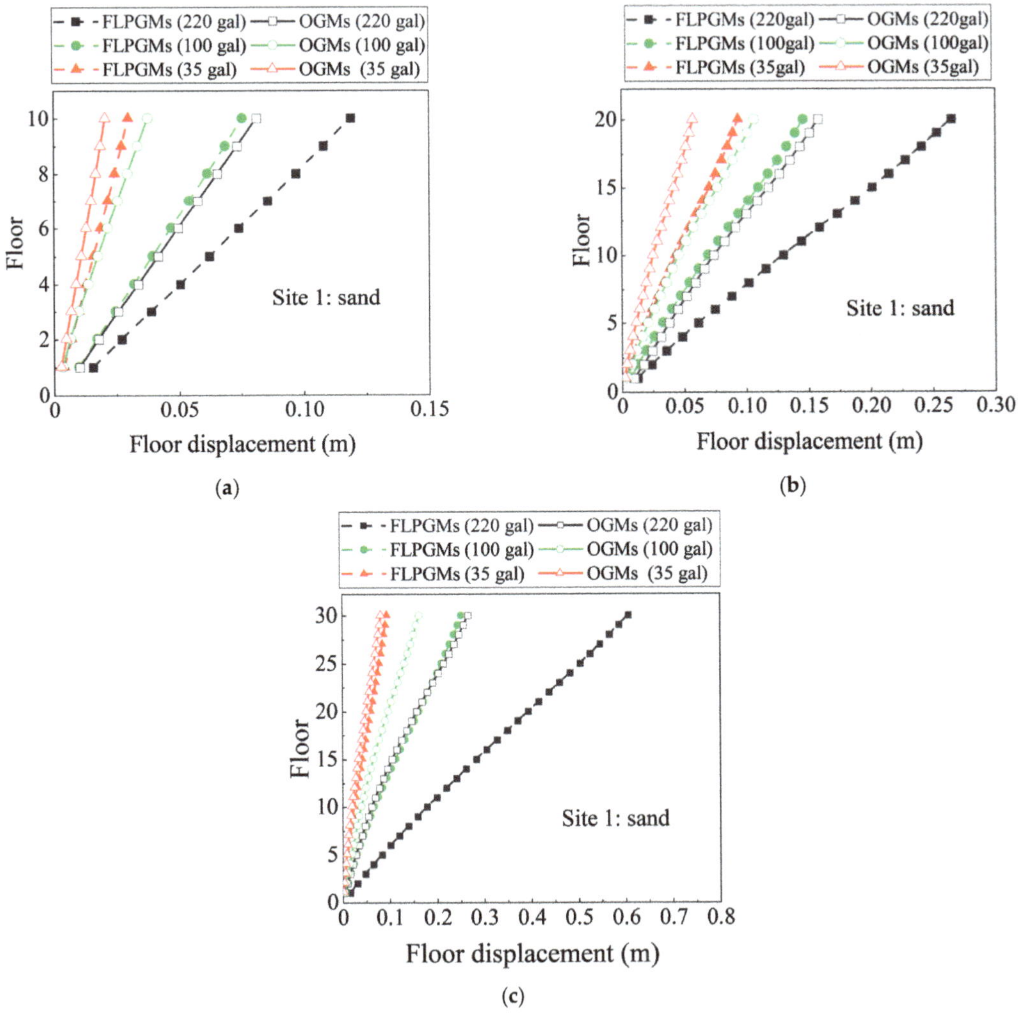

Figure 8. Floor displacement response considering SSI under two types of ground motions for the structures on sand soil (Site 1): (**a**) S10; (**b**) S20; and (**c**) S30.

It can be observed from Figure 8 that, as the natural vibration period of the structure increases, the floor displacement also increases. For the S10 structure, the displacement follows an approximately straight line along the floors, whereas the S30 structure shows a slight bending, although the displacement still increases monotonically under earthquakes. This bending trend in the floor displacement is more pronounced for S30 under far-field long-period ground motion. The floor displacement under far-field long-period ground motions is significantly larger than that under ordinary ground motions at the same PGA level. The unfavorable impact of far-field long-period ground motions on structures with longer periods is more pronounced, both with and without the SSI effect.

Figure 9 shows the floor displacement under two types of ground motion for the structures on loess soil (Site 2).

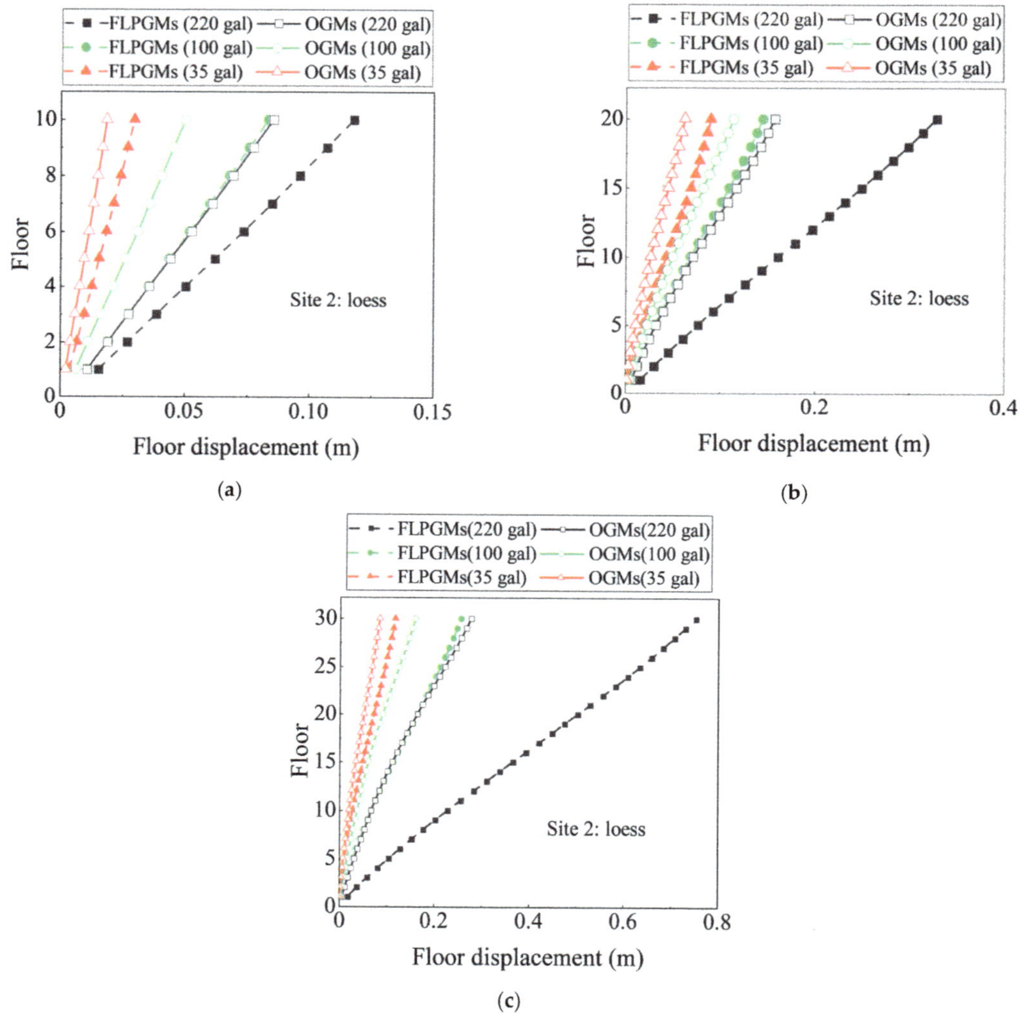

Figure 9. Comparison of structural floor displacement under two types of ground motions for the structures on loess soil (Site 2): (**a**) S10; (**b**) S20; and (**c**) S30.

It can be seen in Figure 9 that the floor displacement increases progressively as the PGA increases. For the S20 and S30 structures, the floor displacement at 220 gal is significantly greater than that at other PGA levels. For high-rise building structures, coupling the effect of far-field long-period ground motions and soil–structure interaction, resonance is prone to occur, resulting in the amplification of the displacement. This can be attributed to the low-frequency characteristics of far-field long-period ground motions and the structural period extension considering SSI. Comparing Figures 8 and 9, it can be seen that the SSI effect is more pronounced on loess soil (Site 2) than on sand soil (Site 1). And the displacement amplification is aggravated, especially for the longer-period structures subjected to far-field long-period ground motions with a high PGA.

3.2. Inter-Story Drift Ratio

The comparison of the inter-story drift ratio of the structures on sand soil (Site 1) considering SSI under two types of ground motions is shown in Figure 10. According to the China Code [21,22], the maximum inter-story drift ratio for RC frame–wall structures should not exceed 1/800 and 1/100 when subjected to seismic action with a PGA of 35 gal and 220 gal, respectively, corresponding to the seismic precautionary objectives of "no-damage" and "no-collapse." The limitation is indicated in Figure 10.

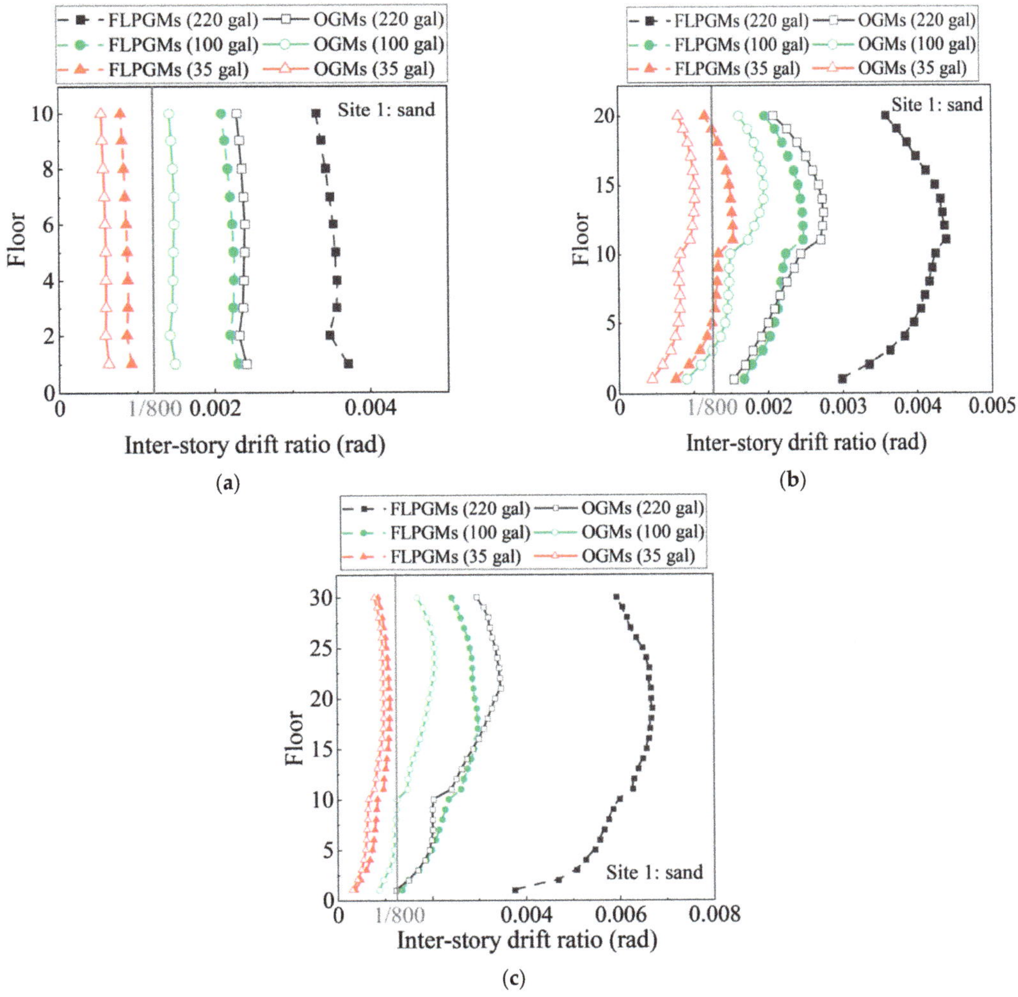

Figure 10. Inter-story drift ratio response considering SSI under two types of ground motions for the structures on sand soil (Site 1): (**a**) S10; (**b**) S20; and (**c**) S30.

It can be seen from Figure 10 that the structural deformation under far-field long-period ground motions is more pronounced than that under ordinary ground motions, with a larger value of the inter-story drift ratio. For the S20 structure on Site 1 under far-field long-period ground motions with a PGA of 35 gal, the maximum inter-story drift ratio exceeds the 1/800 elastic limit, presenting a potential risk for seismic safety. The drift amplification of far-field long-period ground motions is aggravated as the PGA and structural period increases. The SSI makes the inter-story drift ratio curve irregular. As the PGA decreases, the deformation reduces. The inter-story drift ratio curves under ground motions with different PGA levels have a similar shape. The SSI effect amplifies the inter-story drift ratio, and the amplification becomes more significant as the PGA increases.

Figure 11 shows the comparison of the inter-story drift ratio of RC frame–shear wall structures on loess soil (Site 2) under two types of ground motions.

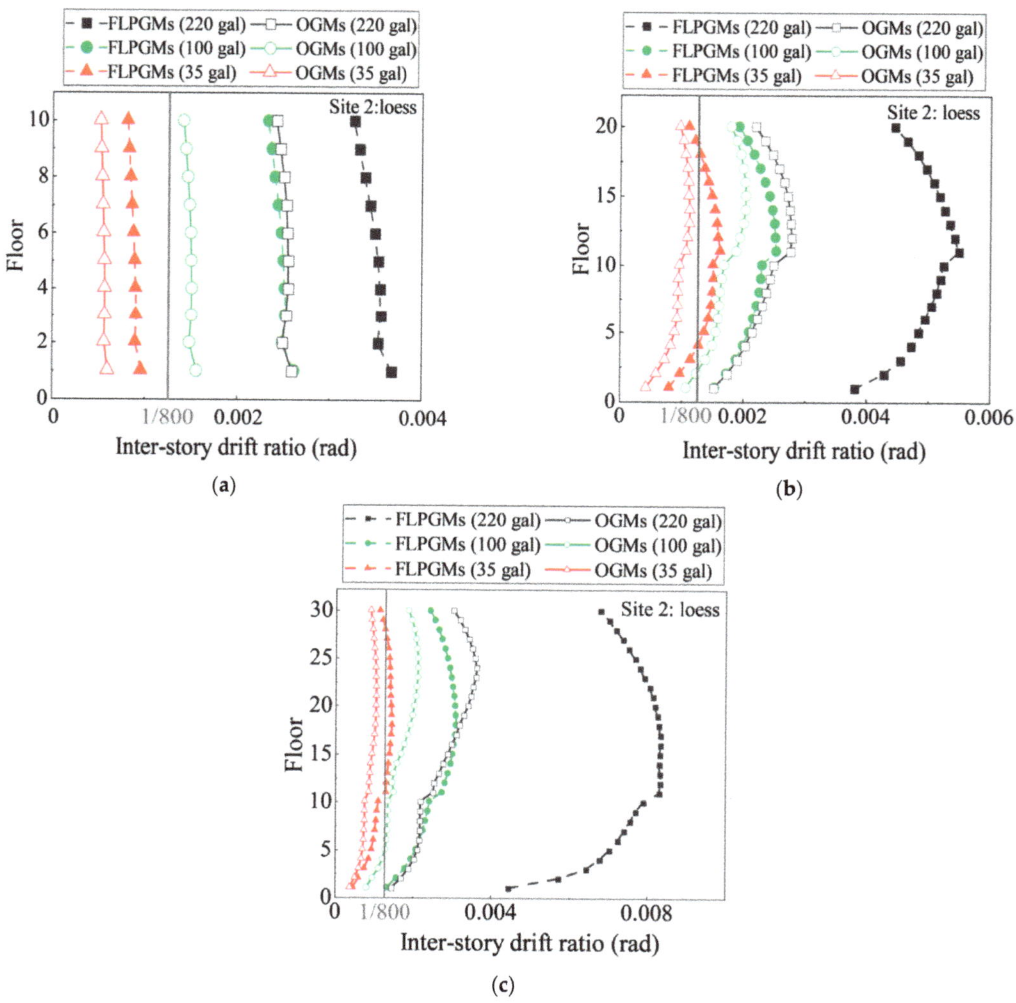

Figure 11. Comparison of inter-story drift ratio under two types of ground motions for the structures on loess soil (Site 2): (**a**) S10; (**b**) S20; and (**c**) S30.

It can be inferred from Figure 11 that, as the PGA and the structural period increase, the maximum inter-story drift ratio also grows. The most significant increase in the inter-story drift ratio occurs at the bottom of the structure when the PGA is 220 gal, with the S30 structure showing a much greater rise compared to the S20 and S10 structures. On Site 2 conditions, the inter-story drift ratio under far-field long-period ground motions is noticeably larger than that under ordinary ground motions, while the maximum inter-story drift ratio of the structure has a smaller value. In the S20 and S30 structures, the inter-story drift ratio at a PGA of 100 gal is even greater than that at 220 gal in the lower parts of the structure. For the S30 structure on Site 2 under far-field long-period ground motions with a PGA of 35 gal, the maximum inter-story drift ratio exceeds the 1/800 elastic limit. Compared with Figure 10c, it can be inferred that the SSI effect on the seismic response to far-field long-period ground motions is more significant for the longer-period structures on the softer soil site.

3.3. Floor Shear

Figure 12 shows the floor shear forces considering SSI under two types of ground motions for the structures on sand soil (Site 1).

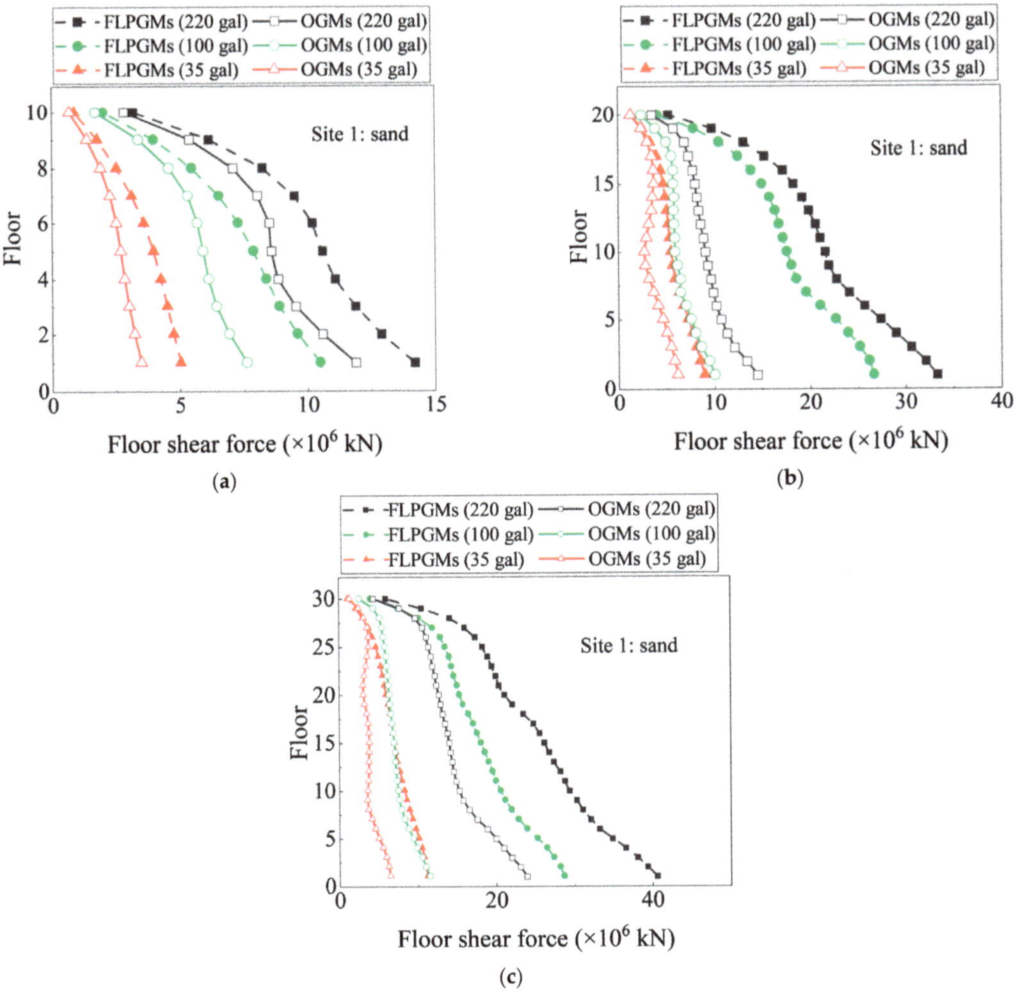

Figure 12. Floor shear force response considering SSI under two types of ground motions for the structures on sand soil (Site 1): (**a**) S10; (**b**) S20; and (**c**) S30.

As shown in Figure 12, the SSI effect has a greater influence on the force response of high-rise buildings with longer periods. Considering the SSI effect, the floor shear force decreases remarkably. The shear force under far-field long-period ground motions is greater than that under ordinary ground motions, especially for long-period structures. When the PGA is 220 gal, the floor shear force of the S20 and S30 structures increases at individual floors, resulting in less smooth curves than those of the S10 structure.

The comparison of the floor shear force under two types of ground motions for the structures on Site 2 is plotted in Figure 13.

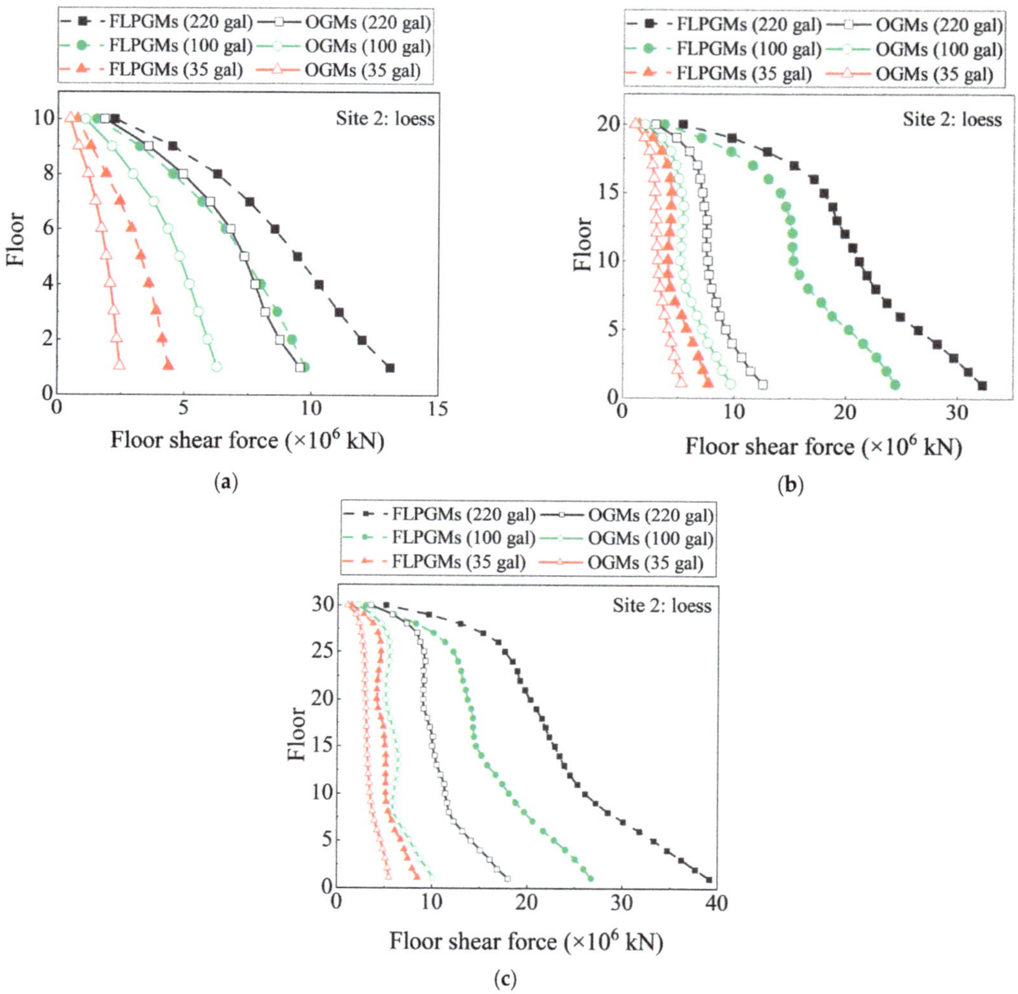

Figure 13. Comparison of structural floor shear force under two types of ground motions for the structures on loess soil (Site 2): (**a**) S10; (**b**) S20; and (**c**) S30.

As shown in Figure 13, the shear force in the lower floors increases progressively with the increase in PGA and the structural period, which shows a "convex up and concave down" trend. The shear force is largest at the bottom floor and smallest at the top. For the S20 and S30 structures under far-field long-period ground motions, the shear force exhibits a remarkable increase when the PGA increases from 35 gal to 100 gal. The shear force at 100 gal is significantly greater than that under ordinary ground motions at 220 gal.

3.4. Base Shear

The base shear response of the structures on sand soil considering SSI under 20 OGMs and 20 FLPGMs is compared in Figure 14.

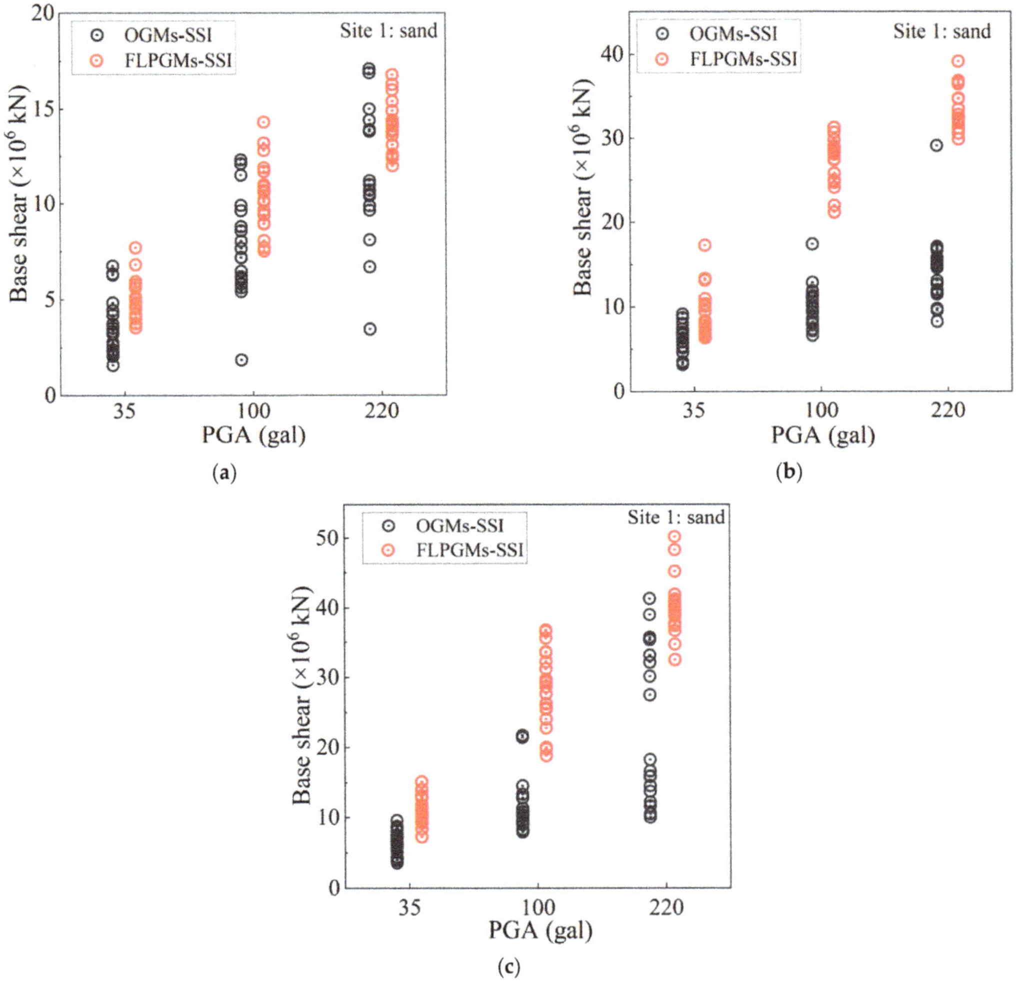

Figure 14. Base shear distribution considering SSI under two types of ground motions for the structures on sand soil (Site 1): (**a**) S10; (**b**) S20; and (**c**) S30.

It can be observed from Figure 14 that the base shear force considering SSI under far-field long-period ground motions is significantly greater than that under ordinary ground motions. The base shear force increases with the increase both in the PGA and the structural period. At a PGA of 220 gal, the base shear force of the S30 structure is more than twice that of the S10 structure. The base shear distribution is more concentrated at the lower PGA level. The distribution differences between the base shear under two types of ground motions increase with PGA.

The base shear force distribution of RC frame–shear wall structures on Site 2 under the selected ground motions is presented in Figure 15.

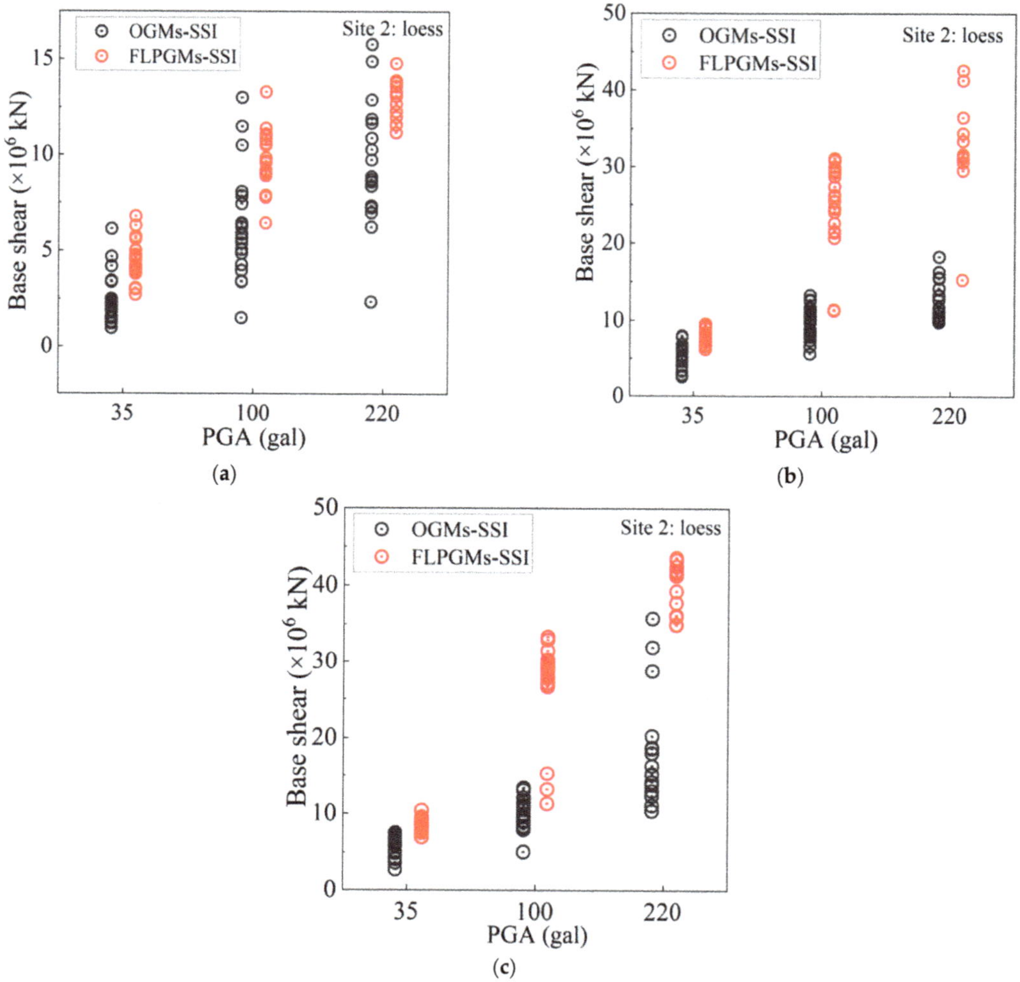

Figure 15. Base shear force distribution under two types of ground motions for the structures on loess soil (Site 2): (**a**) S10; (**b**) S20; and (**c**) S30.

For the S10 structure, the maximum value of the base shear under two types of ground motions shows little difference. For the S20 and S30 structures, far-field long-period ground motions exhibit an obvious increase in the base shear compared to ordinary ground motions. Under ordinary ground motions, the base shear on loess soil has a small value and concentrated distribution than that on sand soil (Figure 14).

3.5. The Influence of Structural Frequency–Feature on Seismic Response Considering SSI Effect

Three site conditions are considered in this study, including the rigid base, sand soil, and loess soil. The frequencies of the S10, S20, and S30 structures on the different site conditions are listed in Table 5. As indicated in Table 3, the first mode is in the x-direction, the second mode is in the y-direction, and the third mode is the torsion.

Table 5. Structural frequency–feature under different site conditions.

Structural Frequency (Hz)	S10 Structure Rigid Base	Site 1	Site 2	S20 Structure Rigid Base	Site 1	Site 2	S30 Structure Rigid Base	Site 1	Site 2
First mode (x-direction)	1.76305	0.66637	0.55244	0.84460	0.57687	0.51871	0.48345	0.32837	0.29859
Second mode (y-direction)	1.76523	0.66641	0.55249	0.90861	0.63525	0.55103	0.53590	0.37670	0.34329
Third mode (torsion)	2.98240	0.66646	0.55354	1.44100	0.66193	0.55136	0.89475	0.66194	0.55103

As shown in Table 5, considering the SSI effect causes a significant decrease in the natural frequency of the structures. The structural frequency reduction on the loess soil (Site 2) is more obvious than that on the sand soil (Site 1). Comparing three structures with different floors, S10 presents the most significant frequency reduction. Take the first mode (x-direction) on Site 1 as an example: S10 presents a 62.2% reduction from 1.73605 Hz to 0.66637 Hz, while S20 and S30 present 31.7% and 32.1% reduction, respectively. Therefore, it can be concluded that the frequency reduction becomes more significant as the number of floors decreases.

In conclusion, the SSI effect can reduce the structural frequency, especially for the structures with lower number of floors on the softer soil site. The decrease in structural frequency results in the increase in structural period. This is unfavorable for the structures under far-field long-period ground motions.

4. Influence Mechanism of SSI on Structural Seismic Response

4.1. Quantitative Analysis of the SSI Effect on Structural Seismic Response

The SSI influence coefficient is proposed in this study to quantify the effect of SSI on the seismic response under far-field long-period ground motions. First, the amplification factor of the SSI effect on the seismic response is obtained; then, the response under two types of ground motions is compared. There are two sets of variable parameters in the coefficient, including whether or not to consider the SSI effect and the types of ground motions. The SSI effect on the seismic response is compared to that on the rigid base, for far-field long-period and ordinary ground motions, respectively. The two are then compared to obtain the influence coefficient. The calculation equations are given below:

$$\alpha_\Delta = \frac{u_{\text{SSI,FLPGM}} / u_{\text{Rigid,FLPGM}}}{u_{\text{SSI,OGM}} / u_{\text{Rigid,OGM}}} \quad (1)$$

$$\alpha_\theta = \frac{\theta_{\text{SSI,FLPGM}} / \theta_{\text{Rigid,FLPGM}}}{\theta_{\text{SSI,OGM}} / \theta_{\text{Rigid,OGM}}} \quad (2)$$

$$\alpha_V = \frac{V_{\text{SSI,FLPGM}} / V_{\text{Rigid,FLPGM}}}{V_{\text{SSI,OGM}} / V_{\text{Rigid,OGM}}} \quad (3)$$

where α_Δ, α_θ, and α_V are the SSI influence coefficient of the floor displacement, inter-story drift ratio, and floor shear, respectively; and u, θ, and V are the lateral displacement, inter-story drift ratio, and shear force of the structures, respectively.

When the SSI influence coefficient is greater than 1, it means the SSI effect on the seismic response under far-field long-period ground motions is more pronounced than that under ordinary ground motions. When the coefficient is less than 1, the opposite is true.

4.1.1. SSI Influence Coefficient of Floor Displacement

Figure 16 shows the SSI influence coefficient of the floor displacement (α_Δ) of the structures on two types of site soil.

It can be seen that, for the structures with different periods, the SSI influence coefficient presents various trends under different PGA levels. For the S10 structure, the SSI coefficient at 35 gal is less than 1, while that at 100 gal and 220 gal is both larger than 1. For the S20 structure on loess soil (Site 2), when the PGA is 220 gal, the SSI influence coefficient of the floor displacement is remarkably greater than 1. A similar phenomenon occurred at a PGA of 100 gal for the S30 structures. In this case, the SSI effect under far-field ground motions is more aggravated than that under ordinary ground motions. The SSI influence coefficient on Site 2 is noticeably larger than that on Site 1, highlighting the significant impact of the foundation soil type on the SSI effect.

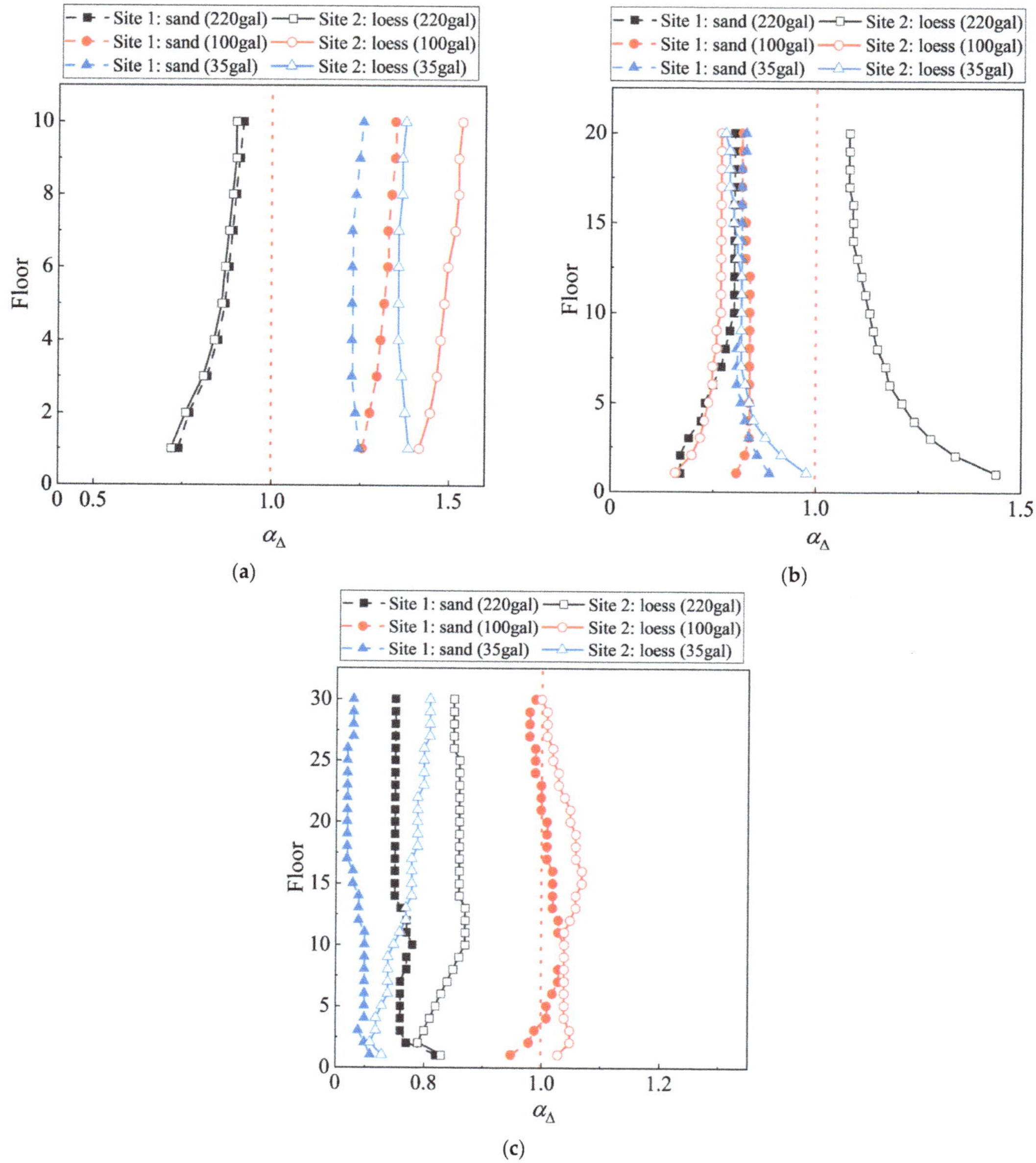

Figure 16. SSI influence coefficient of floor displacement under two types of site condition: (**a**) S10; (**b**) S20; and (**c**) S30.

4.1.2. SSI Influence Coefficient of Inter-Story Drift Ratio

The SSI influence coefficient of the inter-story drift ratio of the structures on two types of site conditions is displayed in Figure 17.

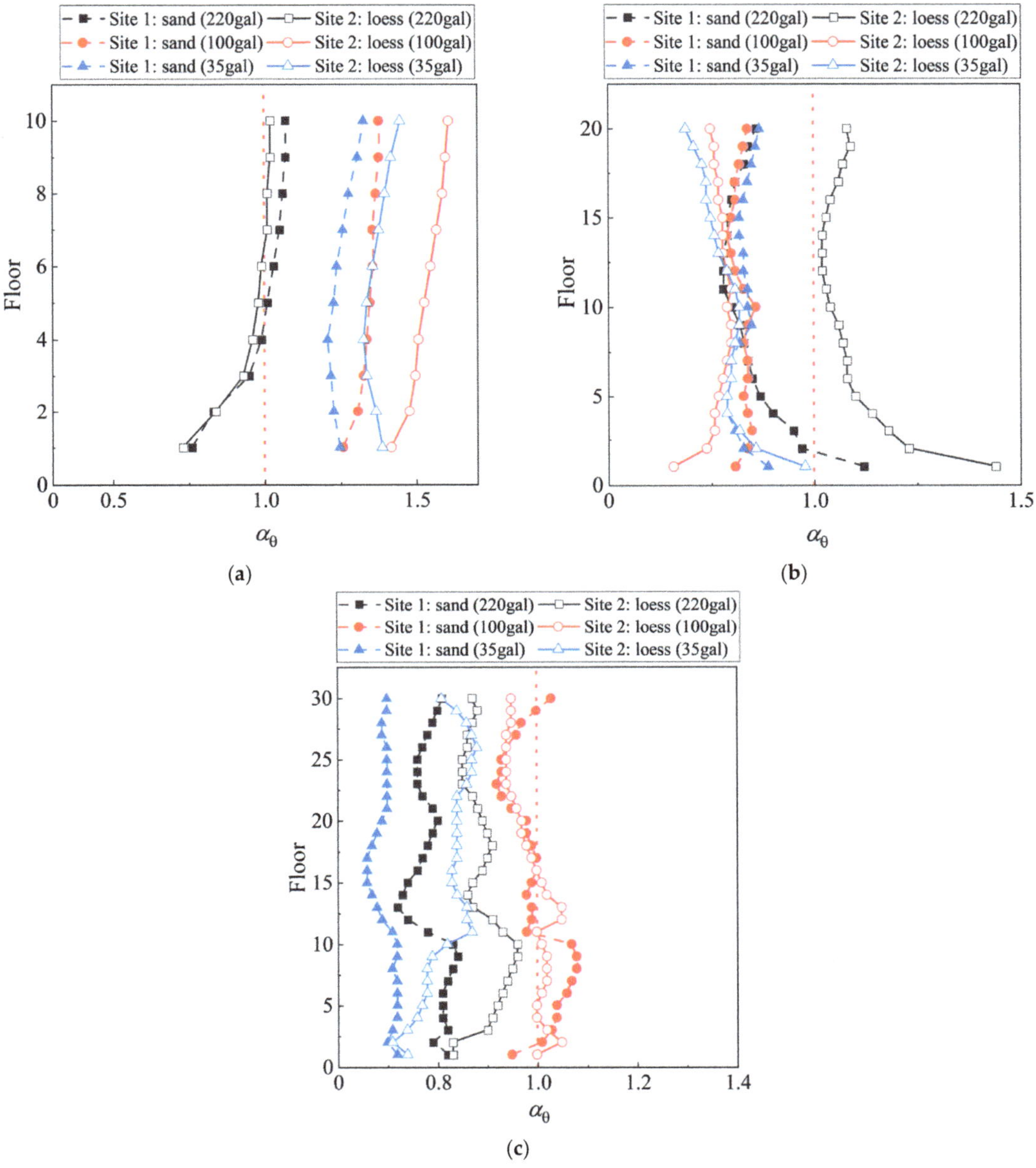

Figure 17. SSI influence coefficient of inter-story drift ratio under two types of site condition: (**a**) S10; (**b**) S20; and (**c**) S30.

It can be observed from Figure 17 that the SSI effect on the inter-story drift ratio is generally consistent with that on floor displacement. The SSI effect on the deformation response of the structures with different periods exhibits distinct trends. The SSI influence coefficient on Site 2 is significantly higher than on Site 1, with the bottom floors of the structure being more affected. This suggests that far-field long-period ground motion has a more pronounced impact on the lower parts of the structure which plays a critical role under earthquakes.

4.1.3. SSI Influence Coefficient of Floor Shear

Figure 18 gives the SSI influence coefficient of the floor shear force (α_V) on two types of site conditions.

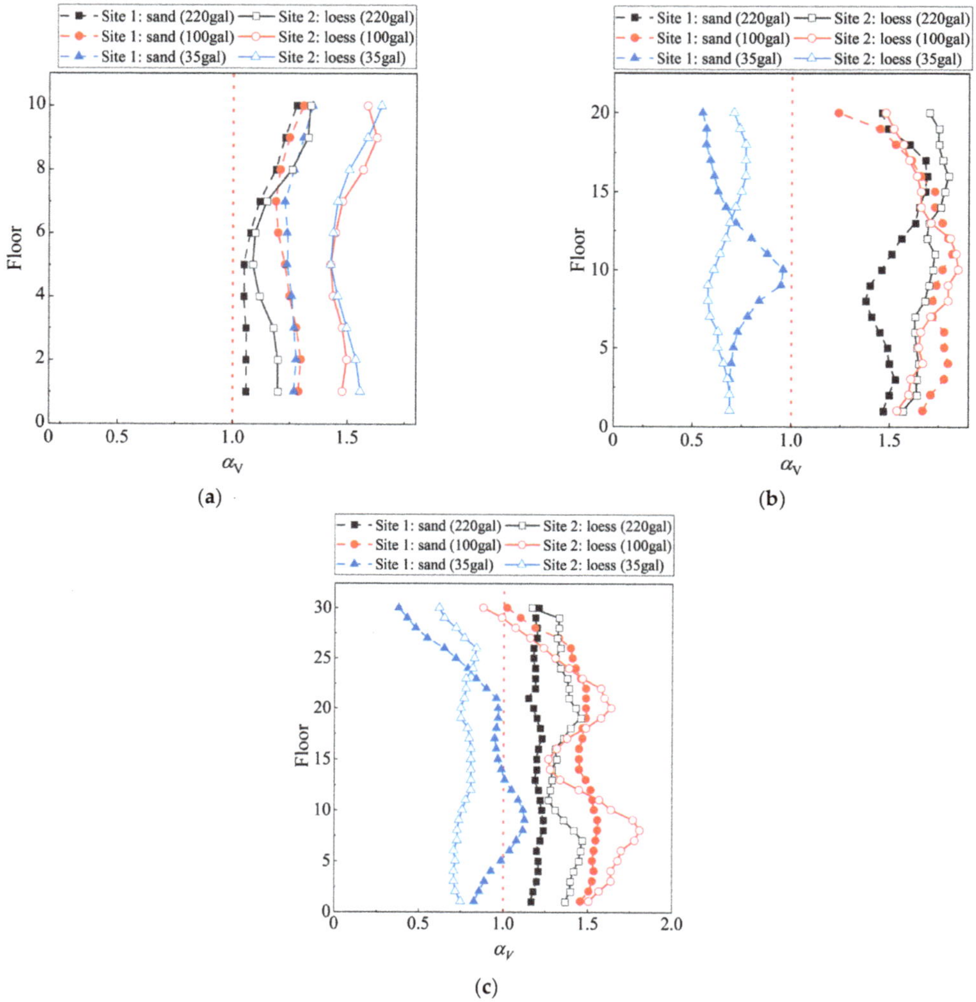

Figure 18. SSI Influence coefficient of floor shear force under two types of site condition: (**a**) S10; (**b**) S20; and (**c**) S30.

For the S10 structures, the SSI influence coefficient of the floor shear is all greater than 1. This indicates that, for the structures with shorter periods, the SSI effect on the floor shear under far-field long-period ground motions is more significant than that under ordinary ground motions. The SSI influence coefficient of the upper structures is greater than that of the middle and bottom parts. For the S20 and S30 structures, the SSI influence coefficient of the floor shear force (α_V) is greater than 1 when the PGA is 100 gal and 220 gal. That is to say, for the structures with longer periods, the SSI effect on the floor shear is aggravated when subjected to the far-field long-period ground motions with higher PGA levels.

4.2. SSI Effect on the Maximum Response Under Far-Field Long-Period Ground Motions

The SSI effect on the maximum seismic response is evaluated by comparing it to the seismic response on the rigid base. It concludes the SSI effect on the base shear, top displacement, and the maximum inter-story drift ratio. The corresponding influence coefficients are defined as β_V, β_Δ, and β_θ, shown as below, respectively:

$$\beta_V = \frac{V_{SSI,max}}{V_{Rigidity,max}} \tag{4}$$

$$\beta_\Delta = \frac{u_{SSI,max}}{u_{Rigidity,max}} \tag{5}$$

$$\beta_\theta = \frac{\theta_{SSI,max}}{\theta_{Rigidity,max}} \tag{6}$$

where β_V, β_Δ, and β_θ are the SSI influence coefficient of the maximum value of structural shear force, top displacement, and inter-story drift ratio, respectively.

It can be observed in Equations (4)–(6) that the SSI coefficient is greater than 1, indicating that the consideration of the SSI effect will result in an increase in seismic response. An SSI coefficient less than 1 indicates that the SSI effect will reduce the seismic response. The SSI influence coefficients of the maximum seismic response on two site conditions under far-field long-period and ordinary ground motions with different PGA levels are plotted in Figure 19.

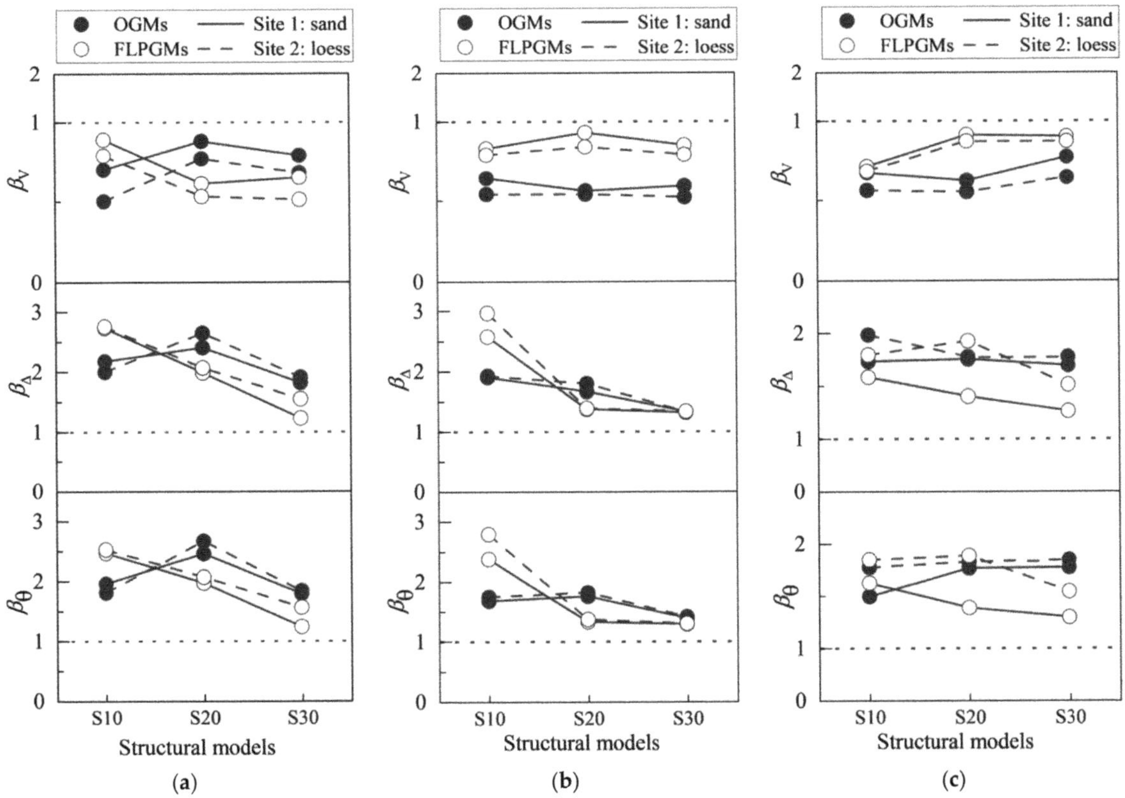

Figure 19. SSI influence coefficient of the maximum structural response under two types of ground motions: (**a**) PGA = 35 gal; (**b**) PGA = 100 gal; and (**c**) PGA = 220 gal.

It is evident in Figure 19 that the SSI coefficients of the base shear are all less than 1, while those of the top displacement and the maximum inter-story drift ratio are all greater than 1 (less than 3). This means that the consideration of SSI causes the decrease in the base shear and the increase in the displacement response. Comparing the results on two site conditions, the SSI influence coefficient of the base shear on the loess site is all less than that on sand soil, while, for the displacement response, the SSI influence coefficient on the loess site is all greater than that on sand soil. The SSI coefficient of the base shear is all less than 1, ranging from 0.5 to 1. Therefore, it can be concluded that loess soil has a more remarkable effect on the seismic response than sand soil for both types of ground motions.

4.3. Influence Mechanism of Far-Field Long-Period Ground Motions on Structural Response

A previous study [38] has pointed out that far-field long-period ground motions have a relatively smaller value of PGA but a higher value of PGV, PGD, PGV/PGA, and PGD/PGA than ordinary ground motions, and exhibit long-duration and low-frequency characteristics in their time history. When the structural period exceeds 1 s, the velocity and displacement spectral values of far-field long-period ground motions are higher than those of ordinary ground motions. The acceleration response spectrum of far-field long-period ground motions presents "bimodal characteristics" with a second peak appearing when the period is between 4 s and 7 s [39], as shown in Figure 20.

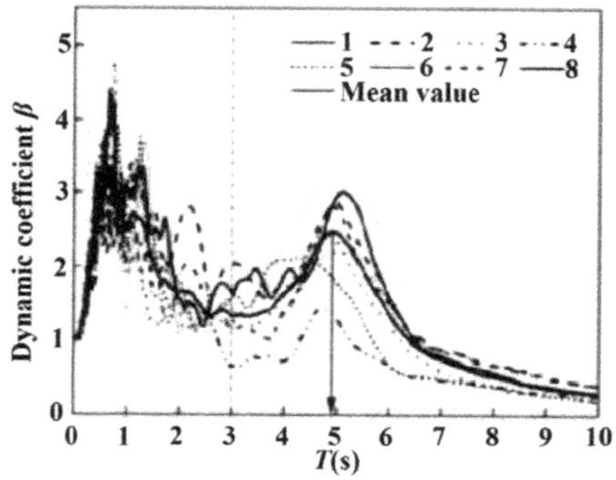

Figure 20. The standard acceleration response spectra of far-field long-period ground motion [38].

As mentioned above, the SSI effect reduces the structural frequency. The site soil filters out the high-frequency components in the seismic wave, leading to an increase in the low-frequency components. The structures considering the SSI effect are prone to resonance with far-field long-period ground motions at the second peak. Coupling long-period ground motions with the SSI effect, high-rise building structures are more prone to aggravated seismic damage.

The simplified model of the acceleration response spectrum for far-field long-period ground motions and ordinary ground motions is shown in Figure 21.

According to the different trends over the period, the acceleration response spectrum can be divided into five parts. When the period is less than 0.1 s, the curves for both types of ground motions are in the rising stage. When the period is in the range of 0.1 s to the structural period T_g, both curves remain at their peak values. In the period range of T_g to T_1, the acceleration response spectra of both ground motions are in the descending stage. Between the periods of T_1 and T_2, the acceleration response of ordinary ground motions continues to decrease, while that of far-field long-period ground motions starts to increase to the second peak. As for the period ranges from T_2 to 6 s, both types of ground motions exhibit a declining trend.

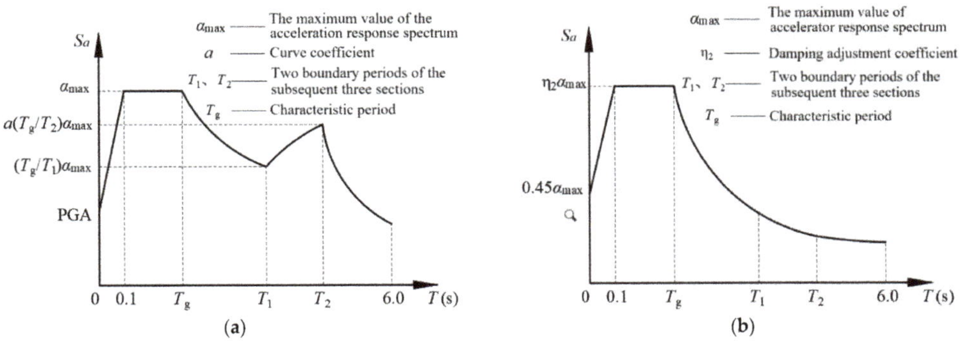

Figure 21. Simplified model of acceleration response spectrum for two types of ground motions: (**a**) far-field long-period ground motions; and (**b**) ordinary ground motions.

When the SSI effect extends the structural period, according to the acceleration response spectrum of far-field long-period and ordinary ground motions, the following may occur:

Case 1: The structural period is located in region I, where the acceleration response spectra for two types of ground motions are both rising, and β_V values are greater than 1, as shown in Figure 22.

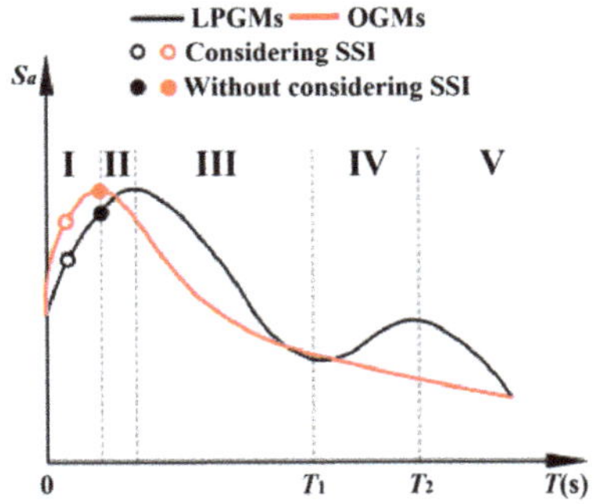

Figure 22. Schematic diagram of Case 1.

Case 2: The structural period is located in region II, where the spectral curves for two types of ground motions show a distinct trend, resulting in the long-period ground motion having β_V values greater than 1, while the ordinary ground motions have β_V values less than 1, as shown in Figure 23.

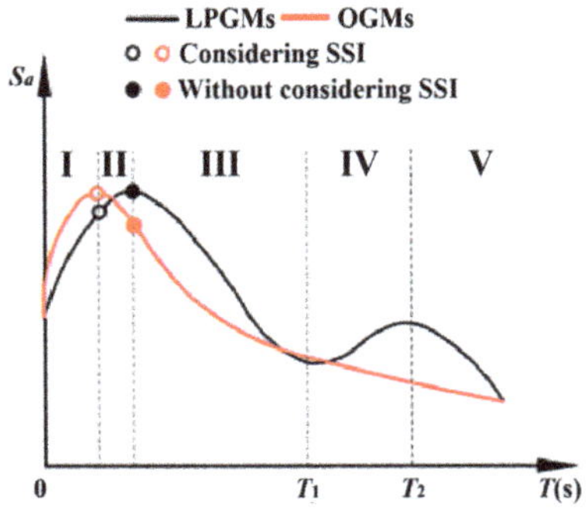

Figure 23. Schematic diagram of Case 2.

Case 3: The structural period is located in region III, where both spectra show a downward trend, and β_V values are less than 1, as shown in Figure 24.

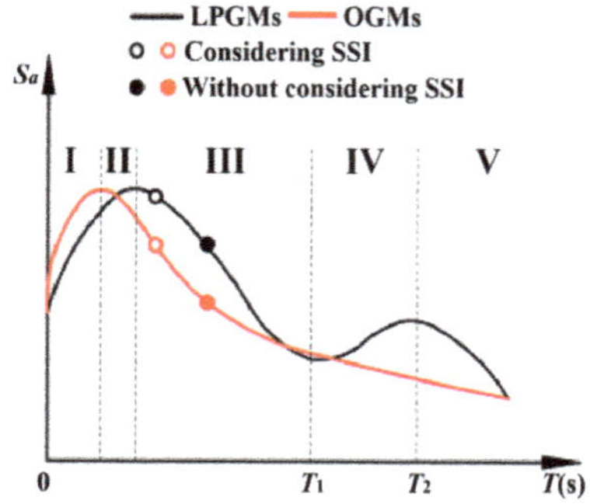

Figure 24. Schematic diagram of Case 3.

Case 4: With the structural period located in region III and the period considering SSI located in region IV, the spectrum curve for ordinary ground motions continues to decline; that for far-field long-period ground motions rises, but the spectrum value is still less than that in region III; β_V values are both less than 1, as shown in Figure 25.

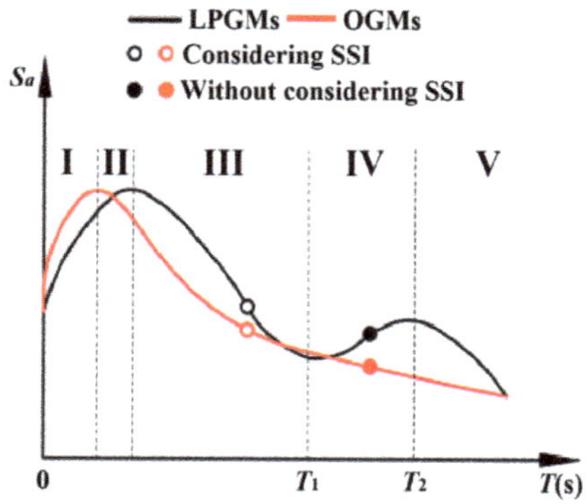

Figure 25. Schematic diagram of Case 4.

Case 5: The structural period is located in region IV, where the curve for long-period ground motions rises with β_V values greater than 1, while that for ordinary ground motions falls with β_V values less than 1, as shown in Figure 26.

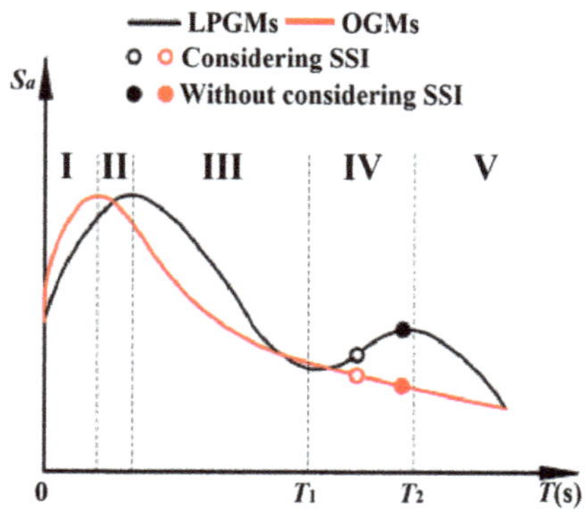

Figure 26. Schematic diagram of Case 5.

Case 6: The structural period is located in region V, where both spectral curves descend, and β_V values are less than 1, as shown in Figure 27.

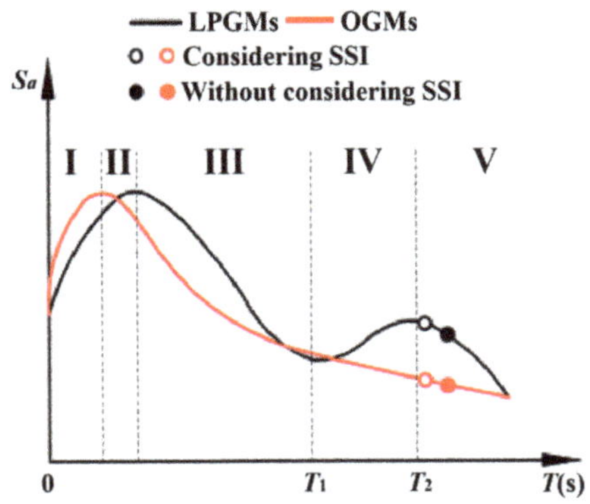

Figure 27. Schematic diagram of Case 6.

For the S10, S20, and S30 structures in this study, the structural periods with and without considering the SSI effect are all located in the descending region of the acceleration response spectra for both far-field long-period and ordinary ground motions, resulting SSI influence coefficients of the base shear (β_V) are less than 1.

5. Conclusions

In this paper, ABAQUS was employed to conduct numerical simulations aimed at studying the influence of SSI on the seismic response of RC frame–shear wall structures under far-field long-period ground motions. Three sets of various parameters are discussed, including ground motion types (far-field long-period and ordinary ground motions); site conditions (rigid base, sand soil, and loess soil); and structural period (the S10, S20, and S30 structures). Then, the mechanism of the SSI effect on the structural response to far-field long-period ground motions is explored based on the ground motion spectrum analysis. The main conclusions are as follows:

(1) The SSI effect increases the floor displacement and the maximum inter-story drift ratio. The displacement amplification is aggravated for the longer-period structures subjected to far-field long-period ground motions with a high PGA. The unfavorable impact of far-field long-period ground motions on structures with longer periods is more pronounced, both with and without the SSI effect. This can be attributed to the resonance motivated by the coupling of far-field long-period ground motions with the SSI effect.

(2) Considering the SSI effect, the floor shear force decreases. The SSI effect has a greater influence on the force response of high-rise buildings with longer periods. In the current code, the SSI effect is considered by reducing the designed shear force on the assumed rigid base. As a result, the structures are designed with reduced resistance. However, the SSI effect amplifies the displacement response, especially for the longer-period structures on softer soil under the strong far-field long-period ground motions. This is unfavorable for the seismic safety and resilience. That is to say, shear force reduction may not be applicable to the structural design considering the SSI effect under far-field long-period ground motions.

(3) Loess soil (Site 2) has a more remarkable effect on the seismic response than sand soil (Site 1) for both far-field long-period and ordinary ground motions. Due to the low-frequency characteristics of far-field long-period ground motions and the structural period extension considering SSI, the resonance at the softer soil site is more pronounced.

(4) The SSI effect can reduce the structural frequency, especially for the structures with a lower number of floors on the softer soil site. The decrease in the structural frequency means an increase in the structural period. This is unfavorable for the structures under far-field long-period ground motions.

(5) The SSI influence coefficients of the base shear on two site conditions under far-field long-period and ordinary ground motions with different PGA levels are all less than 1 (ranging from 0.5 to 1), while those of the top displacement and the maximum inter-story drift ratio are all greater than 1 (less than 3). The SSI has a beneficial effect on the base shear, but an unfavorable effect on the top displacement and the maximum inter-story drift ratio. The "bimodal characteristics" of the acceleration response spectrum for far-field long-period ground motions make it possible for the structure to have an SSI influence coefficient of the base shear β_V greater than 1.

Future investigations are necessary to validate the conclusions that can be generalized to all situations. More structure types, structural plane layout, soil types, ground motion types (i.e., near-fault long-period ground motions), etc. need to be considered in the further analysis.

Author Contributions: Conceptualization, B.W.; formal analysis, Z.Z. and Q.H.; funding acquisition, B.W.; investigation, R.Z.; project administration, B.W.; software, R.Z.; visualization, K.Y., P.C. and B.H.; writing—original draft, K.Y. and P.C.; writing—review and editing, K.Y. All authors have read and agreed to the published version of the manuscript.

Funding: This research was supported by the National Natural Science Foundation of China (NSFC), grant numbers 51978076 and 52478493; and the Fundamental Research Funds for Central Universities, grant number 300102283202.

Data Availability Statement: The raw data supporting the conclusions of this article will be made available by the authors on request.

Conflicts of Interest: Author Bin Hao was employed by the company Shaanxi Architecture Seismic Reduction and Isolation Science Research Institute Co., Ltd. The remaining authors declare that the research was conducted in the absence of any commercial or financial relationships that could be construed as a potential conflict of interest.

References

1. Koketsu, K.; Miyake, H. A seismological overview of long-period ground motion. *J. Seismol.* **2008**, *12*, 133–143. [CrossRef]
2. Beck, J.L.; Hall, J.F. Factors contributing to the catastrophe in Mexico City during the earthquake of September 19, 1985. *Geophys. Res. Lett.* **1986**, *13*, 593–596. [CrossRef]
3. Fletcher, J.B.; Wen, K.L. Strong ground motion in the Taipei basin from the 1999 Chi-Chi, Taiwan, earthquake. *Bull. Seismol. Soc. Am.* **2005**, *95*, 1428–1446. [CrossRef]
4. Koketsu, K.; Hatayama, K.; Furumura, T.; Ikegami, Y.; Akiyama, S. Damaging long-period ground motions from the 2003 Mw 8.3 Tokachi-oki, Japan Earthquake. *Seismol. Res. Lett.* **2005**, *76*, 67–73. [CrossRef]
5. Takewaki, I.; Murakami, S.; Fujita, K.; Yoshitomi, S.; Yoshitomi, M. The 2011 off the Pacific coast of Tohoku earthquake and response of high-rise buildings under long-period ground motions. *Soil Dyn. Earthq. Eng.* **2011**, *31*, 1511–1528. [CrossRef]
6. Furumura, T.; Hayakawa, T. Anomalous propagation of long-period ground motions recorded in Tokyo during the 23 October 2004 Mw 6.6 Niigata-ken Chuetsu, Japan, Earthquake. *Bull. Seismol. Soc. Am.* **2007**, *97*, 863–880. [CrossRef]
7. Liang, X.W.; Dong, Z.P.; Wang, Y.S.; Deng, M.K. Damage to tall buildings in areas with large epicentral distance during M8.0 Wenchuan Earthquake. *J. Earthq. Eng. Eng. Vib.* **2009**, *29*, 24–31. (In Chinese) [CrossRef]
8. Yoshimoto, K.; Takemura, S. A study on the predominant period of long-period ground motions in the Kanto Basin, Japan. *Earth Planets Space* **2014**, *66*, 100. [CrossRef]
9. Zhang, Z.H.; Liu, Q.S. Seismic responses of super high-rise buildings under long-period ground motions. *Eng. Struct.* **2023**, *294*, 116824. [CrossRef]
10. Chen, H.C.; Tsai, V.C.; Niu, F.L. Observations and Modeling of Long-Period Ground-Motion Amplification Across Northeast China. *Geophys. Res. Lett.* **2018**, *45*, 5968–5976. [CrossRef]
11. Furumura, T.; Kennett, B.L.N. Unusual Strong Ground Motion Across Japan From the 680 km Deep 30 May 2015 Ogasawara Islands Earthquake. *J. Geophys. Res. Solid Earth* **2017**, *122*, 8143–8162. [CrossRef]
12. Zhou, F.; Cui, H.; Shigetaka, A.B.E.; Lü, X.; Sun, Y.; Li, Z.; Li, A.; Feng, D.; Li, Y.; Xue, S.; et al. Inspection report of the disaster of the East Japan earthquake by Sino-Japanese joint mission. *Build. Struct.* **2012**, *42*, 1–20. (In Chinese) [CrossRef]
13. Abouelsaad, M.N.; Shaaban, M.; El Bagalaty, S.; El Madawy, M.E. Comparison between seismic analysis of twisting and regular 52-story towers considering soil-structure interaction. *Earthq. Eng. Eng. Vib.* **2024**, *23*, 663–675. [CrossRef]

14. Forcellini, D. Seismic fragility of tall buildings considering soil structure interaction (SSI) effects. *Structures* **2022**, *45*, 999–1011. [CrossRef]
15. Forcellini, D. Analytical fragility curves of shallow-founded structures subjected to Soil-Structure Interaction (SSI) effects. *Soil Dyn. Earthq. Eng.* **2020**, *141*, 106487. [CrossRef]
16. Bolisetti, C.; Whittaker, A.S. Numerical investigations of structure-soil-structure interaction in buildings. *Eng. Struct.* **2020**, *215*, 110709. [CrossRef]
17. Askouni, P.K.; Karabalis, D.L. SSI influence on the seismic response of asymmetrical small, low-rise R/C buildings. *Structures* **2021**, *32*, 1355–1373. [CrossRef]
18. Fatollahpour, A.; Tafakori, E.; Arjmandi, S.A.A. The effects of structure-soil-structure interaction on seismic response of high-rise buildings equipped with optimized tuned mass damper. *Structures* **2023**, *50*, 998–1010. [CrossRef]
19. Dassault Systèmes. *ABAQUS Analysis User's Manual*; Dassault Systèmes: New York, NY, USA, 2016.
20. Ministry of Housing and Urban-Rural Development of China. *Code for Design of Concrete Structures (GB50010-2010) (2015 Edition)*, 1st ed.; China Construction Industry Press: Beijing, China, 2016. (In Chinese)
21. Ministry of Housing and Urban-Rural Development of China. Revised provisions (2024 edition) of the Code for Seismic Design of Buildings (GB 50011-2010) (2016 edition). *Stand. Eng. Constr.* **2024**, *9*, 72–79. (In Chinese) [CrossRef]
22. Ministry of Housing and Urban-Rural Development of China. *Technical Specification for Concrete Structures of Tall Building (JGJ 3-2010)*, 1st ed.; Construction Industry Publishing House: Beijing, China, 2010. (In Chinese)
23. Archer, J.S. Consistent Matrix Formulations for Structural Analysis using Finite-Element Techniques. *Am. Inst. Aeronaut. Astronaut. J.* **1965**, *3*, 1910–1918. [CrossRef]
24. Cowper, R.G. The Shear Coefficient in Timoshenko's Beam Theory. *J. Appl. Mech.* **1966**, *33*, 335–340. [CrossRef]
25. Drucker, D.C.; Prager, W. Soil mechanics and plastic analysis of limit design. *Q. Appl. Math.* **1952**, *10*, 157–165. [CrossRef]
26. Li, P.; Xie, W.L.; Pak, R.Y.S.; Vanapalli, S.K. Microstructural evolution of loess soils from the Loess Plateau of China. *Catena* **2018**, *173*, 276–288. [CrossRef]
27. Luo, Z. Seismic Response Analysis of High-rise Framework Considering Soil-Structure Interaction. Master's Thesis, Huazhong University of Science and Technology, Wuhan, China, January 2012. (In Chinese).
28. Li, Z.Q. Earthquake response analysis of Historical Timber Buildings Considering Soil-Structure Interaction. Master's Thesis, Xi'an University of Architecture and Technology, Xi'an, China, 2008. (In Chinese).
29. Bracci, J.M.; Reinhorn, A.M.; Mander, J.B. *Seismic Resistance of Reinforced Concrete Frame Structures Designed Only for Gravity Loads: Part I—Design and Properties of One-Third Scale Model Structure*; Technical Report NCEER (92-0027); State University of New York at Buffalo, U.S.: Buffalo, NY, USA, 1992.
30. Bracci, J.M.; Reinhorn, A.M.; Mander, J.B. *Seismic Resistance of Reinforced Concrete Frame Structures Designed Only for Gravity Loads: Part III—Experimental Performance and Analytical Study of a Structural Model*; Technical Report NCEER (92-0029); State University of New York at Buffalo, U.S.: Buffalo, NY, USA, 1992.
31. Sargent, R.G. Verification and validation of simulation models. *J. Simul.* **2013**, *7*, 12–24. [CrossRef]
32. Zhang, S.; Lu, X.L.; Zhang, H.M. Experimental and analytical studies on the ultimate displacement of RC shear walls. *China Civ. Eng. J.* **2009**, *42*, 10–16. (In Chinese)
33. Li, Y.S.; Bai, C.T.; Tang, Z.Y. Effects of soil-structure interaction on seismic collapse resistance of mass-irregular steel frames. *China Civ. Eng. J.* **2020**, *53*, 22–27. (In Chinese) [CrossRef]
34. Ghotbi, A.R.; Taciroglu, E. Effects of conditioning criteria for ground motion selection on the probabilistic seismic responses of reinforced concrete buildings. *Earthq. Eng. Struct. Dyn.* **2020**, *50*, 1414–1428. [CrossRef]
35. Sinkovic, N.L.; Dolsek, M.; Zizmond, J. Impact of the type of the target response spectrum for ground motion selection and of the number of ground motions on the pushover-based seismic performance assessment of buildings. *Eng. Struct.* **2018**, *175*, 731–742. [CrossRef]
36. Dhakal, R.P.; Mander, J.B.; Mashiko, N. Identification of critical ground motions for seismic performance assessment of structures. *Earthq. Eng. Struct. Dyn.* **2006**, *35*, 989–1008. [CrossRef]
37. Li, X.; Wang, W.; Wu, D.; Xu, X.; Li, Z.; Li, Y. The bounded method and characteristics analysis for long-period ground motions. *J. Vib. Eng.* **2014**, *27*, 685–692. (In Chinese) [CrossRef]
38. Wang, B.; Liu, B.Q.; Wu, T.; Bai, G.L. Analysis of low-frequency pulse characteristics and response spectrum for long-period ground motions. *J. Earthq. Eng. Eng. Vib.* **2018**, *38*, 142–151. (In Chinese)
39. Cao, G.R.; Wang, B.; Yang, K.; Li, Z.; Liu, B.Q. Analysis of the Bimodal Characteristics of Acceleration Response Spectra of Far-field Harmonic-like Ground Motions. *J. Disaster Prev. Mitig. Eng.* **2022**, *42*, 480–489+498. (In Chinese)

buildings

MDPI

Article

Axial Compressive Behavior and Calculation Model for Axial-Compressive-Load-Carrying Capacity of Locally Corroded RC Short Columns

Xiaojuan Liu [1,2], Baorui He [1], Xueqiong Chen [1], Yang Liu [1,2,*] and Xin Li [1]

1 College of Civil Engineering, Huaqiao University, Xiamen 361021, China; liuxjty@hqu.edu.cn (X.L.); hbr2677071083@stu.hqu.edu.cn (B.H.); 21014086049@stu.hqu.edu.cn (X.C.); 18014086017@stu.hqu.edu.cn (X.L.)

2 Key Laboratory of Structural Engineering and Disaster Prevention of Fujian Province, Xiamen 361021, China

* Correspondence: lyliuyang@hqu.edu.cn

Abstract: The individual effects of the main reinforcement corrosion and stirrup corrosion on the axial compressive behavior of reinforced concrete (RC) columns were evaluated through axial compression tests on 10 full-scale short columns. The primary experimental parameters were the corrosion location and the corrosion ratio of the steel bar. The electrochemical accelerated corrosion method was applied on nine of the columns, including three columns corroded in the main reinforcement, three columns corroded in the stirrup, and three columns corroded in both the main reinforcement and stirrup. The full-field displacement of the column and strain of concrete were evaluated using a non-contact 3D-DIC (digital image correlation) technique. The results indicated that, with the increase in the main reinforcement corrosion ratio, the width of the longitudinal corrosion crack increased. The transverse corrosion cracks appeared when the stirrup corrosion ratio is larger than 8%, and the increase in stirrup corrosion ratio increased the crack number, but had little effect on the crack width. Compared to the non-corroded RC column, the peak load of specimens with main reinforcement corrosion ratios of 8.02%, 9.01%, and 19.27% decreased by 10.53%, 13.56%, and 19.77%, respectively, and that of the specimens with stirrup corrosion ratios of 7.08%, 12.33%, and 24.36% decreased by 11.59%, 12.07%, and 17.15%, respectively. The axial-compressive-load-carrying capacity of RC columns decreased almost linearly as the corrosion ratio of the main reinforcement increases, while it exhibited an approximately bilinear degradation as the corrosion ratio of the stirrups increases. The stirrup corrosion ratio had less effect on the axial compressive loading capacity of the RC column when it was larger than 7.5%. A model for calculating the axial-compressive-load-carrying capacity of the corroded RC short columns was developed based on the impact mechanisms of the corroded main reinforcement and stirrups on the columns' axial compressive behavior. The calculated results closely matched the test data, demonstrating that the proposed model can reliably predict the residual load-carrying capacity of corroded columns.

Keywords: corroded RC column; stirrup corrosion; main reinforcement corrosion; axial compressive behavior; axial-compressive-load-carrying capacity model

1. Introduction

Steel corrosion due to various reasons such as carbonation and chloride attack is the main problem that affects the durability of reinforced concrete (RC) structures [1]. Corrosion of steel not only results in a decrease in its cross-sectional area and mechanical properties, but also induces damage to the concrete cover and deterioration of the bonding behavior between concrete and steel. All of these factors negatively impact the mechanical properties of the RC members [2]. The RC column is the main structural member in building structures, subject to both gravity and lateral forces. The safety of the structure will be substantially threatened once the performance of the column becomes impaired due to steel corrosion. The RC structure with a severely corroded column is more vulnerable to collapse when subjected to a rare earthquake. Consequently, investigating the mechanical characteristics of corroded RC columns is essential from an engineering safety standpoint.

Recently, several studies have focused on the mechanical properties of corroded RC beams and columns. Hou et al. conducted an experimental study on the flexural performance of RC/ultrahigh toughness cementitious composite (RC/UHTCC) beams [3] and proposed a strengthening technique that utilizes the UHTCC to repair and reinforce the corroded RC beams [4]. Goksu et al. [5] and Ge et al. [6] performed experimental and theoretical studies on corroded RC columns, revealing a significant decline in seismic performance as the steel corrosion ratio increases. Similar studies and conclusions can be found in the references [7,8]. According to Dai et al. [7] and Vu et al. [8], the axial compressive ratio is a critical factor influencing the seismic behavior of RC columns. Thus, examining the axial compression behavior of corroded RC columns is essential for accurately assessing their seismic performance.

Understanding the mechanism by which steel corrosion affects the compressive behavior of RC members is crucial for estimating their residual seismic performance, including moment–axial loading capacity, effective stiffness, deformation capacity, and energy capacity. Wu et al. [9] examined the combined effects of sustained load and steel corrosion on the axial compression characteristic of RC columns, discovering that, as the corrosion ratio increases, the peak load declines and is reached at a smaller displacement. Based on the numerical results and theoretical analysis, Campione et al. [10] and Wang [11] proposed simple models for the corroded RC columns to access their axial compressive bearing capacity. In reference [10], the load-carrying capacity of corroded RC columns is considered as the sum of four different strength contributions, termed the cracked cover concrete, cracked core concrete, internal area of core concrete, and longitudinal bars. Compared with the main reinforcement, the stirrup is closer to the concrete surface and smaller in diameter, which results in more severe corrosion than the main reinforcement in the actual structure, arousing more research interests. Joshi et al. [12] observed in the experimental study that, as the stirrup corrosion ratio increases, the confinement effect on the core concrete diminishes, which has a more pronounced impact on deformation than on axial strength. Zhang et al. [13], Vu et al. [14] and Andisheh et al. [15] proposed an updated compressive constitutive model for the confined concrete based on the Mander model [16]. Zhang et al. [17] identified the corrosion ratio, stirrup configuration, and size effect as the primary parameters for investigating the confinement behavior of square reinforced concrete columns based on their test results. They found that, as the corrosion ratio increased, the brittle failure characteristics of the column became more pronounced. In addition, Andisheh et al. [18] examined the influence of stirrup corrosion on the axial properties of the column's cover concrete to evaluate the moment–curvature relationship of the corroded columns. Xia et al. [19] investigated the mechanical properties of corroded RC columns under eccentric compression, discovering that both the stiffness and ultimate compressive strength, as well as deformation behavior, decrease with the increase in corrosion degree. Although some scholars have proposed mechanical models to describe the axial compressive properties of corroded RC columns, most existing calculation models primarily focus on the deteriorating confinement effects of the corroded steel reinforcement on the core concrete [13–15,17] and the cracking of the concrete cover caused by corrosion [18]. However, these models often fail to adequately consider the impact of steel corrosion on the buckling behavior of the main reinforcement, which plays a critical role in the significant degradation of the axial compression performance of RC columns.

Despite this, a substantial amount of research has been conducted on the mechanical behavior of corroded RC columns, particularly on compressive behavior, providing valuable references for evaluating the performance of corrosion damaged columns. However, most studies have primarily concentrated on columns corroded in the stirrup only or in all steel reinforcements, without thoroughly examining the distinct influence mechanisms of the main reinforcement corrosion and stirrup corrosion. In practical engineering, the main reinforcement and stirrups in the same RC component exhibit varying corrosion ratios due to differences in the concrete cover thickness and steel diameter. Furthermore, the mechanisms by which the main reinforcement corrosion and stirrups corrosion affect the mechanical properties of RC members are different. Thus, it is crucial to clarify the different influence mechanism of the main reinforcement corrosion and stirrup corrosion on the mechanical properties of RC columns, and to develop a proper performance evaluation model for corroded RC columns that accurately considers the varying corrosion degrees in the main reinforcement and stirrups.

To this end, this study conducts axial compressive tests on 10 full-scale RC column specimens to examine the impact of the main reinforcement corrosion and stirrup corrosion on the axial compressive performance of RC columns. Through a strain analysis of both the steel reinforcement and concrete, the mechanisms underlying the performance degradation of corroded RC columns are investigated. Additionally, a calculation model for the axial-compressive-load-carrying capacity of corroded RC columns—accounting for the different influence mechanism in the corroded main reinforcement and stirrups—is established, which can be used to assess the residual performance of corroded RC columns.

2. Experimental Program

2.1. Specimen Details

Ten full-scale RC short column specimens were designed and constructed. The corrosion degree of the main reinforcement and stirrup were considered as the main parameters. The specimen information is shown in Table 1. Nine of the specimens were exposed to electrochemically accelerated corrosion, while the other one was the non-corroded reference specimen. The corroded location and corrosion ratios were the main experimental parameters. All the corroded specimens were divided into three groups, including three specimens corroded in the main reinforcement, three specimens corroded in the stirrup, and the other three corroded in all steel reinforcements. The corrosion ratio η is defined as the average mass loss ratio, as shown in Equation (1):

$$\eta = \frac{\Delta m}{m_0} \tag{1}$$

where Δm is the mass loss of the corroded bar, and m_0 is the original mass of the non-corroded bar. In this study, three corrosion levels, i.e., 10%, 20%, and 30%, were considered, as shown in Table 1. All specimens were numbered as RC-η_s-η_l, in which η_s represents the stirrup corrosion ratio, while η_l represents the main reinforcement corrosion ratio.

Table 1. Design parameters of specimens.

Specimen No.	η_l/%	η_s/%	Corrosion Location
RC-0-0	0	0	No corrosion
RC-0-10	10	0	Main reinforcement
RC-0-20	20	0	
RC-0-30	30	0	
RC-10-0	0	10	Stirrup
RC-20-0	0	20	
RC-30-0	0	30	
RC-10-10	10	10	All steel bars
RC-20-20	20	20	
RC-30-30	30	30	

As illustrated in Figure 1, the geometrical dimensions and reinforcement arrangement of all specimens were identical. The RC column featured a square cross-section with a side length of 350 mm and a height of 1200 mm, resulting in an aspect ratio of 3.4. The concrete cover thickness of the column was 20 mm. The hot-rolled ribbed steel bar with a designed yield strength of 400 MPa was adopted for both the main reinforcement and transverse reinforcement. A total of four 20 mm-diameter longitudinal steel bars were configured as the main reinforcement in each RC column, resulting in a main reinforcement ratio of 1.02%. Furthermore, 8 mm-diameter steel bars were used as the stirrup. The stirrup spacing at the middle part of the column is 100 mm, while the stirrup spacing at both ends of the column is 50 mm to provide effective confinement and prevent the column end from crushing first. In addition, a 10 mm-wide carbon fiber sheet was wrapped around both ends of the column to further constrain the column ends.

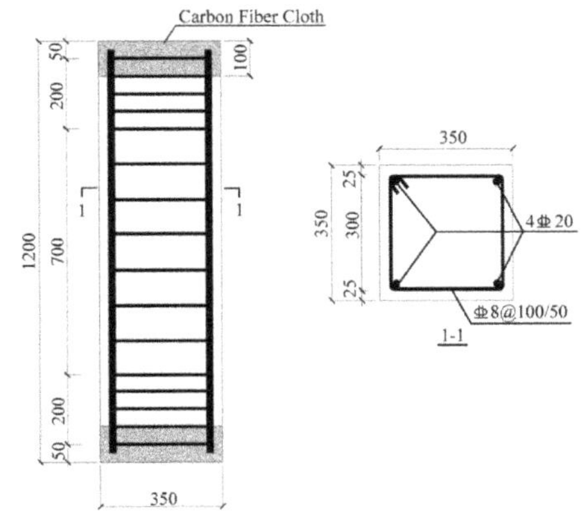

Figure 1. Dimensions and reinforcement details of specimen (unit: mm).

According to the testing method specified in the Chinese standard GB/T50081-2019 "test methods of concrete physical and mechanical properties" [20], the axial compressive strength of the concrete was measured to be 36.6 MPa. The mechanical properties of the steel reinforcement before corrosion were determined according to the test method in the Chinese test standard GB/T228.1-2010 "Metallic materials—Tensile Testing—Part 1: method of test at room temperature" [21]. The mechanical properties of the non-corroded steel reinforcement are illustrated in Table 2.

Table 2. Mechanical properties of steel reinforcement.

d/mm	f_y/MPa	f_u/MPa	$E_s \times 10^5$/MPa	δ/%
8	440	625	1.96	27.5
20	430	624	2.01	25

Note: d is the diameter of the steel bar; f_y is the yield strength; f_u is the ultimate strength; δ is the elongation rate; and E_s is the elastic modulus.

2.2. Electrochemically Accelerated Corrosion of Specimen

All steel bars were numbered and weighed after rust cleaning before the manufacturing of the specimen to accurately determine the corrosion ratio. In the process of binding the steel bars, the intersection position of the longitudinal bars and stirrup bars were isolated by insulation sleeves, and then plastic cable ties were used to bind the steel bars to ensure mutual insulation. After the casting of the concrete and 28 days of standard curing of the specimen, the specimen to be corroded was lifted into a water pool for accelerated corrosion.

The electrochemical corrosion method shown in Figure 2 was employed to accelerate the corrosion. The specimen to be corroded was completely immersed in a 5% sodium chloride solution. The main reinforcement and stirrup reinforcement in the same RC column were connected in parallel with cathodes of separate D.C power supplies. Additionally, copper plates were submerged in the solution and connected to the cathodes of the DC power supplies to establish a current circuit with the steel bars designated for corrosion. Separate current circuits were required for the main reinforcement and the stirrup, respectively.

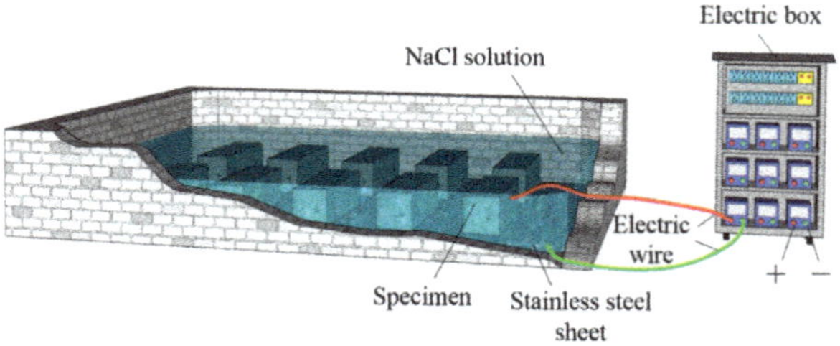

Figure 2. Schematic diagram of accelerated corrosion apparatus.

After the column was immersed in the NaCl solution for 72 h, the DC power supply began to work for accelerated corrosion. Faraday's law was applied to determine the corrosion period of the specimen to achieve its designed corrosion ratio. The corrosion period required to achieved the designed corrosion ratio of the steel reinforcement was listed in Table 3.

Table 3. Corrosion duration.

Corroded Reinforcement	Steel Diameter/mm	Designed Corrosion Ratio/%	Electric Current Density/μA/cm^2	Corrosion Period/Day
Main reinforcement	20	10	400	39
		20		78
		30		117
Stirrup	8	10		16
		20		32
		30		47

2.3. Test Setup and Instrumentation

A 10,000 kN hydraulic jack was used for the test, as shown in Figure 3. The specimen was positioned at the center of the machine, and the axial compression load was directly applied to its base. Three linear variable displacement transducers (LVDTs) were placed along the centerline of three side surfaces of the column to measure the vertical deformation during loading, and the remaining side surface was sprayed with speckles to measure the full-field displacement and strain using the Digital Image Correlation (DIC) technique. Additionally, eight strain gauges were attached on the middle of the side surface to obtain the concrete strain. Twelve additional strain gauges were attached to the main reinforcement and stirrups to measure their strain. Figure 4 presents the layout of the strain gauges. The axial compression load on the specimen was measured using a force sensor integrated into the loading machine. All data, including the load, displacement, and strain, were collected simultaneously through the 3816 N data acquisition system.

The specimen was firstly preloaded to 100 kN to check that both the loading system and data acquisition system were working properly. At the start of the formal loading stage, the specimen was subjected the force control mode, with a loading speed of 50 kN/min until the compression load reached 200 kN. Afterward, the displacement control mode was applied at a rate of 1 mm/min until the end of loading. Loading was stopped when either the average axial displacement reached 15 mm or the axial compression load capacity became unstable.

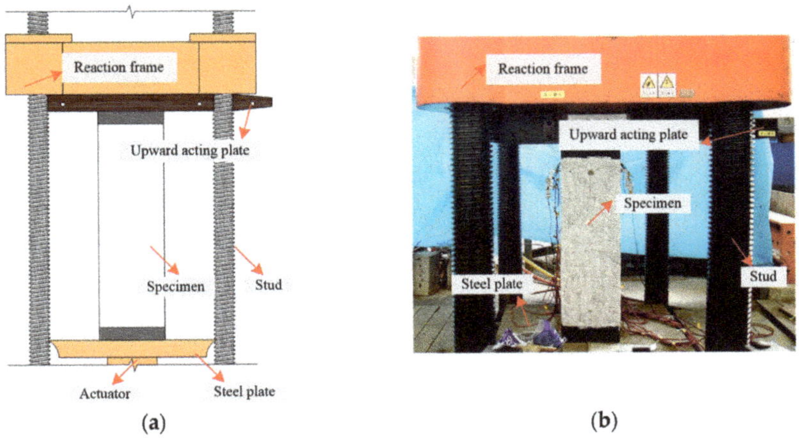

Figure 3. Test setup: (**a**) schematic; and (**b**) actual image.

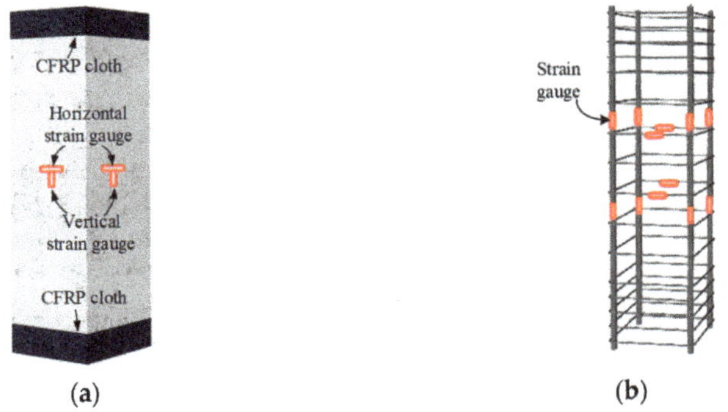

Figure 4. Arrangement of strain gauges: (**a**) strain gauges on concrete; and (**b**) strain gauges on steel reinforcement.

3. Discussion of Test Results

3.1. Observed Corrosion Damage of Specimens

Due to the volume expansion of the corrosion products, cracks gradually developed on the surface of the specimen as the corrosion process progressed. The corrosion damage in the specimen was visually evaluated following the accelerated corrosion process. Figure 5 illustrates the pattern of corrosion cracks. It was observed that, in specimens only corroded in the main reinforcement, vertical corrosion cracks appeared and widened as the corrosion ratio of the main reinforcement increased, as presented in Figure 5a–c. Transverse corrosion cracks appeared on the surface, with an increased number of cracks corresponding to a higher stirrup corrosion ratio. A similar phenomenon was also observed in the study conducted by Zhang et al. [17]. However, the corrosion degree of the stirrup had a minimal impact on the transverse crack width, as illustrated in Figure 5e–f. For specimens RC-10-0, with a stirrup corrosion ratio of 7.08%, no transverse cracks appeared on the surface of the specimen. Similarly, the stirrup corrosion ratios of specimens RC0-10, RC0-20, and RC0-30 were 3.06%, 5.69%, and 6.06%, respectively, and no transverse corrosive cracks were observed on their surfaces either. For specimen RC10-10, with a stirrup corrosion ratio of 8.13%, only one transverse crack was present on the surface. According to the above phenomena, it can be concluded that no significant transverse cracks are likely to develop on the specimen surface when the stirrup corrosion ratio was less than 8%. In specimen RC-30-0, which had a stirrup corrosion ratio of 24.36%, localized longitudinal cracks formed between several transverse cracks, as depicted in Figure 5f. For specimens that experienced corrosion in both the main reinforcement and stirrups, both vertical and transverse cracks developed on the surface, as shown in Figure 5g–i.

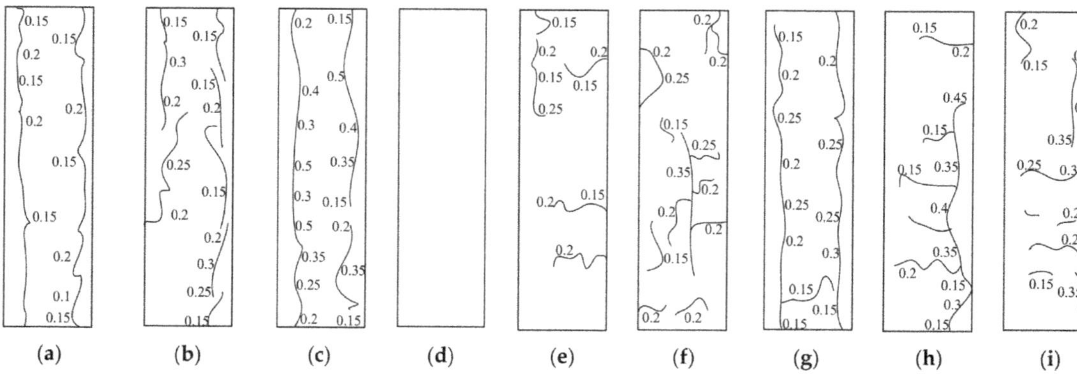

Figure 5. Distribution of corrosion cracks: (**a**) RC-0-10; (**b**) RC-0-20; (**c**) RC-0-30; (**d**) RC-10-0; (**e**) RC-20-0; (**f**) RC-30-0; (**g**) RC-10-10; (**h**) RC-20-20; and (**i**) RC-30-30.

After the loading test, the specimen was dismantled to retrieve the corroded steel bar and measure their residual mass after removing the corrosion rust. The actual corrosion ratios are listed in Table 4. As the corrosion progressed, rust accumulated around the steel bars, gradually obstructing the corrosive cracks and the surrounding concrete pores, thereby hindering the corrosion process. Consequently, the actual corrosion ratios of the specimens were generally lower than the designed values, and the efficiency of the corrosion decreased with an increase in the designed corrosion ratio. For specimens designed to corrode only in the main reinforcement, since no anti-corrosion measures were applied to the stirrups, they experienced slight corrosion during the process. Comparing specimens RC-0-10 and RC-0-20, the latter exhibited a lower corrosion ratio in the main bars but a higher corrosion ratio in the stirrups. This discrepancy might be due to the failure of insulation measures during concrete casting, causing electrical conductivity between the main reinforcement and the stirrups, which not only reduced the corrosion efficiency of the longitudinal bars but also led to more severe stirrup corrosion. For specimens designed to only corrode in the stirrups, as the whole specimen was immersed in the sodium chloride solution, the main reinforcement also experienced slight corrosion. Similar phenomena can be found in the experimental study conducted by Zhang et al. [17]. Since there is no contact between the main reinforcement and the stirrup, the corrosion ratio of the longitudinal reinforcement did not necessarily increase with that of the stirrups, in comparing the specimens RC-10-0, RC-20-0, and RC-30-0. For specimens designed to corrode in all steel reinforcements, due to the smaller diameter and reduced concrete cover thickness of the stirrups, the actual corrosion degree of the stirrups was greater than that of main reinforcement. However, since both the main reinforcement and the stirrup were connected to the power supply, this causes the steel reinforcement to be subject to a complex electric field environment, causing the development of steel corrosion to deviate from expectations.

The corrosion morphology of the steel bars in specific specimens is shown in Figure 6. It can be observed that, when the corrosion ratio of the steel bars is less than 10%, the corrosion of both the longitudinal and transverse reinforcement is relatively uniform. However, as the corrosion ratio increases, the corrosion becomes un-uniform, with noticeable corrosion pits and even the formation of band-shaped pits. For the stirrups, due to their smaller diameter, pitting corrosion is more pronounced, and, in some cases, corrosion-induced fractures occur. As shown in Figure 6c, multiple fracture points are observed on the same stirrup.

Table 4. Actual corrosion ratio of steel reinforcement.

Specimen No.	Design Corrosion Ratio/%		Actual Corrosion Ratio/%	
	η_l	η_s	η_l	η_s
RC-0-0	0	0	0	0
RC-0-10	10	0	9.01	3.06
RC-0-20	20	0	8.20	5.69
RC-0-30	30	0	19.27	6.06
RC-10-0	0	10	1.21	7.08
RC-20-0	0	20	0.88	12.33
RC-30-0	0	30	2.61	24.36
RC-10-10	10	10	3.44	8.13
RC-20-20	20	20	12.07	15.48
RC-30-30	30	30	7.61	20.56

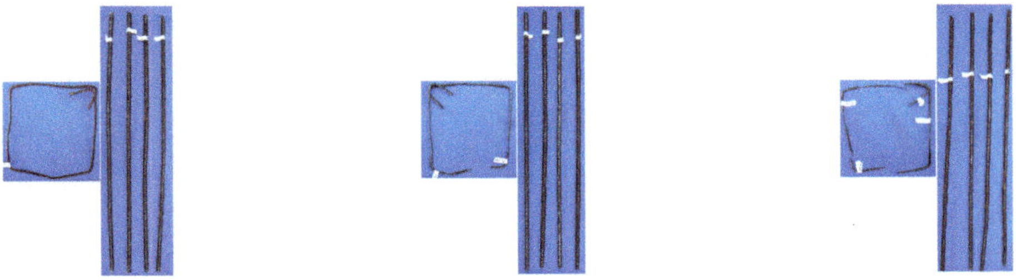

(a) (b) (c)

Figure 6. The corrosion morphology of the steel bars in specific specimens: (**a**) RC-10-10; (**b**) RC-20-20; and (**c**) RC-30-30.

3.2. Damage Evolution of Specimens During Axial Compressive Loading

Throughout the loading process, the specimen exhibited a sequence of failures, including the cracking and spalling of the concrete cover, fracture of the stirrups, buckling of the main reinforcement, and crushing of the core concrete. The full-field strain and displacement of concrete can be obtained by the DIC technique. Figure 7 illustrates the strain cloud diagram of specimens at the peak load and ultimate state, and the ultimate damage modes of specimens. For the non-corroded specimen RC-0-0, noticeable cracks emerged on the concrete surface once the load reached 80% of the peak load. The spalling of the concrete cover was initiated at the peak load. Following the peak load, the loading capacity abruptly decreased to 38% of the peak load. Due the effective confinement provided by the stirrups, the spalling extent of the concrete cover was relatively small, and no significant buckling of the main reinforcement was observed.

In specimens with corrosion in the main reinforcement, the corrosion cracks began to extend further and merge with new cracks when the load reached approximately 50% to 80% of the peak load. At the peak load, the spalling of the concrete cover occurred, and the longitudinal bar buckled (as shown in Figure 7b–d), resulting in a sudden decrease in the axial compressive force.

For specimens corroded in the stirrup, the transverse corrosion crack had little effect on the development of loading cracks. The concrete began to crack and crush when the column is loaded to 56~87% of the peak load. At the peak load, multiple vertical cracks developed on the surface of the specimen, leading to the spalling of the concrete cover. Subsequently, the spalling of a significant area of the concrete cover occurred, and the main reinforcement began to buckle, resulting in a decrease in loading capacity, as shown in Figure 7e–g. In specimen RC-30-0, which exhibited the most severe stirrup corrosion, the stirrup fractured after reaching the peak load, causing the noticeable crushing of the core concrete.

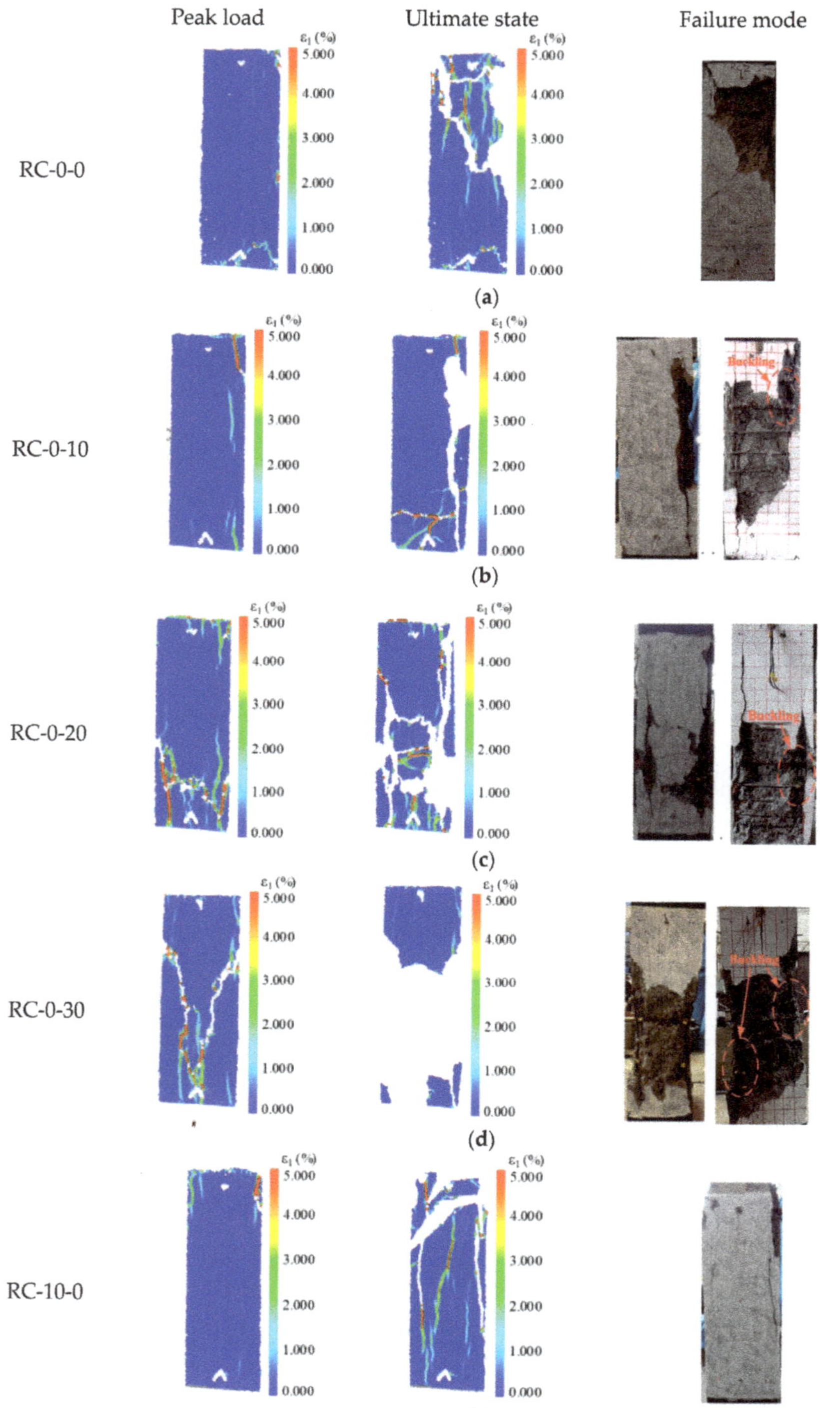

Figure 7. *Cont.*

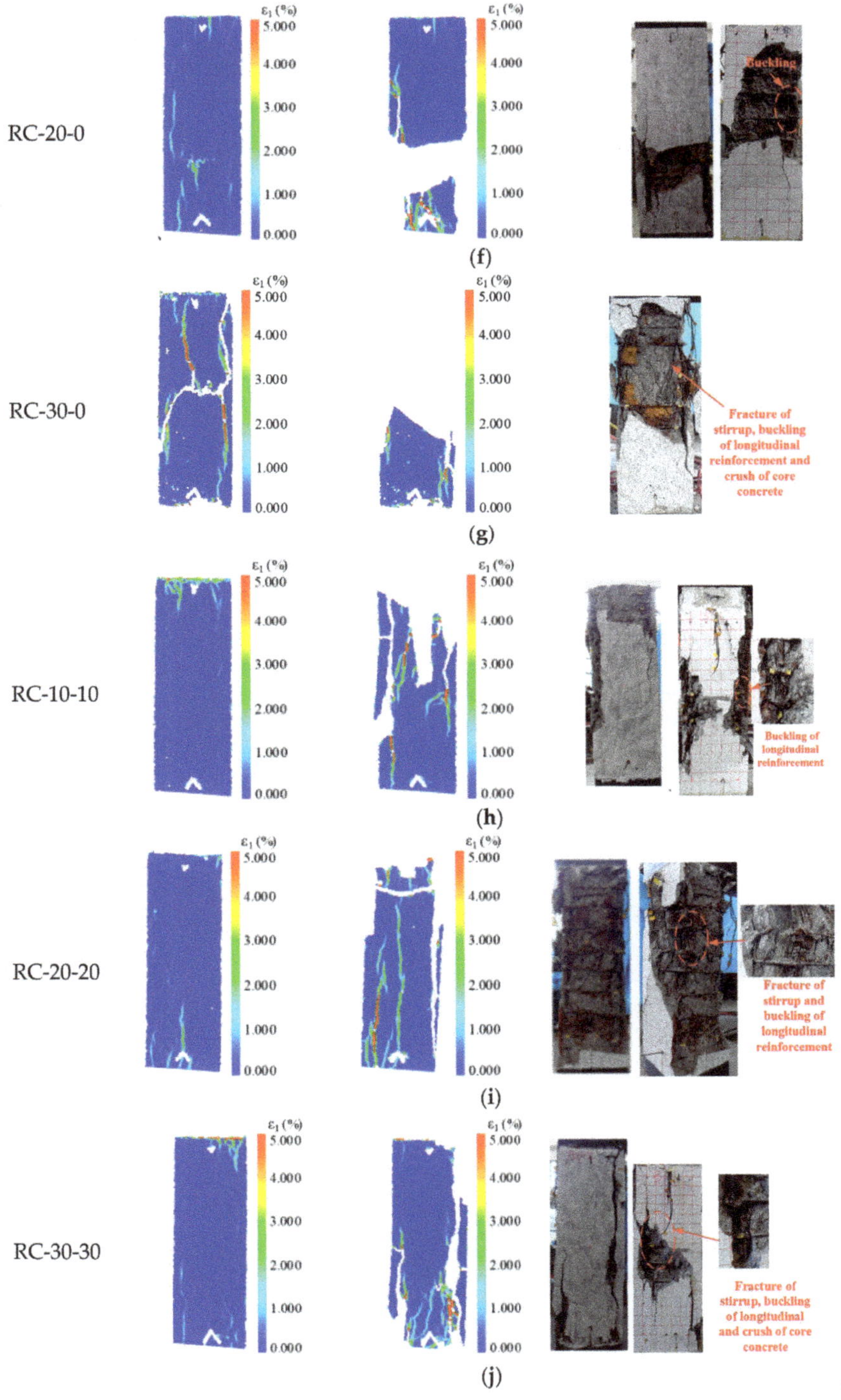

Figure 7. Strain cloud diagram and failure modes of the specimen: (a) RC-0-0; (**b**) RC-0-10; (**c**) RC-0-20; (**d**) RC-0-30; (**e**) RC-10-0; (**f**) RC-20-0; (**g**) RC-30-0; (**h**) RC-10-10; (**i**) RC-20-20; and (**j**) RC-30-30.

For specimens corroded both in the main reinforcement and stirrup, at the peak load, a large range of the concrete cover spalled, resulting in a suddenly drop in loading capacity. In specimen RC-10-10, which exhibited a relatively low stirrup corrosion ratio of 3.44%, the stirrup did not fracture under the peak load. However, for specimen RC-20-20 with a relatively larger corrosion ratio in the stirrup, the actual corrosion ratio of the stirrups is 12.07%, and multiple stirrups fractured under the peak load, resulting in an obvious buckling of the main reinforcement and serious damage of the core concrete.

Based on the observed test phenomena, it can be concluded that the corrosion of both the main reinforcement and stirrups accelerates the compression damage process of the RC columns. The corrosion of the main reinforcement results in vertical corrosive cracks in the concrete cover, which further exacerbates the damage to the concrete cover. Additionally, non-uniform corrosion causes the main reinforcement to buckle prematurely. The corrosion of the stirrups reduces their lateral confinement of the core concrete, which accelerates the buckling of the main reinforcement and the damage of the core concrete, making the brittleness of the specimens more pronounced.

3.3. *Verification of the DIC Measurement*

The accuracy of the DIC measurements was assessed by comparing the displacement/concrete strain obtained by the 3D-DIC and LVDTs/strain gauges during the whole loading process. The displacement readings from DIC were obtained by the virtual extensometer with two reference points positioned at the bottom and top of the columns, corresponding to the LVDT locations. Strain readings from DIC were also obtained by the virtual extensometer with the same position and gauge length with the traditional strain gauge. The comparison of the vertical load versus deformation curves obtained from the DIC and LVDTs/strain gauges is illustrated in Figure 8. It is shown that the DIC deformation readings are in good consistency with the traditional LVDT and strain gauge measurements in the early-loading stage. However, the data obtained by the LVDT and strain gauge are more susceptible to premature failure due to the disturbance caused by the damage of the concrete after the peak load, while the readings recorded from DIC is more stable during the whole loading process. Accordingly, it can be ascertained that both the displacement and strain obtained from the DIC measurement can be used in the following discussion of the test results.

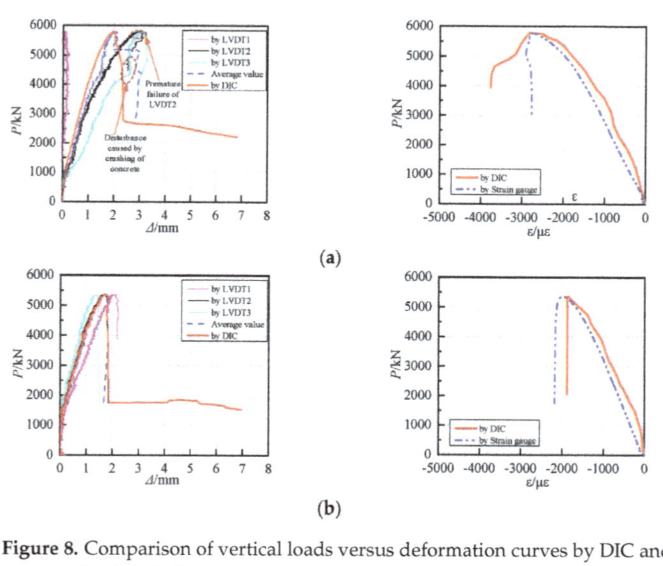

Figure 8. Comparison of vertical loads versus deformation curves by DIC and traditional measurement method: (**a**) RC-0-0; and (**b**) RC-10-10.

3.4. Axial Load–Displacement Curves

Figure 9 presents the axial load versus displacement curves for the specimens. These curves can generally be divided into three sections: an ascending section, a descending section, and a plateau section. As shown in Figure 9, the following observations were made:

1. As the corrosion ratio of the main reinforcement increases, the stiffness of the specimens gradually decreases, primarily due to the vertical corrosion cracks caused by the main reinforcement corrosion. During the loading process, the vertical corrosion cracks expand further, leading to a further reduction in stiffness. In contrast, the stirrup corrosion mainly results in transverse cracks, which have a minimal impact on the early-stage damage evolution of the concrete. As a result, the stiffness of the specimen is minimally impacted by the stirrup corrosion.
2. The bearing capacity of the specimen decreases as the steel corrosion ratio increases. Compared to the non-corroded column, the peak load of the specimens with main reinforcement corrosion ratios of 9.01%, 8.02%, and 19.27% decreased by 13.56%, 10.53%, and 19.77%, respectively. For the stirrup-corroded specimens with actual corrosion ratios of 7.08%, 12.33%, and 24.36%, the load-carrying capacity decreased by 11.59%, 12.07%, and 17.15%, respectively, compared with that of the non-corroded specimen. Figure 10 illustrates the variation in load-carrying capacity with the corrosion ratio. It is evident that the loading capacity of specimens decreases linearly with the increasing corrosion degree of the main reinforcement, while the decrease follows a two-phase degradation trend with the increase in stirrup corrosion degree. The non-uniform corrosion of the main reinforcement reduces its cross-sectional area, leading to a deterioration in the mechanical properties and premature buckling. Additionally, vertical corrosion cracks accelerate the concrete compression damage, resulting in the linear decline in the axial compressive capacity with the increasing corrosion degree of the main reinforcement. For the columns with corroded stirrups, the lateral confinement provided by the stirrups on the core concrete and longitudinal bar diminishes as the stirrup corrosion ratio increases, causing a gradual reduction in the specimen's axial compressive capacity. Once the stirrup corrosion ratio exceeds 7.5%, the stirrups' lateral confinement becomes relatively weak, leading to a reduced influence of stirrup corrosion on the loading capacity.
3. For specimens corroded in all steel reinforcements, the corrosion ratios of the main reinforcement and stirrups exhibit noticeable inconsistencies. Comparing specimens RC-20-20 and RC-30-30 (as shown in Figure 9c), the main reinforcement corrosion ratio of RC-20-20 is 58.6% higher than that of RC-30-30, while the stirrup corrosion ratio is 25% lower. Despite this, the axial compressive bearing capacity of RC-20-20 is 6% lower than that of RC-30-30. This observation highlights the more significant impact of the main reinforcement corrosion on the axial compressive capacity of the RC columns. To further analyze this effect, the degradation of the loading capacity of the RC columns corroded in all steel reinforcements is plotted in Figure 10, using the corrosion ratio of the main reinforcement as the horizontal axis. In this group of specimens, where the stirrup corrosion ratios exceed 8%, the capacity degradation curve closely resembles that of specimens corroded primarily in longitudinal reinforcement, which further confirms that, when the stirrup corrosion ratio exceeds 7.5%, its influence on the loading capacity becomes negligible.
4. Consistent with the conclusions reported in the existing studies [9], in most cases, the displacement corresponding to the peak load decreases with the increase in the steel corrosion ratio. However, there are three specimens in Figure 9a,b (RC-0-20, RC-0-30, and RC-30-0) whose peak displacement is larger than that of the non-corroded specimen. The main reason is that the displacement obtained by the DIC technique is adopted in Figure 9. For the above three specimens, the corrosion degree of the main reinforcement near the DIC measuring surface is relative larger than the other steel bar. When approaching the peak load, the longitudinal bar near the DIC measuring surface

buckled first, resulting in a large vertical deformation on this side, which is larger than the average vertical displacement of the specimen. According to Figure 7c,d,g, it can also be seen that the more serious concrete spalling and longitudinal reinforcement buckling occurred on the DIC measuring surface of these specimens.

5. After reaching the peak load, both the main reinforcement corrosion and stirrup corrosion contribute to the rapid decline in the axial compressive load. In specimens with a stirrup corrosion degree exceeding 15%, the stirrup fractured after reaching the peak load, leading to the buckling of the main reinforcement and significant damage of the core concrete, ultimately resulting in a rapid decline in the loading capacity, as shown in Figure 9b,c.

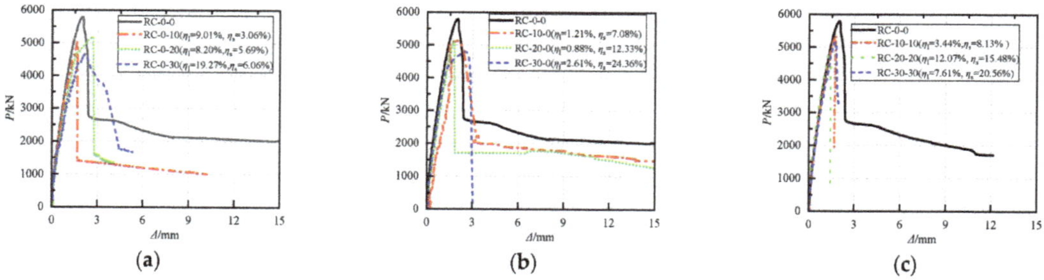

Figure 9. Axial compressive load–displacement curves of specimens: (**a**) main reinforcement corrosion; (**b**) stirrup corrosion; and (**c**) all steel bar corrosion.

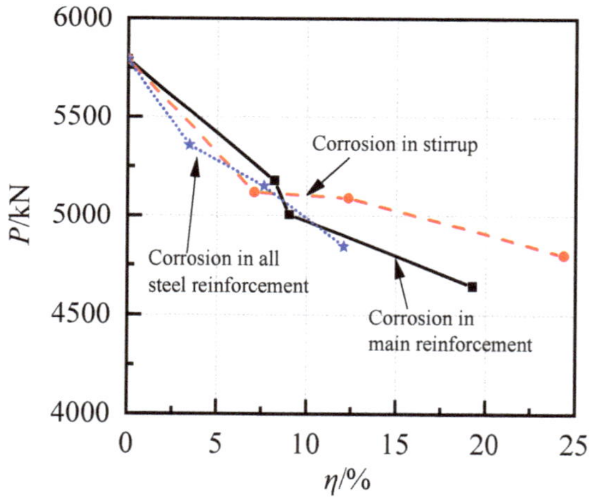

Figure 10. The relationship of ultimate axial-load-carrying capacity versus steel corrosion ratio. Note: For the curve of specimens corroded in all steel reinforcements, the main reinforcement corrosion ratio is taken as the horizontal coordinate value.

3.5. Discussion of Material Strain

3.5.1. Strain of Concrete

Figure 11 presents the relationship between the axial compressive load and the vertical strain of the concrete. It is indicated that the deformation performance of the non-corroded specimen RC-0-0 is enhanced due to the lateral confinement provided by the stirrups and the collaborative action of the main reinforcement. At the peak load, the concrete compressive strain reaches 2800 με (as shown in Figure 11a), which is significantly greater than the plain concrete strain corresponding to the peak stress. In the column with the corroded longitudinal bar, the vertical strain of the concrete at the peak load gradually decreases with the increasing corrosion degree, which indicates the gradually deceasing deformation capacity of the concrete. Nevertheless, the concrete compressive strain at the peak load is still above 2000 με. For the specimens corroded in the stirrup, the maximum vertical strain of the concrete is about 2000 με with little plastic deformation. For the specimen corroded both in the main reinforcement and stirrup, the maximum vertical strain of the concrete is less than 2000 με due to the serious damage of the concrete cover.

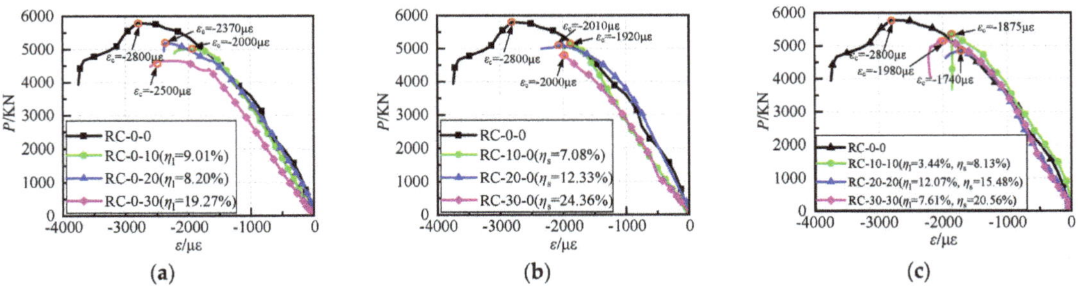

Figure 11. Axial compressive load-vertical strain of concrete curves: (**a**) main reinforcement corrosion; (**b**) stirrup corrosion; and (**c**) all steel bar corrosion.

3.5.2. Strain of Main Reinforcement

The relationship of the axial load versus main reinforcement strain relationships of the specimen is presented in Figure 12. As illustrated in the figure, the development of the specimen's longitudinal steel strain can be roughly divided into two distinct stages. During the initial loading stage, the strain in the longitudinal steel bar exhibited an approximately linear increase as the axial compressive load increased. After the yield of the longitudinal bar, the strain developed more rapidly than the load, as indicated in the smaller slope on the load–strain curve. For the non-corroded specimen RC-0-0, the main reinforcement did not buckle due to the good transverse constraint of the stirrup, and an obvious plastic strain appeared in the main reinforcement. In the specimens corroded in the main reinforcement, the premature buckling of the longitudinal bars occurred at the peak load due to the extensive spalling of the concrete cover and the reduction in the steel cross-sectional area. Consequently, the plastic strain in the corroded main reinforcement was limited. The plastic strain at the peak load decreased with the increasing corrosion degree, as shown in Figure 12a. In specimens with corroded stirrups, the reduced confinement provided by the stirrups led to the buckling of the main reinforcement. The maximum strain in the main reinforcement decreased as the stirrup corrosion ratio increased, as illustrated in Figure 12b. For specimens with corrosion in both the longitudinal bars and stirrups, the strain in the longitudinal bars was approximately 2000 με, lower than in the other specimens, due to the combined effects of the main reinforcement and stirrup corrosion.

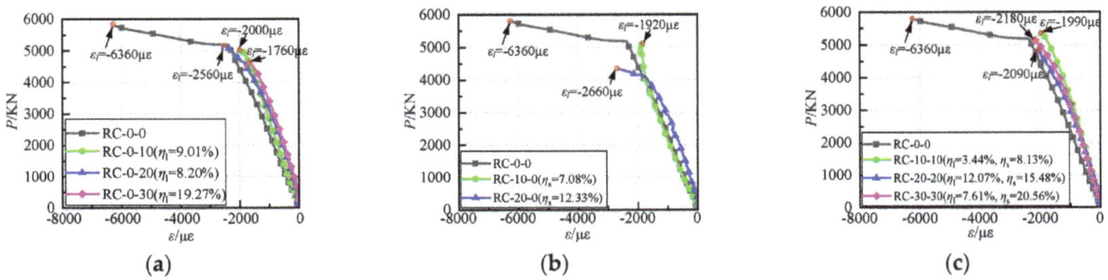

Figure 12. Axial compressive load-strain of main reinforcement curves: (**a**) main reinforcement corrosion; (**b**) stirrup corrosion; and (**c**) all steel bar corrosion.

3.5.3. Strain of Stirrup

The relationship between the vertical load and the stirrup strain of the specimen is illustrated in Figure 13. At the initial loading stage, both the core concrete and the longitudinal bars were in the elastic state with a small deformation; thus, the tensile strain of the stirrup developed slowly. In the uncorroded specimen and those with minimal stirrup corrosion, as the load neared its peak, the transverse deformation of the core concrete became more pronounced, extruding the stirrup and resulting in the rapid growth of the stirrup strain. The maximum stirrup strain of the specimens with a low corrosion ratio in the stirrup can reach 1000 με, as shown in Figure 13a. For the specimen corroded in the stirrup or in all reinforcements, the maximum strain of the stirrup was about 500 με, as indicated in Figure 13b,c, indicating that the lateral confinement of the corroded stirrup to the core concrete and the main reinforcements is significantly diminished, leading to a decline in both the loading and deformation capacity of the RC column.

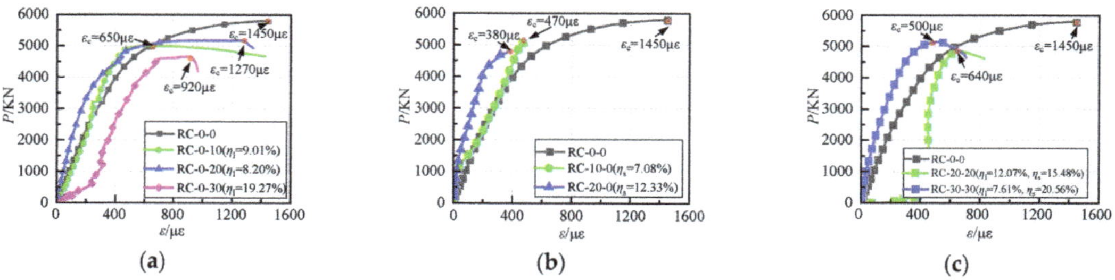

Figure 13. Axial compressive load–strain of stirrup curves: (**a**) main reinforcement corrosion; (**b**) stirrup corrosion; and (**c**) all steel bar corrosion.

4. Calculation Model of Axial-Compressive-Load-Carrying Capacity of Corroded RC Columns

Based on the results discussed above, it can be concluded that the corrosion of the main reinforcement reduces its cross-sectional area, and increases its slenderness ratio, which may lead to premature compression buckling of the corroded longitudinal bar and affect its compressive loading capacity. Meanwhile, the corrosion products of the main reinforcement led to longitudinal cracks, accelerating the damage progression of the concrete cover. Stirrup corrosion reduces the lateral confinements of both the core concrete and the longitudinal bar, thereby impacting their mechanical performance. As a result, the axial-compressive-load-carrying capacity of the corroded RC columns can be calculated by the following calculation model:

$$N_{uc} = f'_{cc}A_{cor} + f'_{\text{cov}}A_{\text{cov}} + f'_{yc}A'_s \tag{2}$$

where N_{uc} represents the axial compressive load of the corroded columns; f'_{cc} denotes the axial compressive strength of the core concrete, accounting for the confinement provided by the steel reinforcement; A_{cor} is the area of the core concrete; f'_{cov} and A_{cov} refer to the axial compressive strength and sectional area of the unconfined cover concrete, respectively; and f'_{yc} and A'_s represent the compressive strength and cross-sectional area of the main reinforcement, respectively.

4.1. *The Axial Compressive Strength of Core Concrete Confined by Corroded Steel Reinforcement*

Considering the influence of the stirrup corrosion, Vu et al. [14] proposed the stress–strain relationship of the concrete confined by corroded stirrups. Accordingly, further taking into account the effect of the corroded main reinforcement, the concrete axial compressive strength considering confinement by both corroded longitudinal bars and stirrups can be expressed as follows:

$$f'_{cc} = (1 - \alpha\eta_s) f_c(-1.254 + 2.254\sqrt{1 + \frac{7.94 f'_l}{f_c}} - 2\frac{f'_l}{f_c}) \tag{3}$$

where η_s represents the stirrup corrosion ratio; and α is the coefficient that accounts for the impact of stirrup corrosion-0.19 for rectangular stirrups, 0.40 for composite stirrups, and 0.51 for spiral stirrups [14]. f_c denotes the axial compressive strength of the unconfined concrete, while f'_l is the effective lateral confinement stress, which considers the detrimental influence of steel corrosion. The value of f'_l can be determined using the following formulae:

$$f'_l = \frac{1}{2} k_e \rho_{sc} f_{ysc} \tag{4}$$

$$k_e = \frac{(1 - \sum_{i=1}^{n} \frac{(w'_i)^2}{6 b_c d_c})(1 - \frac{s'}{2b_c})(1 - \frac{s'}{2d_c})}{1 - \rho_{lc}} \tag{5}$$

$$\rho_{sc} = (1 - \eta_s)\rho_s \tag{6}$$

$$f_{ysc} = (1 - \alpha_s \eta_s) f_{ys} \tag{7}$$

$$\rho_{lc} = (1 - \eta_l)\rho_l \tag{8}$$

where η_l represents the corrosion ratio of the longitudinal bar; k_e is the effective confinement coefficient; ρ_s and ρ_{sc} represent the volume stirrup ratio before and after corrosion, respectively; f_{ys} and f_{ysc} are the yield strength of the stirrup before and after stirrup corrosion, respectively; α_s is the strength degradation coefficient of the corroded stirrup, taken as 0.005 [22]; ρ_l and ρ_{lc} are the reinforcement ratios of the longitudinal bars before and after corrosion, respectively; w'_i is the clear distance between the longitudinal bars; b_c and d_c are the center-to-center distances between the stirrup legs along the two orthogonal directions of the cross-section; and s' is the clear spacing of the stirrup.

4.2. The Axial Compressive Strength of Corrosion-Damaged Cover Concrete

The axial compressive strength of the corrosion-damaged cover concrete can be calculated according to the following equation [23]:

$$f'_{cov} = \frac{f_c}{1 + K\varepsilon_1/\varepsilon_{c0}} \tag{9}$$

where K is a coefficient relying on the physical properties of the steel reinforcement, with a value of 0.1 [24]. ε_{c0} refers to the compressive strain at peak stress, with a value of 0.002, while ε_1 represents the average transverse tensile strain at the onset of concrete cracking, which can be determined using Equation (10):

$$\varepsilon_1 = (b_f - b_0)/b_0 \tag{10}$$

where b_0 represents the initial width of the specimen's cross-section before cracking, and b_f is the cross-sectional width of the specimen after cracking due to corrosion expansion, which can be calculated as follows:

$$b_f = b_0 + n_{bars} \omega_{cr} \tag{11}$$

where n_{bars} is the number of compression steel bars; and ω_{cr} is the total width of the corrosion expansion cracks, which can be calculated according to the formula suggested by Molina et al. [25], as follows:

$$\omega_{cr} = \sum_{i=1}^{n} u_{icorr} = 2\pi(v_{rs} - 1)\chi \tag{12}$$

where u_{icorr} is the width of the ith corrosion crack; v_{rs} is the volume expansion coefficient of the corrosion product, which is taken as 2 according to Molina et al. [25]; and χ represents the pitting corrosion depth of the longitudinal bar.

4.3. Compression Strength of Corroded Main Reinforcement

The non-uniform corrosion of the steel reinforcement results in a reduction in its cross-sectional area and nominal yield strength. The cross-sectional area reduction increases the slenderness ratio, alters the buckling mode, and further diminishes the compressive capacity of the reinforcement. Taking into account the above detrimental effects, Li et al. [26] developed a calculation model for the compressive capacity of the corroded steel bars, as follows:

$$f_{yc1}' = \begin{cases} (1 - 1.602\eta_l)f_y & (s/D \leq 15) \\ 0.969^{s/D-15}(1 - 1.602\eta_l)f_y & (s/D > 15) \end{cases} \tag{13}$$

where f_y and f_{yc1}' represent the nominal compression strength of the steel bars before and after corrosion, respectively; s is the center-to-center spacing of the stirrup; and D denotes the steel diameter.

The lateral confinement provided by the corroded stirrups to the longitudinal bar diminishes as the stirrup corrosion degree increases, which, in turn, impacts the buckling performance of the main reinforcement. The lateral confinement of the stirrup on the main reinforcement can be equivalent to a transverse distribution spring constraint, as shown in Figure 14 [27]. Based on the elasticity theory, the nominal strength of the longitudinal bar at buckling can be determined as follows [27]:

$$f_{yc2}' = \frac{N_{l,buck}}{A_{lc}} = \frac{3.46\sqrt{E_r \times J_l \times k}}{A_{lc}} \tag{14}$$

$$E_r = \frac{4 \times E_s \times E_p}{\left(\sqrt{E_s} + \sqrt{E_p}\right)^2} \tag{15}$$

$$k = \frac{k_{eq}}{s} \tag{16}$$

where f_{yc2}' and $N_{l,buck}$ represent the critical stress and bearing capacity of the longitudinal steel bar under buckling, respectively; A_{lc} is the cross-sectional area of the corroded bar; E_r denotes a reduction coefficient of the steel deformation modulus; J_l is the inertia moment of the corroded bar; k is the confinement coefficient of the stirrup on the main reinforcement; and k_{eq} denotes the stirrup stiffness, which is dependent on its position [27].

To account for the combined effects of both the main reinforcement corrosion and stirrup corrosion, the compressive strength f_{yc}' of the main reinforcement is taken as the smaller value calculated according to Equations (13) and (14).

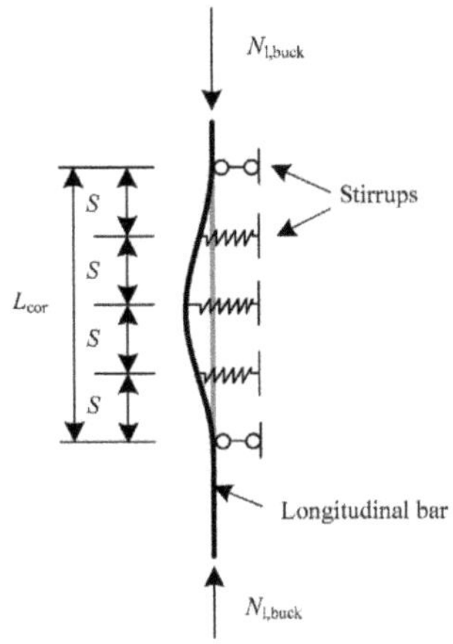

Figure 14. Schematic diagram of the buckling model of main reinforcement.

4.4. Validation of the Axial Compressive Loading Capacity Model for Corroded RC Columns

A total of seven groups of test results regarding the axial compressive bearing capacity of the RC columns were collected for model validation. The comparison of the test and calculated results is presented in Table 5. The ratio of the calculated results to test results ranges from 0.85 to 1.28, with an average value of 1.02 and a standard deviation of 0.08. Therefore, it can be concluded that the calculated results are in good agreement with the experimental findings, confirming the validity of the calculation model.

Table 5. Comparison of calculating results with test results.

Data Sources	Specimen No.	b/mm	h/mm	ρ_s/%	ρ_l/%	f_{ck}/MP	f_{ys}/MPa	f_{yl}/MPa	s/mm	η_s/%	η_l/%	$N_{u,exp}$/kN	$N_{u,pre}$/kN	$N_{u,pre}$/$N_{u,exp}$	μ	CV
This research	RC-0-0	350	1200	0.67	1.03	36.6	440.0	430.0	100	0	0	5791	5168	0.89	0.91	0.05
	RC-0-10									3.1	9.0	5006	4561	0.91		
	RC-0-20									5.7	8.2	5181	4559	0.88		
	RC-0-30									6.1	19.3	4646	4264	0.92		
	RC-10-0									7.1	1.2	5120	4959	0.97		
	RC-20-0									12.3	0.9	5092	4925	0.97		
	RC-30-0									24.4	2.61	4798	4632	0.97		
	RC-10-10									8.1	3.44	5356	4772	0.89		
	RC-20-20									15.5	12.1	4842	4308	0.89		
	RC-30-30									20.6	7.6	5154	4400	0.85		
Wu [9]	A-0-0	250	600	4.23	1.29	40.53	427.2	427.2	100	0	0	3400	3570	1.05	1.12	0.07
	A-0-5									5.5	3	3000	3246	1.08		
	A-0-10									10.8	7.6	2650	2952	1.11		
	A-0-15									16.7	9.6	2200	2812	1.28		
	A-1-5									5.2	3.1	3100	3243	1.05		
	A-1-10									9.9	7.5	2700	2967	1.10		
	A-1-15									15.2	10.1	2350	2815	1.20		
	A-2-5									4.5	2.5	3150	3293	1.05		
	A-2-10									9.2	6.3	2800	3024	1.08		
	A-2-15									13.2	8.8	2450	2880	1.18		
	B-0-0	250	600	4.23	1.29	40.53	468.7	468.7	100	0	0	3450	3652	1.06	1.00	0.04
	B-0-5									1	4.3	3350	3303	0.99		
	B-0-10									1.4	9.2	3250	3091	0.95		
	B-0-15									2.5	13.5	3050	2962	0.97		
	B-1-5									0.8	4.9	3300	3273	0.99		
	B-1-10									1.2	9.7	3150	3078	0.98		
	B-1-15									2.1	14.9	3000	2938	0.98		
	B-2-5									0.9	5.8	3250	3228	0.99		
	B-2-10									1.1	12	3100	3016	0.97		
	B-2-15									2	7.4	2900	3146	1.08		

Table 5. *Cont.*

Data Sources	Specimen No.	b/mm	h/mm	ρ_s/%	ρ_l/%	f_{ck}/MP	f_{ys}/MPa	f_{yl}/MPa	s/mm	η_s/%	η_l/%	$N_{u,exp}$/kN	$N_{u,pre}$/kN	$N_{u,pre}$/$N_{u,exp}$	μ	CV
Zhang [17]	RC1S4	150	300	1.46	2.01	19.8	377.8	511.2	125	0	0	619	541	0.87	0.92	0.05
	RC1S4R2									9.2	1.6	519	496	0.95		
	RC1S4R4									21.3	3.4	480	455	0.95		
	RC2S1	250	500	1.91	0.72	19.8	377.8	511.2	50	0	0	1737	1907	1.10	1.10	0.01
	RC2S1R1									5.1	0.2	1694	1862	1.10		
	RC2S1R2									9.5	1.8	1635	1787	1.09		
	RC2S1R3									13.8	2.2	1605	1745	1.09		
	RC2S1R4									20.3	2.5	1535	1689	1.10		
	RC2S2	250	500	1.37	0.72	19.8	377.8	511.2	70	0	0	1632	1717	1.05	1.04	0.01
	RC2S2CR1									6.4	3.6	1530	1595	1.04		
	RC2S2R1									8.6	0.3	1585	1657	1.05		
	RC2S2R2									12.1	2.4	1548	1585	1.02		
	RC2S2CR2									12.8	5.8	1496	1520	1.02		
	RC2S2R3									15.8	3.2	1503	1546	1.03		
	RC2S2CR3									17.3	6.5	1423	1482	1.04		
	RC2S2R4									21.6	3.1	1477	1513	1.02		
	RC2S2CR4									22.2	8.9	1337	1423	1.06		
	RC2S3	250	500	0.91	0.72	19.8	377.8	511.2	105	0	0	1513	1540	1.02	1.03	0.03
	RC2S3CR1									7.2	2.7	1402	1442	1.03		
	RC2S3R1									8.2	1.2	1478	1472	1.00		
	RC2S3R2									12.8	2.2	1425	1429	1.00		
	RC2S3CR2									13.2	7.1	1323	1344	1.02		
	RC2S3R3									15.3	1.6	1386	1432	1.03		
	RC2S3CR3									15.6	8.7	1309	1314	1.00		
	RC2S3R4									18.6	1.9	1318	1411	1.07		
	RC2S3CR4									19.5	9.3	1207	1290	1.07		

Note: μ and CV are the mean value and coefficient of variation of calculation error for each date set, respectively.

5. Conclusions

Axial compression tests on corroded RC short columns were performed to examine the effects of the corrosion location and corrosion ratio of the steel reinforcement on the damage evolution and axial compressive capacity of RC columns. The main conclusions are as follows:

1. Longitudinal corrosion cracks developed and widened as the corrosion ratio of the main reinforcement increased. The transverse corrosive cracks appeared when the stirrup corrosion ratio is larger than 8%, and the increase in the stirrup corrosion ratio significantly raises the number of transverse corrosive cracks, although it has a minimal impact on the width of these cracks.
2. Both the main reinforcement corrosion and stirrup corrosion accelerate the compression damage process of RC columns. Specifically, the corrosion of the main reinforcement primarily affects the damage to the concrete cover and the buckling behavior of the main reinforcement before reaching the peak load, while the stirrup corrosion mainly influences the damage process after the peak load, hastening the buckling of the main reinforcement and the damage of the core concrete, which leads to a more pronounced brittle characteristic in the RC columns.
3. The axial compressive capacity of the RC column decreases approximately linearly as the corrosion ratio of the main reinforcement increases, while it exhibits a bi-fold degradation relationship with the rising stirrup corrosion ratio. Additionally, the impact of the stirrup corrosion on the axial compressive capacity of the RC column diminishes when the corrosion ratio exceeds 7.5%.
4. Taking into account the mechanisms through which the longitudinal and stirrup corrosion affects the axial compression performance of RC columns, a calculation model for the axial-compressive-load-carrying capacity of corroded RC columns has been developed. The theoretical results from this model are in good agreement with the experimental results, demonstrating that the proposed model can effectively evaluate the axial compression performance of RC short columns.
5. This research focuses on the individual effects of the main reinforcement corrosion and stirrup corrosion on the axial compressive behavior of the RC short column. The additional eccentricity induced by the premature buckling of the localized longitudinal steel bar is not considered, which will significantly influence the compressive

behavior of the RC long columns. Therefore, further research is required in order to investigate the compressive behavior of non-uniformly corroded RC long columns.

Author Contributions: Conceptualization, X.L. (Xiaojuan Liu) and Y.L.; methodology, X.L. (Xiaojuan Liu); investigation, B.H., X.C. and X.L. (Xin Li); writing—original draft preparation, B.H. and X.C.; writing—review and editing, X.L. (Xiaojuan Liu); supervision, X.L. (Xiaojuan Liu) and Y.L.; project administration, X.L. (Xiaojuan Liu) and Y.L.; funding acquisition, X.L. (Xiaojuan Liu). All authors have read and agreed to the published version of the manuscript.

Funding: This research was funded by the NATIONAL NATURAL SCIENCE FOUNDATION OF CHINA, grant number 51808236, and the SCIENCE AND TECHNOLOGY PLAN PROJECT OF QUANZHOU CITY, grant number 2024NS013.

Data Availability Statement: The original contributions presented in the study are included in the article.

Conflicts of Interest: The authors declare no conflicts of interest.

References

1. Chen, H.P.; Jiang, Y.; Markou, G. Structural performance deterioration of corroding reinforced concrete columns in marine environment. *Ocean. Eng.* **2022**, *262*, 112155. [CrossRef]
2. Liu, X.F.; Li, Y. Static bearing capacity of partially corrosion-damaged reinforced concrete structures strengthened with PET FRP composites. *Constr. Build. Mater.* **2019**, *211*, 33–43. [CrossRef]
3. Hou, L.J.; Wang, J.; Huang, T.; Shen, C.; Aslani, F.; Chen, D. Flexural behavior of corroded reinforced concrete beams repaired with ultra-high toughness cementitious composite. *Constr. Build. Mater.* **2019**, *211*, 1127–1137. [CrossRef]
4. Hou, L.; Zhou, B.; Guo, S.; Aslani, F.; Chen, D. Corrosion behavior and flexural performance of reinforced concrete/ultrahigh toughness cementitious composite (RC/UHTCC) beams under sustained loading and shrinkage cracking. *Constr. Build. Mater.* **2019**, *198*, 278–287. [CrossRef]
5. Goksu, C.; Ilki, A. Seismic behavior of reinforced concrete columns with corroded deformed reinforcing bars. *ACI Struct. J.* **2016**, *113*, 1053–1064. [CrossRef]
6. Ge, X.; Dietz, M.S.; Alexander, N.A.; Kashani, M.M. Nonlinear dynamic behavior of severely corroded reinforced concrete columns: Shaking table study. *B Earthq. Eng.* **2020**, *18*, 1417–1443. [CrossRef]
7. Dai, K.Y.; Lu, D.G.; Yu, X.H. Experimental investigation on the seismic performance of corroded reinforced concrete columns designed with low and high axial load ratios. *J. Build. Eng.* **2021**, *44*, 102615. [CrossRef]
8. Vu, N.S.; Yu, B.; Li, B. Prediction of strength and drift capacity of corroded reinforced concrete columns. *Constr. Build. Mater.* **2016**, *115*, 304–318. [CrossRef]
9. Wu, X.; Chen, L.; Li, H.; Xu, J. Experimental study of the mechanical properties of reinforced concrete compression members under the combined action of sustained load and corrosion. *Constr. Build. Mater.* **2019**, *202*, 11–22. [CrossRef]
10. Campione, G.; Cannella, F.; Minafò, G. A simple model for the calculation of the axial load-carrying capacity of corroded RC columns. *Mater. Struct.* **2016**, *49*, 1935–1945. [CrossRef]
11. Wang, G.Y.; Pan, Z.; Yu, W. Bearing capacity of the corrosion reinforced concrete axial compression members. *Adv. Mater. Res.* **2012**, *430–432*, 1830–1833. [CrossRef]
12. Joshi, J.; Arora, H.C.; Sharma, U.K. Structural performance of differently confined and strengthened corroding reinforced concrete columns. *Constr. Build. Mater.* **2015**, *82*, 287–295. [CrossRef]
13. Zhang, G.; Cao, X.; Fu, Q. Experimental study on residual strength of concrete confined with corroded stirrups. *Can. J. Civil. Eng.* **2016**, *43*, 583–590. [CrossRef]
14. Vu, N.S.; Yu, B.; Li, B. Stress-strain model for confined concrete with corroded transverse reinforcement. *Eng. Struct.* **2017**, *151*, 472–487. [CrossRef]
15. Andisheh, K.; Scott, A.; Palermo, A. Effects of corrosion on stress-strain behavior of confined concrete. *J. Struct. Eng.* **2021**, *147*, 04021087. [CrossRef]
16. Mander, J.B.; Priestley, M.J.N.; Park, R. Theoretical stress-strain model for confined concrete. *J. Struct. Eng.* **1988**, *114*, 1804–1826. [CrossRef]
17. Zhang, Q.; Zheng, N.-H.; Gu, X.-L.; Wei, Z.-Y.; Zhang, Z. Study of the confinement performance and stress-strain response of RC columns with corroded stirrups. *Eng. Struct.* **2022**, *266*, 114476. [CrossRef]
18. Andisheh, K.; Scott, A.; Palermo, A. Experimental evaluation of the residual compression strength and ultimate strain of chloride corrosion-induced damage concrete. *Struct. Concrete* **2019**, *20*, 296–306. [CrossRef]
19. Xia, J.; Jin, W.; Li, L. Performance of Corroded Reinforced Concrete Columns under the Action of Eccentric Loads. *J. Mater. Civil. Eng.* **2016**, *28*, 04015087. [CrossRef]
20. *GB/T50081-2019*; Ministry of Housing and Urban-Rural Development of the People's Republic of China. Standard for Test Methods of Concrete Physical and Mechanical Properties. China Industrial Press: Beijing, China, 2019. (In Chinese)

21. *GB/T228.1-2010*; State General Administration of the People's Republic of China for Quality Supervision and Inspection and Quarantine. Metallic Materials -Tensile Testing-Part 1: Method of Test at Room Temperature. Standards Press of China: Beijing, China, 2011. (In Chinese)
22. Du, Y.G.; Clark, L.A.; Chan, A.H.C. Residual capacity of corroded reinforcing bars. *Mag. Concrete Res.* **2005**, *57*, 135–147. [CrossRef]
23. Vecchio, F.J.; Collins, M.P. The modified compression-field theory for reinforced concrete elements subjected to shear. *ACI J. Proceeding* **1986**, *83*, 219–231.
24. Cape, M. *Residual Service-Life Assessment of Existing R/C Structures*; Chalmers University of Technology: Gotborg, Sweden, 1999; pp. 20–65.
25. Molina, F.J.; Alnonso, C.; Andrade, C. Cover cracking as a function of rebar corrosion: Part 2: Numerical model. *Mater. Struct.* **1993**, *26*, 532–548. [CrossRef]
26. Li, S.; Li, W.; Tang, W.; Lu, Y. Study on constitutive model for compressive stress-strain relationship of corroded steel bars. *J. Huazhong Univ. of Sci. Tech. (Nat. Sci. Ed.)* **2021**, *49*, 98–102. (In Chinese)
27. Campione, G.; Minafò, G. Compressive behavior of short high-strength concrete columns. *Eng. Struct.* **2010**, *32*, 2755–2766. [CrossRef]

buildings

Article

Eccentric Compression Behavior of RCFST Columns with Web Perforated T-Shaped Stiffeners

Qifan Jiang [1], Jun Zhang [1], Lvhong Sun [2], Yanchao Wu [2] and Tuo Lei [1,*]

[1] School of Civil Engineering, Chang'an University, Xi'an 710061, China; 2022128004@chd.edu.cn (Q.J.)
[2] CCCC First Highway Northwest Engineering Co., Ltd., Xi'an 710075, China
* Correspondence: leituo616@163.com

Abstract: Based on the previous axial compression tests by the authors, this study investigates the eccentric compression behavior of rectangular concrete-filled steel tubular (RCFST) columns with web-perforated T-shaped stiffeners through finite element analysis. Finite element models (FEM) of RCFST columns were established and validated against failure modes and load-displacement curves obtained from axial compression tests. Subsequently, a detailed parametric analysis of RCFST columns under eccentric compression was conducted, including the stiffening details (i.e., material strength, the number, and dimension of stiffeners). Results show that the stiffening method enhances the RCFST ultimate bearing capacity. Increasing the number and dimensions of stiffeners, as well as the strengths of steel and concrete, leads to higher peak loads, whereas an increase in the slenderness ratio results in a reduction in ultimate capacity. Finally, an *N-M* (axial force-moment) interaction curve calculation method for RCFST columns with web-perforated T-shaped stiffeners was proposed, simplifying the calculation process for stiffeners and incorporating the correction factors for eccentric columns. The results indicate that this method can achieve higher accuracy while ensuring safety, whereas the calculations based on CECS 28 are overly conservative.

Keywords: RCFST column; eccentric compression; finite element method; ultimate strength; web perforated T-shaped stiffeners

1. Introduction

Concrete-filled steel tubular (CFST) columns are extensively utilized in engineering structures, including bridges, heavy-duty workshops, high-rise buildings, and subway stations [1]. Depending on their cross-sectional shapes, CFST columns can be categorized into circular, square, rectangular, and polygonal types, among others [2–4]. Due to the advantages of convenient joint construction, a more flexible structural layout, and higher flexural stiffness, rectangular CFST (RCFST) columns are increasingly favored by engineers. However, compared to conventional circular CFST columns, RCFST columns exhibit more complex mechanical behavior due to the non-uniform confinements exerted by the external steel tubes on the core concrete, making them more susceptible to local buckling, particularly under eccentric loadings [5].

Extensive studies on RCFST columns [6–10] have indicated that the collapse of most columns is caused by local buckling of the steel tubes due to the weak interactions between concrete and steel tubes. To further improve the bearing capacity and ductility of RCFST columns, scholars have tried various methods to enhance these interactions. Bhartiya and

Sahoo [11] found that adding reinforcement cages does not improve the local buckling characteristics of square CFST columns through experimental studies. Ding et al. [12,13] conducted axial load tests on RCFST columns under spiral hoops and steel reinforcement and found that welding reinforcing bars on the steel tube can suppress local buckling. Du et al. [14] adopted fiber-reinforced polymer (FRP) to confine the steel tube and concrete and found that the local buckling of the specimens confined by FRP lagged behind those without confinement, indicating that FRP inhibits the buckling of steel tubes. Zhou and Gan et al. [15,16] proposed a strengthening scheme of welding diagonal stiffeners in steel tubes and subsequently compared the axial compression and bending performance of RCFST components with different stiffening forms. The results show that the performance of the inclined stiffening specimens is the best under the same material strength and steel consumption. Ge and Usami [17] tested thin-walled square CFST short columns with longitudinal plate stiffeners and found that longitudinal stiffeners delayed the occurrence of local buckling, but the enhancement effect on peak load was not significant. Cai and Long [18] conducted a theoretical study on the elastic local buckling of steel tubes in RCFST columns with binding bars under axial compression. Their findings indicate that the presence of binding bars increases the local buckling coefficient, thereby enhancing the load-bearing capacity of the steel tube. Chen et al. [19] conducted experiments on 12 RCFST columns with T-shaped stiffeners. The results show that T-shaped stiffeners can suppress premature local buckling of the tube wall. As long as the T-shaped stiffeners have sufficient stiffness, even the local buckling of the steel pipe before yielding can be avoided. However, T-shaped stiffeners have little effect on improving the ductility. To further improve the mechanical behavior of the interface between steel and concrete, the PBL-type stiffener was proposed [20]. In this way, the PBL stiffener not only plays a role in stiffening the steel tube wall, but the concrete-filled-in PBL apertures can act as a "tenon" or shear connector, which prevents the separation of the steel tube and concrete. Liu et al. [21] carried out bending and eccentric compression tests and finite element analysis on PBL stiffened square CFST members. The results show that the strengthening effect of PBL stiffeners on the bearing capacity of the specimens is not significant, but the flexural stiffness of the specimens is increased by 20%. Furthermore, Guo and Diao [22] conducted axial compression tests on 12 square stub columns and compared the mechanical performance of CFST columns with straight ribs, single PBL, double PBL, PBL diagonal binding ribs, and CFST columns without stiffeners. The results indicate that the strength index of columns with double longitudinal PBL stiffeners remains relatively stable, accompanying slight improvements in both ultimate strength and ductility. In contrast, columns with PBL diagonal binding ribs exhibit significant enhancements in ultimate strength and ductility while mitigating the risk of a sudden ductility drop.

Among the existing stiffening methods for RCFST columns, it is evident that different stiffening forms lead to varying mechanical behaviors, as well as differences in steel consumption and construction costs. How to find a simple, effective, and economical stiffening method to delay or avoid local buckling is still an open question. Unfortunately, existing main specifications, such as EN1994 [23], AISC 360 [24], AS/NZS 2327 [25], and Chinese standards [26–28], have differences in theoretical basis when calculating the bearing capacity of RCFST columns, and do not give clear suggestions on stiffening methods. This situation has hindered the engineering application of stiffened RCFST columns.

Existing studies on PBL-stiffened RCFST columns have demonstrated that the perforations of ribs can enhance the interaction between the core concrete and the steel tubes. However, their contribution to improving load-carrying capacity remains limited, as the perforations weaken the compressive resistance of the stiffeners. In contrast, T-type stiffeners can improve the bearing capacity of RCFST columns but have little effect on the ductility. Based on previous studies, the authors [29] proposed a novel web-perforated T-shaped stiffener, as shown in Figure 1, and conducted axial compression tests on eight RCFST columns. The results indicate that longitudinal stiffeners provide a slight improvement in load-carrying capacity compared to the findings of Chen et al. [19]. However, the presence of concrete "tenons" helps alleviate the risk of a sharp decline in ductility. Obviously, compared with the PBL and T-shaped stiffeners, this new stiffening form has comprehensive advantages. To promote the use of this stiffening form, more research is still needed.

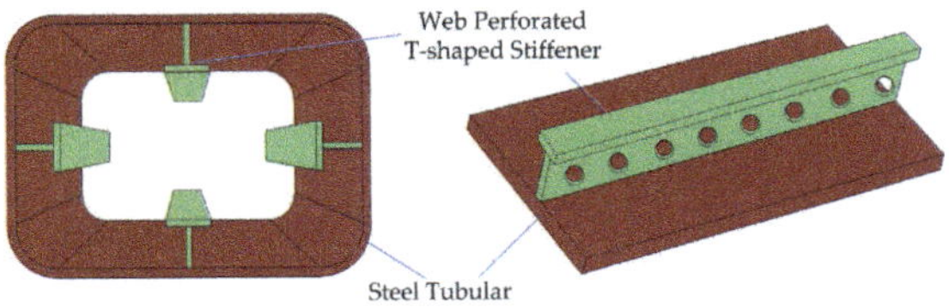

Figure 1. RCFST with web-perforated T-shaped stiffeners.

To examine the eccentric compression behavior of RCFST columns with web-perforated T-shaped stiffeners, this study conducts a numerical simulation based on previous tests. Firstly, FEM of RCFST columns with web-perforated T-shaped stiffeners was developed and validated using the axial compression test results by the authors. Subsequently, a comprehensive parametric analysis of RCFST columns under eccentric compression was performed, considering key stiffening parameters such as the number and size of stiffeners and material strength. Finally, based on the findings, a predictive formula for the ultimate bearing capacity was proposed.

2. Establishment of FEM

2.1. Constitutive Model of Material

2.1.1. Steel

The classical elastic-plastic model [30] was used to describe the behavior of the steel material. The stress–strain curve was presented in Figure 2a. Similar to Zhang et al. [31], the stress–strain relationship is divided into four stages: 1. Elastic Segment (OA): This is a straight line with a slope equal to the elastic modulus; 2. Yield Segment (AB): This is a horizontal line where the stress remains constant while the strain increases from ε_1 to ε_2, where $\varepsilon_2 = 10\varepsilon_1$; 3. Strengthening Segment (BC): This is an oblique straight line where the stress and strain increase from f_y and ε_2 to f_u and ε_3, where $\varepsilon_3 = 100\varepsilon_1$; 4. Secondary Plastic Flow Segment (After C): This is a horizontal line where the stress remains constant while the strain continues to increase.

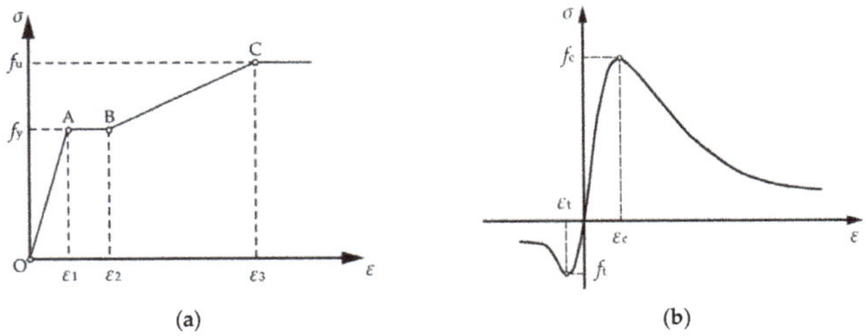

Figure 2. Stress–strain curves: (**a**) Steel; (**b**) Concrete.

2.1.2. Concrete

In this paper, the concrete damage plasticity model implemented in ABAQUS 2023 [32] software was used to describe the behavior of the concrete material. To account for the confinement effect of the steel tube on the concrete, the uniaxial stress–strain relationship curve proposed by Ding et al. [33] was selected. The stress–strain curve of the concrete is shown in Figure 2b. Under compression, the stress–strain relation was given by

$$y = \begin{cases} \frac{A_1 x + (B_1 - 1)x^2}{1 + (A_1 - 2)x + B_1 x^2} & x < 1 \\ \frac{x}{\alpha_1 (x-1)^2 + x} & x > 1 \end{cases} \tag{1}$$

where $y = \sigma / f_c$, $x = \varepsilon / \varepsilon_c$; f_c and ε_c represent the uniaxial compressive strength of the concrete and the peak strain under compression. $A_1 = 9.1 f_{cu}^{-4/9}$ is the ratio of the elastic modulus of concrete to the peak secant modulus, where f_{cu} is the cubic compressive strength of the concrete; $B_1 = 1.6(A_1 - 1)^2$ is the physical quantity that controls the degree of decay of the elastic modulus in the ascending segment of the curve; α_1 is the parameter for the descending segment of the uniaxial compressive stress–strain curve of concrete. Considering the confining effect that concrete experiences in practical situations, it is uniformly taken as 0.15 for calculations.

The tensile behavior of concrete was described as follows:

$$y = \begin{cases} \frac{A_2 x + (B_2 - 1)x^2}{1 + (A_2 - 2)x + B_2 x^2} & x \leq 1 \\ \frac{x}{\alpha_2 (x-1)^2 + x} & x > 1 \end{cases} \tag{2}$$

where $y = \sigma / f_t$, $x = \varepsilon / \varepsilon_t$; f_t represents the uniaxial tensile strength of the concrete; ε_t is the peak strain under tension; A_2 is the parameter for the ascending segment, taken as 1.306 for calculations, $B_2 = 5(A_2 - 1)^2/3 = 0.15$ and $\alpha_2 = 1 + 3 \times 10^{-4} f_{cu}^2$ is the parameter for the descending segment.

The material properties of steel and concrete are obtained from the tests [29], as shown in Table 1.

Table 1. Material properties.

Material	Thickness (mm)	Yield Strength (MPa)	Ultimate Strength (MPa)	Elasticity Modulus
Q355	8	480.67	558.00	1.93×10^5
Q355	10	355.67	523.33	2.13×10^5
Q235	8	304.50	416.50	1.98×10^5
Concrete	/	/	31.20	3.03×10^4

2.2. Element Type and Meshing

The RCFST columns with web-perforated T-shaped stiffeners discussed here is composed of three main components: the steel tube, concrete, and web-perforated T-shaped stiffeners. To accurately analyze the stress states of each part during the compression process of the specimens, solid elements (C3D8R) are used to model the steel tube, concrete, and stiffeners. The steel tube and stiffeners are combined into a single component for calculations to reduce the complexity of contact settings, improve computational efficiency, and avoid issues related to over-constraining that could lead to calculation interruptions. The mesh size of steel tube and concrete is 20 mm, and the mesh size of stiffener is 10 mm. During the meshing process, the geometries of the stiffeners and concrete are complex, necessitating the combination of hexahedral and wedge elements to create an appropriate mesh. Especially in the chamfer areas, a sweeping mesh generation method is employed to handle irregular shapes, ensuring high-quality mesh generation in these regions. Other regular components are meshed using a structured meshing approach, as illustrated in Figure 3. In order to enhance the simplicity and computational efficiency of the FEM, it is essential to minimize any errors that may arise from complex models. Therefore, the influence of the end plates and the height of the spherical hinges on the compression behavior of the specimens is disregarded.

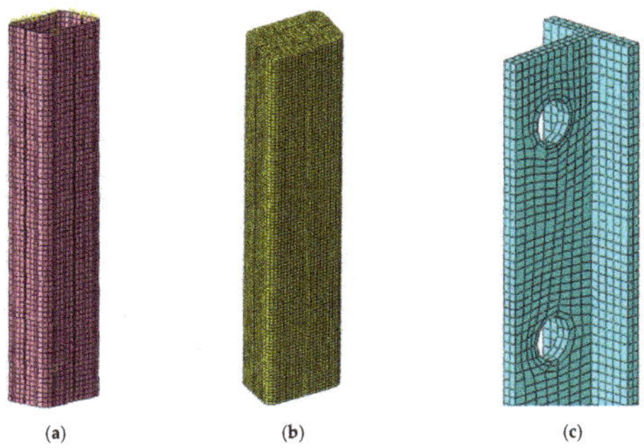

Figure 3. The mesh of the FE model: (**a**) Steel tube; (**b**) Concrete; (**c**) Stiffener.

2.3. Definition of Interaction and Boundary Conditions

The contact surfaces between the steel tube and concrete were defined using a surface-to-surface contact model, with the steel tube acting as the master surface and the concrete as the slave surface. The normal contact property was defined as 'hard' contact, while the tangential contact was modeled using a 'penalty' approach, with a friction coefficient of 0.25 [34]. Given that the concrete "tenon" within the openings of the web-perforated T-shaped stiffeners, the relative slip between the stiffeners and the concrete was neglected. Instead, the stiffeners were embedded in the concrete for analysis using the built-in region settings. Reference points were established at the top and bottom surfaces of the specimen, with the top surface coupled to reference point RP1 and the bottom surface coupled to reference point RP2. Constraints and displacements were applied at these reference points to simulate the boundary conditions and load application, as observed in the actual experiments. Finally, reference points RP1 and RP2 were offset to the loading line at the knife edge, as shown in Figure 4. At reference point RP1, horizontal displacements (U1 and U2) were constrained, while vertical displacement (U3) was allowed to vary to apply displacement loading. No constraints were imposed on the rotation angles (UR1, UR2, and UR3) in all three directions to simulate hinged boundary conditions. At reference point, RP2, displacements in all three directions (U1, U2, and U3) were constrained, while no constraints were imposed on the rotation angles (UR1, UR2, and UR3).

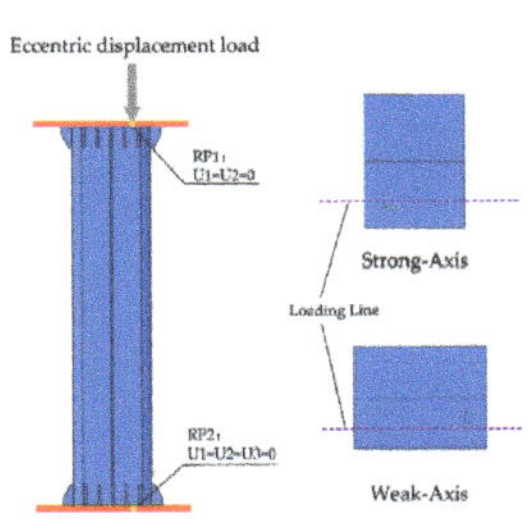

Figure 4. The load and boundary conditions of FE models.

2.4. Validation of FEM

The geometric dimension of the columns is shown in Figure 5. In the figure, B and H represent the width and height of the rectangular cross-section, respectively. b_{s1} and b_{s2} denote the height of the stiffener web and the width of the stiffener flange, respectively. t and t_s represent the thickness of the steel tube and the thickness of the stiffener, respectively. D is the diameter of the openings in the stiffeners.

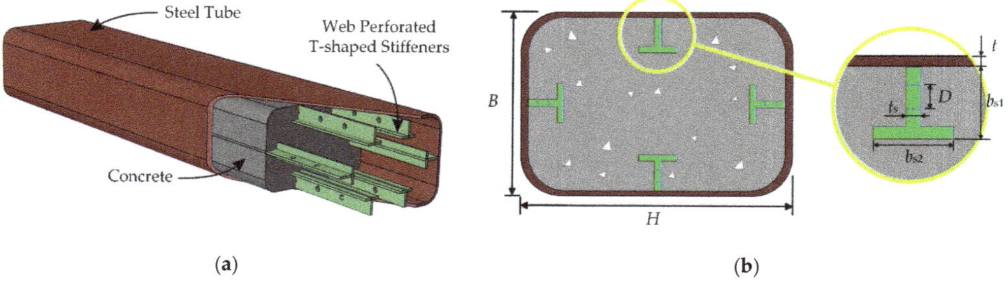

Figure 5. Schematic diagram of specimens: (**a**) 3D diagram; (**b**) Cross-sectional view.

Currently, there are no test studies on the eccentric compression performance of RCFST columns with web-perforated T-shaped stiffeners. To validate the rationality and accuracy of the FEM established, our axial compression test results were used for verification. It is worth noting that in the verification model, the reference points RP1 and RP2 must be moved to the center. The material properties, element selection, mesh generation, and contact settings are exactly the same as in eccentric compression modeling. The load–displacement curve is illustrated in Figure 6. The FEM exhibits more ideal stress states in various parts of the specimen. However, differences in stiffness and the descending segments of the load–displacement curves are observed between the FEM and the experimental results. These discrepancies may arise from initial defects, residual stresses from specimen fabrication, inherent deformation and measurement errors, etc. Table 2 presents a comparison of the load-carrying capacities, showing that the discrepancy between the FEM results and the experimental results remains within 10%, with a maximum error of 5.36%.

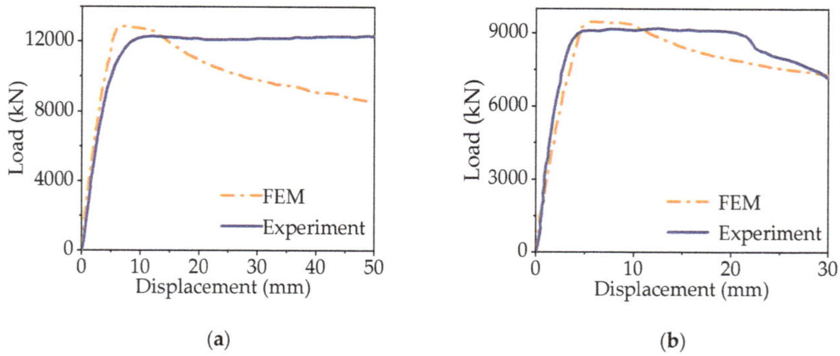

Figure 6. Comparison between experimental results: (**a**) T-CFST-1; (**b**) T-CFST-5.

Table 2. Comparison of experimental and FEM-bearing capacity.

No.	$L \times H \times B$ (mm)	t (mm)	b_{s1} (mm)	b_{s2} (mm)	t_{s1} (mm)	t_{s2} (mm)	D (mm)	y (mm)	N_{exp} (kN)	N_{FEM} (kN)	Error (%)
T-CFST-1	2200 × 480 × 320	10	68	60	8	8	20	120	11,502.49	12,118.80	5.36
T-CFST-2	2200 × 480 × 320	10	68	60	8	8	20	120	12,327.30	12,865.70	4.37
T-CFST-3	1800 × 400 × 266	8	75	60	8	8	25	120	9185.70	9456.80	2.95
T-CFST-4	1800 × 400 × 266	8	58	60	8	8	20	120	8945.40	9393.61	5.01
T-CFST-5	800 × 400 × 266	8	75	40	8	8	25	120	9285.29	9339.66	0.59

Note: L is the length of the column. N_{exp} is the bearing capacity obtained by the test and N_{FEM} is the bearing capacity obtained from FEM; Error is calculated by the formula $(N_{FEM} - N_{exp})/N_{exp} \times 100\%$.

As shown in Figure 7, both the components in the experiment and the model exhibited buckling at the midpoint of the column, accompanied by overall instability failure. The FEM reflects the failure mode observed in the experiment.

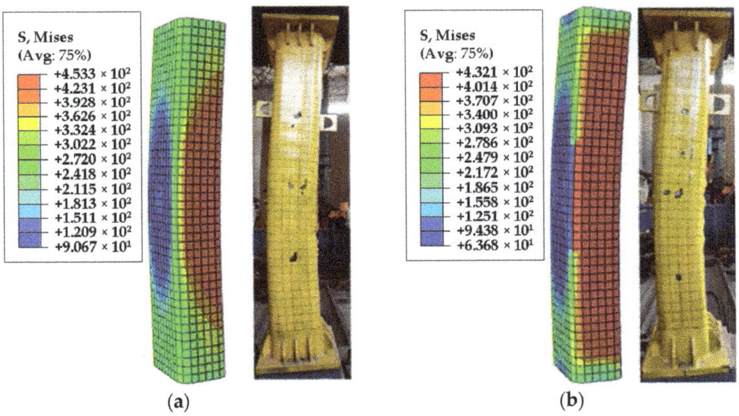

Figure 7. Comparison of failure modes between FE model and test: (**a**) T-CFST-1; (**b**) T-CFST-5.

3. Analysis of Eccentric Compression Behavior Mechanism

Specimen 1 (See Table 3) was analyzed in detail to intuitively understand the process of the specimen from elasticity to yield and final failure.

Table 3. Main parameters of FEM.

No.	L (mm)	n_s	b_{s1} (mm)	b_{s2} (mm)	t_s (mm)	D (mm)	y (mm)	f_{cu} (MPa)	f_y (Mpa)
specimen 1	2200	4	60	60	10	20	120	40	345
L800	800	4	60	60	8	20	120	40	345
L2200	2200	4	60	60	8	20	120	40	345
L3600	3600	4	60	60	8	20	120	40	345
L5000	5000	4	60	60	8	20	120	40	345
NS0	2200	0	/	/	/	/	/	40	345
NS2+2	2200	4	60	60	8	20	120	40	345
NS4+2	2200	6	60	60	8	20	120	40	345
NS4+4	2200	8	60	60	8	20	120	40	345
BS1-40	2200	4	40	60	8	20	120	40	345
BS1-60	2200	4	60	60	8	20	120	40	345
BS1-80	2200	4	80	60	8	20	120	40	345
BS1-100	2200	4	100	60	8	20	120	40	345
BS2-0	2200	4	60	0	8	20	120	40	345
BS2-30	2200	4	60	30	8	20	120	40	345
BS2-60	2200	4	60	60	8	20	120	40	345
BS2-90	2200	4	60	90	8	20	120	40	345
TS4	2200	4	60	60	4	20	120	40	345
TS6	2200	4	60	60	6	20	120	40	345
TS8	2200	4	60	60	8	20	120	40	345

Table 3. *Cont.*

No.	L (mm)	n_s	b_{s1} (mm)	b_{s2} (mm)	t_s (mm)	D (mm)	y (mm)	f_{cu} (MPa)	f_y (Mpa)
TS10	2200	4	60	60	10	20	120	40	345
C30	2200	4	60	60	8	20	120	30	345
C40	2200	4	60	60	8	20	120	40	345
C50	2200	4	60	60	8	20	120	50	345
C60	2200	4	60	60	8	20	120	60	345
Q235	2200	4	60	60	8	20	120	40	235
Q345	2200	4	60	60	8	20	120	40	345
Q390	2200	4	60	60	8	20	120	40	390
Q420	2200	4	60	60	8	20	120	40	420

Note: f_{cu} and f_y represent the compressive strength of concrete poured into steel tubes and the yield limit of steel, respectively.

3.1. Strain Distribution of Steel Tube and Stiffeners

The strain variation of each component in the web-perforated T-shaped stiffened RCFST column during the eccentric loading process is analyzed using equivalent plastic strain (PEEQ) values. A PEEQ value greater than 0 indicates that the material in that region has yielded and entered the plastic stage. For concrete, a PEEQ value exceeding 0.0033 signifies that the concrete in that area has been crushed.

The strain distribution of the steel tube and stiffeners is illustrated in Figures 8 and 9. When the load reaches 84.85% of the peak bearing capacity, the steel plate on the compression side of the steel tube yields first, whereas the steel plate adjacent to the stiffeners remains elastic due to the constraining effect of the stiffeners. Following the yielding of the compression side steel plate, the stiffeners also yield when the load attains 88.32% of the peak bearing capacity, with yielding predominantly occurring on both sides of the openings in the stiffener. Upon reaching its maximum load-bearing capacity, the steel tube and the stiffeners transition into the plastic region. This transition remains primarily concentrated on the compression side in the mid-span of the specimen. The stiffeners have a significant limiting effect on the strain increase in the steel tube wall.

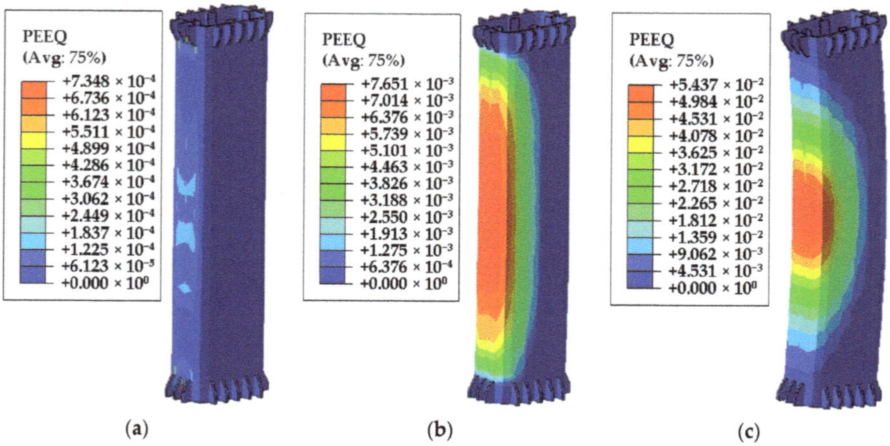

Figure 8. Strain development of steel tube: (**a**) Initial yield; (**b**) Ultimate load capacity; (**c**) Failure.

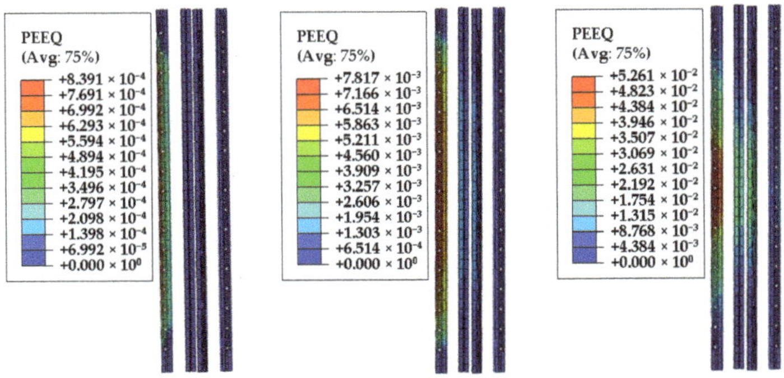

Figure 9. Strain development of stiffeners: (**a**) Initial yield; (**b**) Ultimate load capacity; (**c**) Failure.

3.2. Stress Distribution of Steel Tube and Stiffeners

Figures 10 and 11 show the stress development and distribution of the steel tube and stiffeners along the strong axis, respectively. The stress distribution and development in the steel tube are relatively uniform, with a notable reduction in stress in the steel tube wall near the stiffeners. The involvement of the stiffeners alleviates the overall stress on the steel tube. Once the load reaches the ultimate bearing capacity, the steel tubes and stiffeners on the compression side experience significant damage, while the steel on the tension side begins to engage in load-bearing.

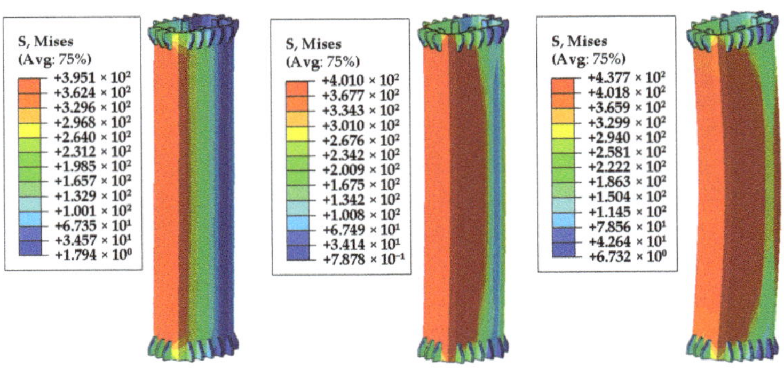

Figure 10. Stress development of steel tube: (**a**) Initial yield; (**b**) Ultimate load capacity; (**c**) Failure.

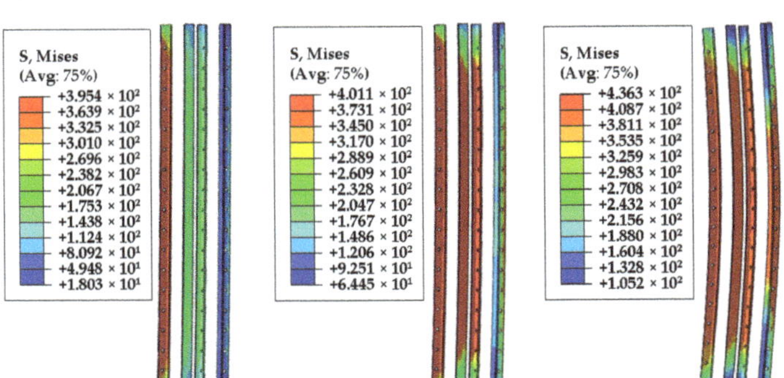

Figure 11. Stress development of stiffener: (**a**) Initial yield; (**b**) Ultimate load capacity; (**c**) Failure.

3.3. Strain Distribution of Concrete

Figure 12 delineates the progressive failure mechanisms of concrete under eccentric compression. At 42.83% of the peak bearing capacity, localized plastic deformation initiates at the steel–concrete interface near the perforated T-stiffeners. As the applied force escalates, the confinement effect induced by the stiffeners significantly mitigates lateral expansion of the concrete matrix, as evidenced by restrained strain accumulation in adjacent regions. Notably, when the loading level approaches 98.83%, compressive crushing initiates preferentially in the unconfined concrete zones on the compression face, whereas the stiffener–proximal regions maintain structural integrity due to enhanced triaxial stress states. Post-peak behavior is characterized by rapid crushing propagation in the compression zone, contrasting with limited damage evolution near the stiffeners—a phenomenon attributable to stress redistribution facilitated by the perforated reinforcement system.

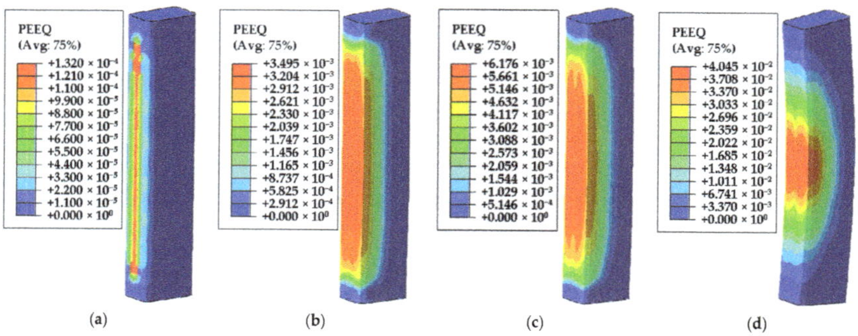

Figure 12. Strain development of concrete: (**a**) 42.83%N_u; (**b**) 98.83%N_u; (**c**) 100%N_u; (**d**) Failure.

3.4. Stress Distribution of Concrete

The vertical stress profiles along the strong axis, depicted in Figure 13, reveal distinct mechanical responses in concrete under eccentric loading. Tensile and compressive stresses are differentiated through positive and negative scalar fields, respectively. Analysis of the pre-yield phase indicates subdued stress escalation in stiffener-adjacent regions, attributable to localized confinement effects delaying plasticity initiation. At peak load, the confinement efficacy of stiffeners is amplified, as evidenced by sustained high compressive stresses within reinforced zones—even surpassing stress concentrations at steel–concrete interfacial corners. Conversely, tensile-dominated regions exhibit negligible stress modulation, aligning with concrete's inherent tensile weakness, which renders stiffener interventions ineffective in mitigating tensile cracking.

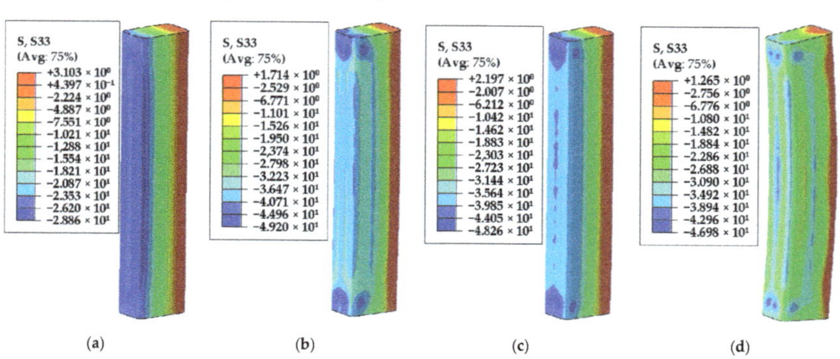

Figure 13. Stress development of concrete: (**a**) 42.83%N_u; (**b**) 98.83%N_u; (**c**) 100%N_u; (**d**) Failure.

4. Parameter Analysis

4.1. Selection of Eccentricity

To facilitate subsequent parameter analysis, FEM were developed for specimen 1, considering both strong axis eccentric compression and weak axis eccentric compression. The eccentricities analyzed were 0.15, 0.3, 0.45, 0.6, 1.0, 1.5, 2.0, 5.0, and 10.0. The resulting *N-M* interaction curves are presented in Figure 14. During the process of increasing eccentricity from 0 to 0.6, the *N-M* interaction curve of specimen 1 is almost a straight line. When the eccentricity is 5.0, the specimen basically reaches the maximum bending moment it can withstand. Therefore, during parameter analysis, eccentricities of 0, 0.6, 1.0, and 5.0 were selected, and a total of 29 models were established. Detailed parameters of all models were summarized in Table 3.

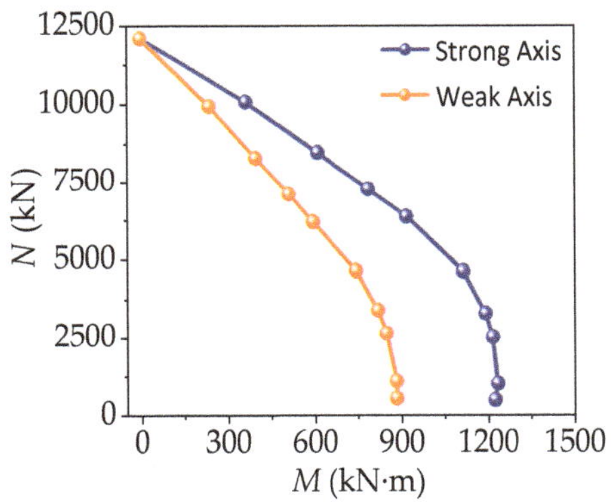

Figure 14. *N-M* interaction curve of specimen 1.

4.2. Slenderness Ratio

The slenderness ratio is a critical index in column design. It plays an important role in the strength, stiffness, and stability of columns. The calculation formula is as follows:

$$\lambda = 2\sqrt{3}L/H, \tag{3}$$

where λ represents the slenderness ratio, L is the length of column, and H is the width of section.

As shown in Figure 15, the larger the slenderness ratio of the column, the greater the influence of eccentricity. When the eccentricity increases, the maximum load and ultimate bending moment it can withstand decrease more significantly. As the slenderness ratio of the specimen increases, the *N-M* interaction curve tends to approximate a straight line. This indicates that the eccentric compression performance of the specimen becomes more similar to that of a pure steel column and is increasingly influenced by compressive stability. Taking the strong axis as an example, at a load eccentricity ratio of 0.6, the peak axial load increased by 41.3% as the slenderness ratio varied from 5.8 to 36.1, compared to an improvement of 36.4% at a load eccentricity ratio of 1. Therefore, when calculating the bearing capacity and moment of the specimen under eccentric loading, it is necessary to consider the reductions for different slenderness ratios and eccentric ratios.

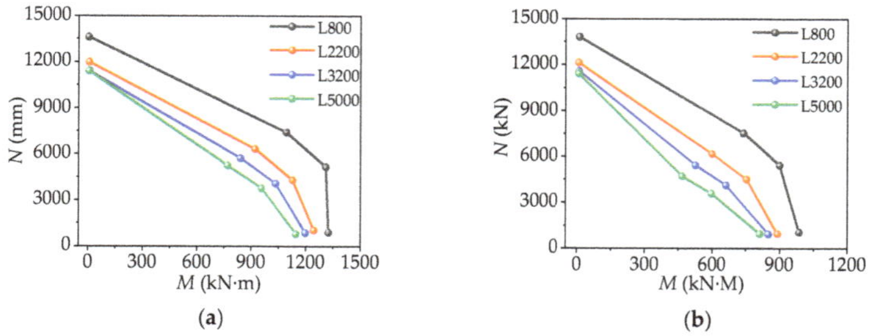

Figure 15. *N-M* interaction curves of specimens with different slenderness ratio: (**a**) Strong Axis; (**b**) Weak Axis.

4.3. Number and Size of Stiffeners

Compared with ordinary CFST columns without stiffeners, the RCFST columns with stiffeners have improved axial compression and bending capacities, as shown in Figure 16. Taking the strong axis and 0.6 eccentricity rate as an example, compared with the NS0 specimen, the N_u of the NS2+2 specimen with four stiffeners increased by 21.1%, and the N_u of the NS4+4 specimen with eight stiffeners increased by 36.8%. The research indicates that the addition of each stiffener increases the bearing capacity by approximately 5%. However, the degree of improvement in bearing capacity tends to diminish as the number of stiffeners increases. Under strong-axis eccentric loading, increasing the number of stiffeners on the short side of the specimen enhances its bearing capacity. In contrast, under weak-axis eccentric loading, adding stiffeners to the long side of the specimen leads to a more substantial improvement in bearing capacity. This indicates that the enhancement of the bending performance of the specimen provided by the stiffeners comes from the flange plates, while the openings in the web affect the mechanical properties of the steel plates. Compared to the *N-M* curve under strong-axis eccentric compression, the increase of the number of stiffeners shows a smaller improvement in bearing capacity under weak-axis eccentric loading. This suggests that small cross-sectional widths can impact the performance of stiffeners.

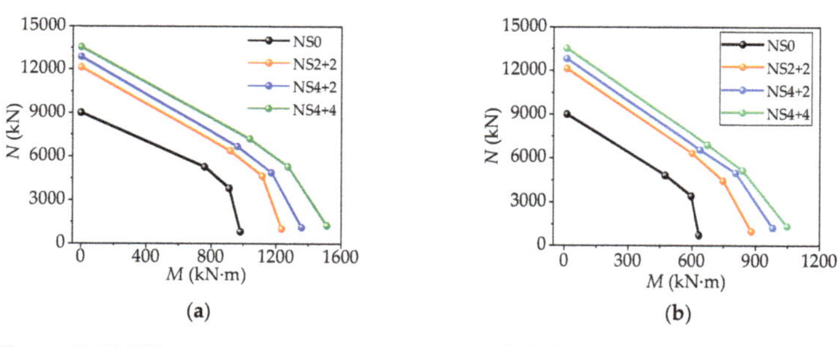

Figure 16. *N-M* interaction curves of specimens with different number of stiffeners: (**a**) Strong Axis; (**b**) Weak Axis.

Figures 17–19 present the *N-M* interaction curves of RCFST columns under compression, with varying stiffener heights, widths, and thicknesses. Taking the cross-sectional area of the stiffeners in specimen 1 as a reference, the change rates in specimens BS1-80 and BS1-100 are +17.86% and +35.71%, respectively. At a strong axis eccentricity ratio of 0.6, the ultimate bearing capacity increases by +1.99% and +3.95%, respectively. The change rate of the cross-sectional area in specimen BS2-90 is +26.79%, with the ultimate bearing capacity increasing by +3.27%. For specimen TS10, the cross-sectional area increases by +22.77%, with a +2.80% increase in ultimate bearing capacity. It can be deduced that increasing the height of the stiffeners by 1% leads to a 0.111% increase in the ultimate bearing capacity. For every 1% increase in cross-sectional area from changes in width or thickness, the ultimate bearing capacity increases by 0.122% and 0.123%, respectively. Compared to increasing the height, enhancing the width and thickness of the stiffeners has a more significant effect on improving the compressive performance of the columns. Increasing the thickness of steel components remains one of the most effective methods to improve the mechanical performance of steel–concrete composite structures. Moreover, increasing the width of the stiffeners results in a performance improvement similar to that achieved by increasing their thickness. This is due to the wider stiffener increasing the contact area with the concrete during bending, thereby enhancing the restraining effect.

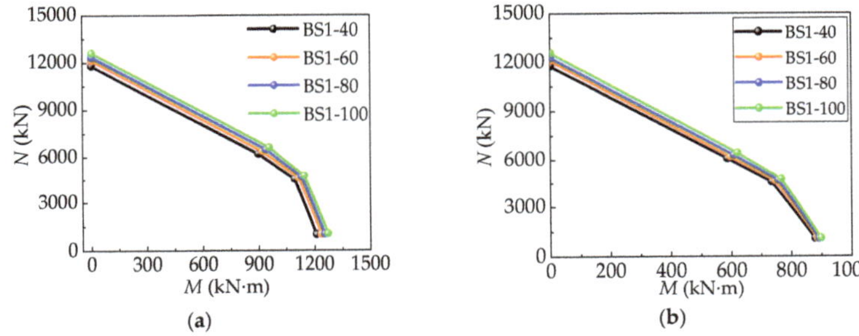

Figure 17. *N-M* interaction curves of specimens with different stiffening heights: (**a**) Strong Axis; (**b**) Weak Axis.

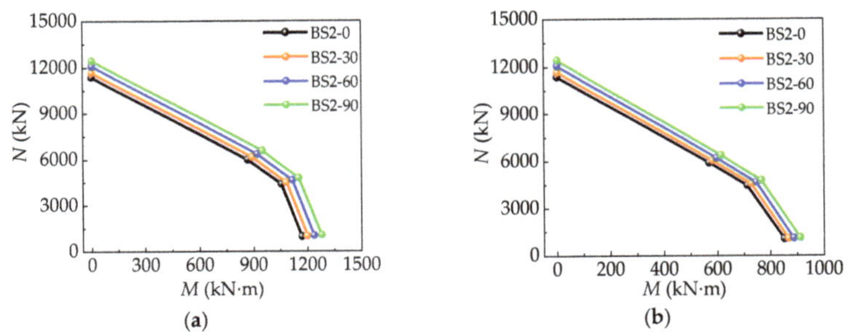

Figure 18. *N-M* interaction curves of specimens with different stiffening widths: (**a**) Strong Axis; (**b**) Weak Axis.

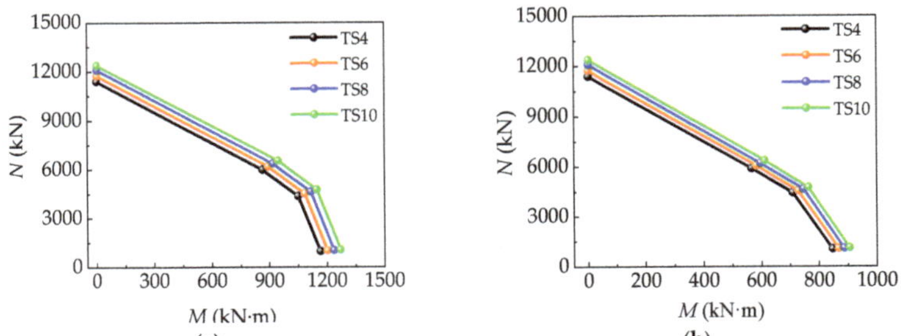

Figure 19. *N-M* interaction curves of specimens with different stiffener thicknesses: (**a**) Strong Axis; (**b**) Weak Axis.

4.4. Material Strength

Figures 20 and 21 show the *N-M* interaction curves of stiffened RCFST columns with different concrete strengths and steel strengths. It is obvious that changing the strength grade of the material is the most significant factor affecting the bearing capacity of the specimen. As shown in Figure 20, with the increase of concrete strength grade, the bending degree of the *N-M* interaction curve of the specimen becomes greater, indicating that the eccentric compression performance of the specimen is closer to that of the concrete column. On the contrary, in Figure 21, with the increase of steel strength grade, the *N-M* interaction curve of the specimen tends towards a straight line, indicating that the eccentric compression performance of the specimen is closer to that of a pure steel column. Therefore, it is essential to choose the strength of steel and concrete reasonably in the design. The strength of the steel pipe is not the higher the better, and the strength of the concrete and the design of the stiffeners should also be considered.

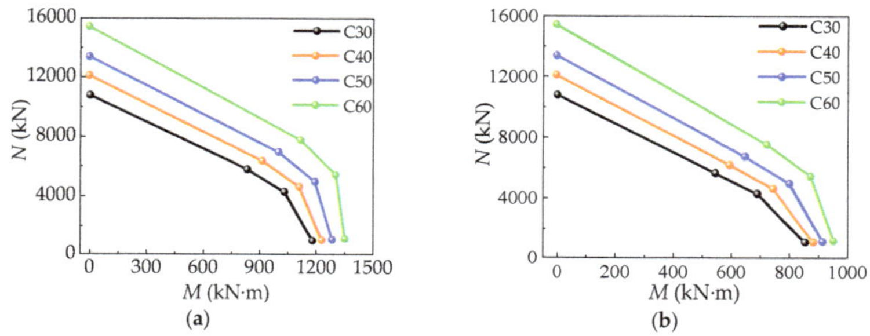

Figure 20. *N-M* interaction curves of specimens with different concrete strengths: (**a**) Strong Axis; (**b**) Weak Axis.

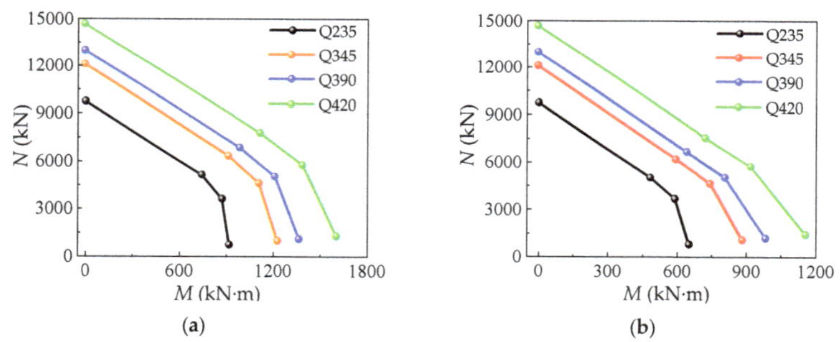

Figure 21. *N-M* interaction curves of specimens with different steel strengths: (**a**) Strong Axis; (**b**) Weak Axis.

4.5. The Load (N)-Midspan Deflection (Δ) Curves

The *N*-Δ curves provide a detailed depiction of the stiffness, yielding behavior, and post-peak load-carrying capacity degradation throughout the entire eccentric compression process of the specimens. Figure 22 illustrates the *N*-Δ curves for specimens with different parameters. The analysis was performed under strong-axis loading conditions with an eccentricity ratio of 0.6. The column length is 2200 mm. As shown in Figure 22a, compared to CFST columns without stiffeners, the use of stiffeners improves the compressive capacity of the specimens and reduces the rate of capacity degradation after reaching the peak load, thereby enhancing their deformation capacity. Figure 22b,c indicates that altering the height and width of the stiffeners results in similar trends in their influence on eccentric compression performance. However, increasing the width of stiffener has a more pronounced effect on improving the performance of the specimens, consistent with the conclusions drawn from the analysis of the N-M interaction curves. Figure 22d shows that changes in stiffener thickness also have a significant impact on the eccentric compression performance of the specimens, while influencing their stiffness to a certain extent. Figure 22e,f reveals that increasing the strength of the concrete and steel has a noticeable effect on enhancing the eccentric compression performance of the specimens. However, higher concrete strength leads to a more rapid decline in load-carrying capacity after reaching the peak load, reducing the ductility and deformation capacity. In contrast, higher steel strength results in a gentler decline in the N-Δ curves, indicating improved ductility and deformation capacity.

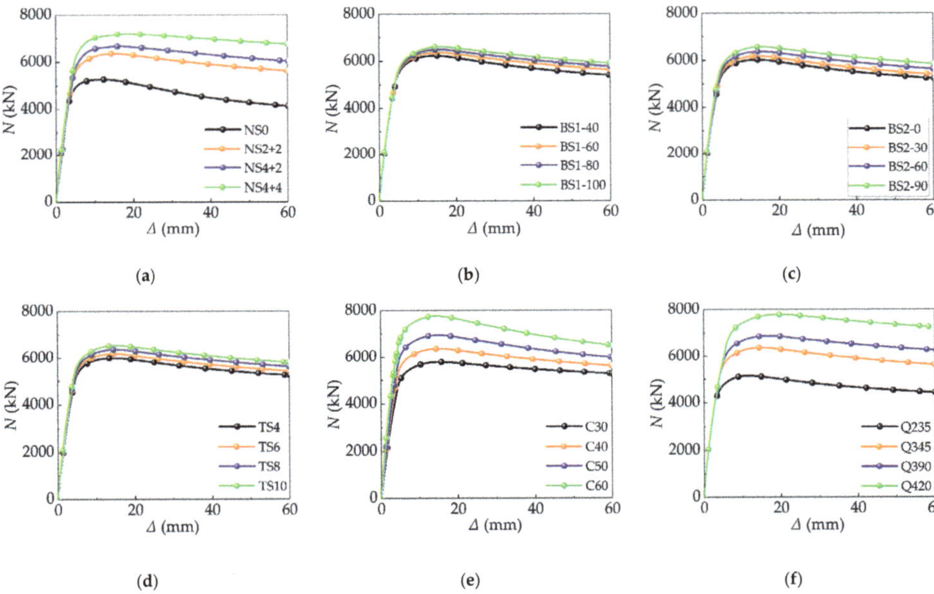

Figure 22. N-Δ curves: (**a**) n_s; (**b**) b_{s1}; (**c**) b_{s2}; (**d**) t_s; (**e**) f_{cu}; (**f**) f_y.

5. Calculation of Eccentric Bearing Capacity

5.1. Contribution of Concrete

The FEM results demonstrate that the perforated T-stiffeners enhance the confinement effects on the infilled concrete within the steel tube. The concrete inside the steel tube can be divided into strong constraint zones and partial constraint zones, as shown in Figure 23. Taking the stiffeners and steel tube walls as equivalent H-beams embedded in concrete, and b_w is the web width of the equivalent H-beam. The majority of the area within the equivalent H-beam flange is a concrete strong constraint zone, with a quadratic parabolic boundary and the vertex height b_{pw} of 0.25 times the height b_w of the web plate.

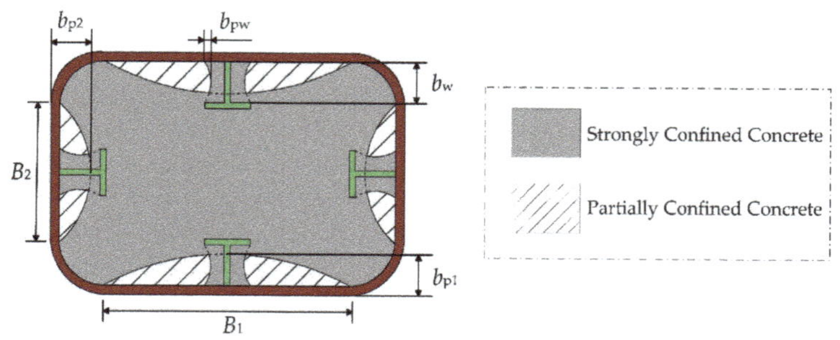

Figure 23. Core concrete effective confinement zone.

The load-bearing capacity of concrete within the stronger constrained zone is determined by the enhanced compressive strength of confined concrete f_{cc}, whereas regions under moderate confinement utilize the confined concrete strength f_c. Following the Eurocode 8 [35] provisions, the confined concrete strength can be calculated as:

$$f_{cc} = \begin{cases} f_c\left(1.0 + 5.0k_e\frac{f_1}{f_c}\right) & \frac{f_1}{f_c} < 0.05 \\ f_c\left(1.125 + 2.5k_e\frac{f_1}{f_c}\right) & \frac{f_1}{f_c} \geq 0.05 \end{cases} \tag{4}$$

$$f_1 = k_e(tf_y/B_1 + tf_y/B_2), \tag{5}$$

$$k_e = \frac{A_h}{A_c}, \tag{6}$$

where f_{cc} is the compressive strength of confined concrete; f_1 is the equivalent uniformly distributed transverse compressive stress exerted by the steel tube on the core concrete; f_c and f_y are the compressive strength of ordinary concrete and the yield strength of steel, respectively; t is the thickness of the steel tube wall; B_1 and B_2 are the width and height of the inner wall of the steel tube, respectively; k_e is the effective confinement coefficient of concrete; A_h and A_c are the area of strongly restrained concrete and the total area of concrete, respectively.

When calculating the *N-M* interaction curve of web perforated T-shaped stiffened RCFST, it is necessary to consider the bearing capacity at different neutral axis positions of the specimen section separately. To simplify the calculation, the concrete strength adopts uniform strengthening strength:

$$f_{ca} = N_c/A_c\,, \tag{7}$$

$$N_c = 0.85(f_cA_p + f_{cc}A_h)\,, \tag{8}$$

where f_{ca} is the uniform strengthening strength of concrete; N_c is the axial compressive bearing capacity provided by concrete; A_p is the concrete area of some confined areas.

5.2. Contribution of Steel Tube and Stiffeners

Adding stiffeners inside the steel tubes is an effective structural measure to prevent or limit local bulging deformation tubes in CFST columns. When the stiffeners possess sufficient rigidity or exhibit strong interaction with the concrete, they remain unaffected by the bulging and deformation of the steel tube walls, thereby providing significant restraint against local buckling.

In terms of the strength reduction caused by local bulging deformation in the middle of the tube wall during compression of CFST columns, the Design Specification AS 4100 [36] provides a formula for calculating the local buckling stress under compression related to the buckling coefficient of the steel plate:

$$f_{\mathrm{ol}} = \frac{\pi^2 E_{\mathrm{s}}}{12(1-v_{\mathrm{s}}^2)} \frac{k_0}{(b/t)^2}, \tag{9}$$

where f_{o1} is the local buckling stress of the steel tube wall; E_{S} is the elastic modulus of steel; v_{s} is the Poisson's ratio of steel, which can be calculated as 0.30; k_0 is the buckling coefficient of the steel plate without stiffeners; *b/t* is the width to thickness ratio of the steel plate, where the width *b* of the steel plate needs to be reduced according to the number of stiffeners.

Considering the effective width of the steel tube wall of the CFST column, the local buckling stress of the steel plate, and the regularized width to thickness ratio, the effective distribution width coefficient ρ [37] of the single-sided stiffened steel plate that distinguishes the difference in ultimate stress states under elasticity and elastoplastic can be taken as follows:

$$\rho = \begin{cases} 1 & \lambda \le 0.5 \\ \frac{0.66}{\lambda^{0.6}} & 0.5 < \lambda \le 1.348, \\ \frac{0.64}{\lambda^{0.5}} & \lambda > 1.348 \end{cases} \tag{10}$$

$$\lambda_{\mathrm{p}} = \sqrt{\frac{f_{\mathrm{y}}}{f_{\mathrm{ol}}}}, \tag{11}$$

where λ_{p} is the regularized width to thickness ratio of the steel plate; f_{y} is the yield strength of steel.

After obtaining the effective distribution width coefficient of the steel plate, the strength of the steel is reduced by calculating the ratio of the effective width area of the steel tube to the total cross-sectional area:

$$f_{\mathrm{ye}} = f_{\mathrm{y}} \frac{A_{\mathrm{se}}}{A_{\mathrm{s}}}, \tag{12}$$

where f_{ye} is the effective strength of steel; $A_{\mathrm{se}} = \rho b_{\mathrm{e}} t$ is the effective width area of the steel plate; As is the total cross-section area of the steel plate.

Calculate the equivalent of web perforated T-shaped stiffeners as a straight shaped stiffener according to the principle of equivalent moment of inertia, as shown in Figure 24. Considering that the opening of the web plate will affect the mechanical properties of the stiffeners, the thickness of the web plate will be reduced before the stiffeners become equivalent as a whole. The equivalent thickness of stiffeners can be calculated using the following formula:

$$t'_{\mathrm{s}} = \gamma_{\mathrm{s}} t_{\mathrm{s}} + t_{\mathrm{E}}, \tag{13}$$

$$\gamma_{\mathrm{s}} = \frac{b_{\mathrm{s1}} - d}{b_{\mathrm{s1}}}, \tag{14}$$

where t_{s}' is the equivalent thickness of web perforated T-shaped stiffeners that are equivalent to straight shaped stiffeners; γ_{s} is the opening reduction coefficient; t_{E} is the thickness of the stiffener flange plate spread flat to the web plate according to the principle of equivalent moment of inertia; *d* is the diameter of the stiffener opening.

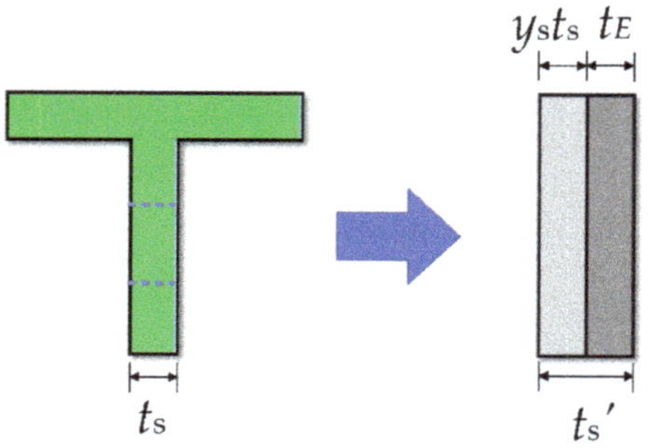

Figure 24. Equivalent principle of web perforated T-shaped stiffener.

According to the analysis results in Section 4.3, when the web perforated T-shaped stiffeners are eccentrically compressed, changing the number of stiffeners on the steel tube web has little influence on the eccentric compression performance. To simplify the calculation and consider the influence of the flange plate of the stiffener on the bending performance of the specimen, the stiffener on the steel tube web is spread flat to the steel tube web for calculation, as shown in Figure 25. There are two methods to spread the web perforated T-shaped stiffeners flat on the steel tube web: (1) Spread the cross-sectional area of the stiffeners evenly over the steel tube web plate; (2) Use the principle of equivalent moment of inertia, the stiffeners are equivalent to the steel tube web. When calculating the *N-M* interaction curve of web perforated T-shaped stiffened RCFST columns, the thickness of the steel tube web plate is calculated using the flattened thickness of the stiffeners, while the thickness of the steel tube remains unchanged.

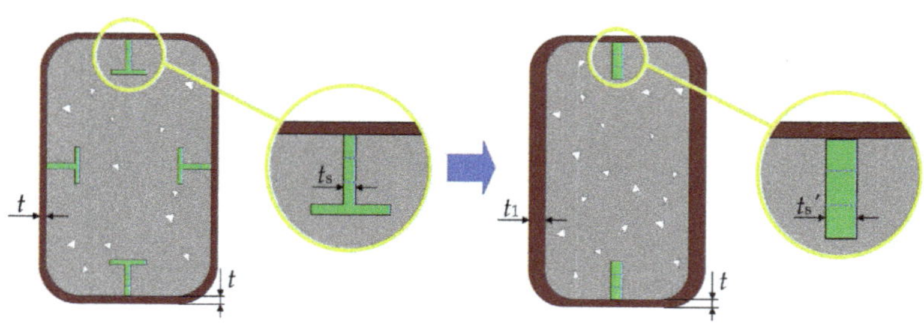

Figure 25. Simplified calculation model of web perforated T-shaped stiffener.

5.3. Derivation Formula for Eccentric Compression

The *N-M* interaction curves of the strong axis and weak axis eccentric compression of web perforated T-shaped stiffened RCFST columns are calculated using the same formula, and are divided into the following three situations based on the different positions of the neutral axis:

(1) The neutral axis is located in the concrete ($0 \le x < \frac{h}{2} - b_{s1} - t$)

When the neutral axis of the specimen is located in the core concrete area, the stress distribution of the web perforated T-shaped stiffened RCFST column section is shown in Figure 26a. At this time, a large part of the concrete is in the tensile zone. If the stress on this part of the concrete is ignored, the axial force and bending moment can be determined using the following formula:

$$N = 4f_{ya}xt_1 + f_{ca}\left[(b-2t)\left(\frac{h}{2} - t + x\right) - n_s t'_s b_{s1}\right], \tag{15}$$

$$M = f_{ya}\left\{ 2t_1\left[\left(\frac{h}{2}-t\right)^2 - x^2\right] + bt(h-t) + 2n_s b_{s1} t'_s\left(\frac{h}{2}-t-\frac{b_{s1}}{2}\right) \right\} + \frac{1}{2} f_{ca}(b-2t)\left[\left(\frac{h}{2}-t\right)^2 - x^2\right]. \quad (16)$$

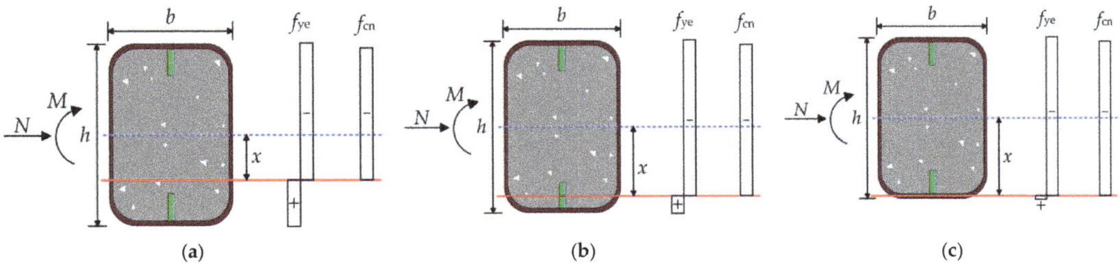

Figure 26. Force diagram of section: (**a**) $0 \le x < \frac{h}{2} - b_{s1} - t$; (**b**) $\frac{h}{2} - b_{s1} - t < x \le \frac{h}{2} - t$; (**c**) $\frac{h}{2} - t < x \le \frac{h}{2}$.

(2) The neutral axis is located in the stiffener ($\frac{h}{2} - b_{s1} - t < x \le \frac{h}{2} - t$)

When the neutral axis of the specimen is located in the stiffener area, the stress distribution of the web perforated T-shaped stiffened RCFST column section is shown in Figure 26b. At this point, the area of the core concrete compression zone decreases, and a portion of the stiffener on one side of the specimen's flange is in the compression zone and another portion is in the tension zone. The axial force and bending moment can be determined using the following formula:

$$N = f_{ya}[4t_1 x + n_s t'_s(2b_{s1} - h + 2t + 2x)] + f_{ca}\left[(b-2t)\left(\frac{h}{2}-t+x\right) - n_s t'_s\left(2b_{s1} + x - \frac{h}{2} + t\right) \right], \quad (17)$$

$$M = f_{ya}\left\{ 2t_1\left[\left(\frac{h}{2}-t\right)^2 - x^2\right] + bt(h-t) + n_s t'_s\left[\left(\frac{h}{2}-t\right)^2 - x^2\right] \right\} + \frac{1}{2} f_{ca}(b-2t)\left[\left(\frac{h}{2}-t\right)^2 - x^2\right]. \quad (18)$$

(3) The neutral axis is located at the flange of the steel tube ($\frac{h}{2} - t < x \le \frac{h}{2}$)

When the neutral axis of the specimen is located inside the flange of the steel tube, the stress distribution of the web perforated T-shaped stiffened RCFST column section is shown in Figure 26c. At this point, except for a small portion of the steel tube flange being under compression, most of the cross-section of the specimen is under tension. The axial force and bending moment can be determined using the following formula:

$$N = f_{ya}\left[2t_1(h-2t) + b(2x-h+2t) + 2n_s t_s b_{s1} \right] + f_{ca}\left[(b-2t)(h-2t) - 2n_s t'_s b_{s1}\right], \quad (19)$$

$$M = f_{ya} b\left[\left(\frac{h}{2}\right)^2 - x^2\right]. \quad (20)$$

Based on the analysis results in Section 4.2, as the slenderness ratio increases, the shape of the *N-M* curve for the stiffened RCFST column specimens gradually approaches a straight line, and the variation in performance amplitude also changes with increasing eccentricity. When the slenderness ratio is large, neither the steel nor the concrete can develop its strength, leading to a decline in the overall mechanical performance of the specimen. Therefore, it is crucial to consider the effects of different slenderness ratios and eccentricities on the eccentric compression behavior of the specimens. Based on the FEM results of the eccentric compression performance of the stiffened RCFST columns with different slenderness ratios under different eccentricities in this paper, a correction factor for the eccentric compression bearing capacity is derived for cases where the eccentricity e is greater than or equal to 0.6.

$$\eta = \frac{22.1}{\lambda^2} + 0.034e + 0.763, \quad (21)$$

wherein λ is the slenderness ratio; e is the eccentric ratio.

5.4. Comparison and Verification of Calculation Methods

The eccentric compression bearing capacity of RCFST columns with web perforated T-shaped stiffeners was compared using the theoretical calculation method proposed in this study and the empirical coefficient calculation method of CECS28 [38], as illustrated in Figure 27. The results of the theoretical method proposed in this paper are more accurate than those of the empirical coefficient method, and overall, its calculation results are lower than experimental and simulation results, leaving a certain margin of safety, making it more suitable for practical engineering design.

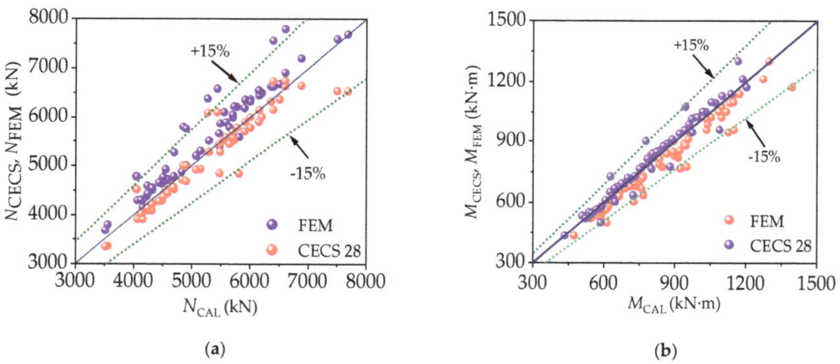

Figure 27. Comparison of calculated and simulated values: (**a**) Eccentric compression bearing capacity; (**b**) Bending moment.

6. Conclusions

(1) A finite element analysis model for web-perforated T-shaped stiffened RCFST columns was developed in this study. The model was validated through comparisons with existing test results, showing a difference within 10% and a maximum error of 5.36%. The outcomes demonstrate the precision of the FEM presented in this research for predicting the mechanical performance of web perforated T-shaped stiffened RCFST columns.

(2) Web-perforated T-shaped stiffeners can restrain the growth and development of stress and strain in various parts of the specimen during eccentric compression. They also help delay the damage of steel tube and concrete while enhancing the compressive strength of concrete in certain areas.

(3) The effects of L, n_s, b_{s1}, b_{s2}, t_s, f_{cu}, and f_y on the eccentrically loaded capacity of web-perforated T-shaped stiffened RCFST columns were analyzed. An increase in L reduces the overall stability of the member, making its mechanical behavior closer to that of a pure steel column. Enhancing f_{cu} and f_y improves the load-bearing capacity; however, increasing f_{cu} reduces the ductility and deformability of the columns. The number of stiffeners has a more significant influence on the bearing capacity of members under strong-axis eccentric loading, with a smaller impact on the weak axis.

Among the stiffener size parameters, b_{s2} and t_s exhibit a more significant influence on the load-bearing capacity compared to b_{s1}.

(4) A novel method is proposed for predicting the ultimate strength of RCFST columns with web-perforated T-shaped stiffeners. This method considers the joint load-bearing contribution of the web-perforated T-shaped stiffeners and the steel tube while accounting for the impact of local buckling of the steel tube on the ultimate strength. The results indicate that this method can achieve higher accuracy while ensuring safety, whereas the calculations based on CECS 28 is conservative.

Author Contributions: Writing—original draft, Q.J.; Writing—review & editing, J.Z. and T.L.; Supervision, L.S. and Y.W. All authors have read and agreed to the published version of the manuscript.

Funding: This research received no external funding.

Data Availability Statement: Data are contained within the article.

Conflicts of Interest: Authors Lvhong Sun and Yanchao Wu were employed by the company CCCC First Highway Northwest Engineering Co., Ltd. The remaining authors declare that the research was conducted in the absence of any commercial or financial relationships that could be construed as a potential conflict of interest.

References

1. Han, L.H.; Li, W.; Bjorhovde, R. Developments and advanced applications of concrete-filled steel tubular (CFST) structures: Members. *J. Constr. Steel Res.* **2014**, *100*, 211–228. [CrossRef]
2. Thomas, J.; Sandeep, T.N. Capacity of short circular CFST columns with inner vertical plates welded intermittently. *J. Constr. Steel Res.* **2020**, *165*, 105840. [CrossRef]
3. Liao, J.J.; Zeng, J.J.; Long, Y.L.; Cai, J.; Ouyang, Y. Behavior of square and rectangular concrete-filled steel tube (CFST) columns with horizontal reinforcing bars under eccentric compression. *Eng. Struct.* **2022**, *271*, 114899. [CrossRef]
4. Hassanein, M.F.; Patel, V.I.; Elchalakani, M.; Thai, H.T. Finite element analysis of large diameter high strength octagonal CFST short columns. *Thin-Walled Struct.* **2018**, *123*, 467–482. [CrossRef]
5. Young, B.; Ellobody, E. Experimental investigation of concrete-filled cold-formed high strength stainless steel tube columns. *Constr. Steel Res.* **2006**, *62*, 484–492. [CrossRef]
6. Alatshan, F.; Osman, S.A.; Altlomate, A.; Alkair, M.; Hamid, R.; Mashiri, F. Design Model of Rectangular Concrete-Filled Steel Tubular Stub Columns under Axial Compression. *Buildings* **2023**, *13*, 128. [CrossRef]
7. Fujinaga, T. Ultimate flexural strength of rectangular concrete-filled steel tubular beam-columns using high-strength materials. *Jpn. Archit. Rev.* **2023**, *6*, e12336. [CrossRef]
8. Li, H.D.; Guo, L.H.; Gao, S.; Elchalakani, M. Analysis and design of wide rectangular concrete-filled steel tubular columns under axial compression. *Structures* **2024**, *70*, 107540. [CrossRef]
9. Wang, Z.B.; Dai, H.W.; Ke, C.L. Test and design of rectangular steel-reinforced concrete-filled stainless steel tubular short columns. *Eng. Struct.* **2023**, *293*, 116689. [CrossRef]
10. Zhou, Z.; Denavit, M.D.; Zhou, X.H. New cross-sectional slenderness limits for stiffened rectangular concrete-filled steel tubes. *Eng. Struct.* **2023**, *280*, 115689. [CrossRef]
11. Bhartiya, R.; Sahoo, D.R. Prediction of axial compression behavior of rectangular RCFST columns with confining ties. *J. Constr. Steel Res.* **2021**, *186*, 106920. [CrossRef]
12. Ding, F.X.; Fang, C.J.; Bai, Y.; Gong, Y.Z. Mechanical performance of stirrup-confined concrete-filled steel tubular stub columns under axial loading. *Constr. Steel Res.* **2014**, *98*, 146–157. [CrossRef]
13. Ding, F.X.; Fu, L.; Liu, X.M.; Liu, J. Mechanical performances of track-shaped rebar stiffened concrete-filled steel tubular (SCFRT) stub columns under axial compression. *Thin-Walled Struct.* **2016**, *99*, 168–181. [CrossRef]
14. Du, Y.S.; Gao, D.H.; Chen, Z.H.; Zheng, Z.H.; Wang, X.D. Behaviors of FRP confined rectangular concrete-filled thin-walled steel tubular stub columns using high-strength materials under axial load. *Compos. Struct.* **2022**, *280*, 114915. [CrossRef]
15. Zhou, Z.; Zhou, X.H.; Gan, D.; Liu, Y.J. Comparison and design of stiffened rectangular concrete-filled steel tubular members. *J. Constr. Steel Res.* **2023**, *208*, 108037. [CrossRef]
16. Gan, D.; Zhou, Z.; Zhou, X.H. Improved composite effect of square concrete-filled steel tubes with diagonal binding ribs. *J. Struct. Eng.* **2019**, *145*, 04019112.

17. Ge, H.; Usami, T. Strength of concrete-filled thin-walled steel box columns: Experiment. *J. Struct. Eng.* **1992**, *118*, 3036–3054. [CrossRef]
18. Cai, J.; Long, Y. Local buckling of steel plates in rectangular CFT columns with binding bars. *J. Constr. Steel Res.* **2009**, *65*, 965–972. [CrossRef]
19. Chen, S.Z.; Liu, Y.; Luo, J.H.; Gao, S.Y. Experimental and numerical analysis on rectangular concrete-filled steel tubular columns with T-shaped stiffeners. *J. Build. Eng.* **2022**, *45*, 103510. [CrossRef]
20. Liu, Y.J.; Zhang, N.; Zhang, J.G. Mechanical behavior of concrete-filled square steel tube stiffened with PBL. *J. Archit. Civ. Eng.* **2012**, *29*, 13–17. (In Chinese)
21. Liu, J.P.; Yang, J.; Chen, B.C.; Zhou, Z.Y. Mechanical performance of concrete-filled square steel tube stiffened with PBL subjected to eccentric compressive loads: Experimental study and numerical simulation. *Thin-Walled Struct.* **2020**, *149*, 106617. [CrossRef]
22. Guo, J.H.; Diao, Y. Experimental behaviors of square concrete filled steel tubular columns with PBL stiffeners. *Structures* **2022**, *38*, 1556–1569. [CrossRef]
23. *CEN, EN 1994-1-1*; Design of Composite Steel and Concrete Structures—Part 1: General Rules and Rules for Buildings. European Standards: London, UK, 2005.
24. *AISC 360-2022*; American Steel Construction Institute Standard, Specification for Structural Steel Buildings. American Institute of Steel Construction, Inc.: Chicago, IL, USA, 2022.
25. *AS/NZS 2327*; Composite Steel-Concrete Construction for Buildings. Australian/New Zealand Standards: Wellignton, New Zealand, 2020.
26. *GB50936-2014*; Technical Code for Concrete Filled Steel Tubular Structures. China Architecture & Building Press Beijing: Beijing, China, 2014. (In Chinese)
27. *JGJ 138—2016*; Code for Design of Composite Structures. China Architecture and Building Press: Beijing, China, 2016. (In Chinese)
28. *CECS 159:2004*; Technical Specification for Structures with Concrete-Filled Rectangular Steel Tube Members. China Architecture and Building Press: Beijing, China, 2004. (In Chinese)
29. Lei, T.; Jiang, Q.F.; Liang, S.F.; Wu, Z.M.; Xu, J.X. *Experimental Investigation on Axial Compressive Performance of Rectangular CFST Columns with Web Perforated T-Shaped Stiffeners*; Chang'an University: Xi'an, China, 2024. (In Chinese)
30. Akhaveissy, A.H. Seismic Retrofit of Defective RC Beam-Column Joints. *Period. Polytech. Civ. Eng.* **2018**, *62*, 596–611. [CrossRef]
31. Zhang, J.G. Research on the Mechanical Performance of Concrete-Filled Square Steel Tube Stiffened with PBL under Axial Com-Pression Load. Ph.D. Thesis, Chang'an University, Xi'an, China, 2012. (In Chinese)
32. ABAQUS. *ABAQUS Standard User's Manual, version 6.12*; Dassault Systems Simulia Corporation: Providence, RI, USA, 2012.
33. Ding, F.X.; Zhou, L.C.; Yu, Z.W. Nonlinear finite element analysis of axially loaded concrete-filled steel tubular stub columns. *China Sci.* **2009**, *4*, 472–479. (In Chinese)
34. Ellobody, E.; Young, B. Nonlinear analysis of concrete-filled steel SHS and RHS columns. *Thin-Walled Struct.* **2006**, *44*, 919–930. [CrossRef]
35. *EN 1998-1:2004*; Eurocode 8: Design of Structures for Earthquake Resistance—Part 1: General Rules, Seismic Actions and Rules for Buildings. British Standard: London, UK, 2005.
36. *AS4100-2020*; Steel Structures. Australian Standards: Sydney, Australia, 2020.
37. Wang, H.T.; Liu, Y.J.; Sun, L.P. Calculation method of eccentric bearing capacity of RCFST based on post-buckling strength of panels. *J. Archit. Civ. Eng.* **2022**, *39*, 41–52.
38. *CECS 28:2012*; Technical Specification for Concrete-Filled Steel Tubular Structures. China Planning Press: Beijing, China, 2012.

buildings

MDPI

Article

Experimental Study on Dynamic Response Characteristics of Rural Residential Buildings Subjected to Blast-Induced Vibrations

Jingmin Pan [1,2,3], Dongli Zhang [1,3,*], Zhenghua Zhou [4,5], Jiacong He [4,5], Long Zhang [4,5], Yi Han [4,5], Cheng Peng [4,5] and Sishun Wang [6]

1 Guangdong Provincial Key Laboratory of Geodynamics and Geohazards, School of Earth Sciences and Engineering, Sun Yat-sen University, Guangzhou 510275, China; panjm26@mail2.sysu.edu.cn
2 Department of Intelligent Architecture, Zhejiang College of Security Technology, Wenzhou 325016, China
3 Southern Marine Science and Engineering Guangdong Laboratory (Zhuhai), Zhuhai 519082, China
4 College of Transportation Engineering, Nanjing Tech University, Nanjing 211816, China; bjsmoc@163.com (Z.Z.); cierhjc@163.com (J.H.); hy0328@njtech.edu.cn (Y.H.)
5 Engineering Mechanics Institute, Nanjing Tech University, Nanjing 211816, China
6 China Railway 24th Bureau Group Co., Ltd., Shanghai 200433, China
* Correspondence: zhangdongli@mail.sysu.edu.cn

Abstract: Numerous rural residential buildings exhibit inadequate seismic performance when subjected to blast-induced vibrations, which poses potential threats to their overall stability and structural integrity when in proximity to blasting project sites. The investigation conducted in conjunction with the Qianshi Mountain blasting operations along the Wenzhou segment of the Hangzhou–Wenzhou High-Speed Railway integrates household field surveys and empirical measurements to perform modal analysis of rural residential buildings through finite element simulation. Adhering to the principle of stratified arrangement and composite measurement point configuration, an effective and reasonable experimental observation framework was established. In this investigation, the seven-story rural residential building in adjacent villages was selected as the research object. Strong-motion seismographs were strategically positioned adjacent to frame columns on critical stories (ground, fourth, seventh, and top floors) within the observational system to acquire test data. Methodical signal processing techniques, including effective signal extraction, baseline correction, and schedule conversion, were employed to derive temporal dynamic characteristics for each story. Combined with the Fourier transform, the frequency–domain distribution patterns of different floors are subsequently obtained. Leveraging the structural dynamic theory, time–domain records were mathematically converted to establish the structure's maximum response spectra under blast-induced loading conditions. Through the analysis of characteristic curves, including floor acceleration response spectra, dynamic amplification coefficients, and spectral ratios, the dynamic response patterns of rural residential buildings subjected to blast-induced vibrations have been elucidated. Following the normalization of peak acceleration and velocity parameters, the mechanisms underlying differential floor-specific dynamic responses were examined, and the layout principles of measurement points were subsequently formulated and summarized. These findings offer valuable insights for enhancing the seismic resilience and structural safety of rural residential buildings exposed to blast-induced vibrations, with implications for both theoretical advancements and practical engineering applications.

Keywords: blast-induced vibrations; rural residential buildings; strong vibration observation; dynamic response; experimental investigation

1. Introduction

As a critical tool in construction projects, engineering blasting technology finds extensive application across tunnel operations, mining activities, and the development of hydro-power infrastructure. Its advantages lie in rapid execution, cost efficiency, and operational adaptability. However, the adverse effects of blasting operations on adjacent structures are equally concerning, particularly through destructive phenomena such as building damage or catastrophic collapse, tunnel deformation, fractures in civil and industrial structures, and slope instability in open excavations [1]. Observations from numerous blasting projects reveal that rural residential buildings are frequently widespread in the adjacent areas. Predominantly constructed in the absence of stringent regulatory frameworks and failing to align with contemporary seismic design codes, these residential structures inherently demonstrate structural deficiencies stemming from an emphasis on architectural aesthetics at the expense of engineering robustness, compounded by construction methodologies adapted to localized contextual constraints. Consequently, the dynamic behavior of such vulnerable dwellings under blast-induced excitations necessitates meticulous scientific investigation to establish empirically grounded assessment protocols for vibration mitigation and structural reinforcement strategies [2–7].

In the existing research, considerable attention has been directed toward investigating the characteristics of blast-induced vibrations. Scholars have systematically analyzed the propagation patterns, spectral signatures, and contributing factors. Empirical evidence demonstrates that blast-induced vibrations may induce a spectrum of adverse effects on adjacent rural residential buildings, manifesting from superficial hairline cracks to catastrophic structural failures [8]. The severity of impacts arises from the synergistic interaction of multiple variables, encompassing blasting operational parameters (such as explosive charge weight, detonation sequencing) [9–11], geophysical transmission media (including subsurface geology and soil typology) [12], and inherent architectural attributes (such as material resilience and construction methodology) [13]. Notably, peak particle velocity (PPV) emerges as the primary metric for assessing blast-induced vibrations severity; frequency is also an important parameter because rural residential buildings respond differently to vibrations of different frequencies [14,15]. The frequency sensitivity stems from the structural particularities of rural residential buildings, which often lack modern seismic fortification measures. Further research indicates that the foundational typology, the distance of blast sources, and the structural conditions significantly affect the seismic performance of rural residential buildings [16,17]. For instance, shallow foundations common in rural residential buildings demonstrate heightened susceptibility to resonant amplification effects at specific frequency ranges, while age-related material degradation exacerbates susceptibility to cumulative damage mechanisms. Concurrently, architectural variables such as irregular mass distribution, non-engineered construction joints, and substandard reinforcement detailing, prevalent in vernacular building practices, contribute to stress concentration under dynamic loading regimes such as blast-induced vibrations.

In impact analysis, the elevation amplification effect is one of the key factors to be considered when assessing the threats. Blast-induced vibrations exhibit a height-directional amplification phenomenon within building structures, whereby taller constructions demonstrate progressively greater vibrational intensity at upper levels [18–20]. Through an actual engineering blasting test, three-directional peak vibration velocities in high-rise buildings show marked amplification trends at elevated positions, with intensity increasing proportionally to height [21]. Similarly, the analysis of the blast-induced vibration monitoring data of a 33-story high-rise building by Faramarzi et al. indicates that both the vertical vibration velocity and the main vibration frequency have elevation amplification effects, which are more obvious in the middle and top of the building [18]. Notably, these dynamic responses demonstrate greater magnitude in mid-story and upper-story locations, suggesting complex interaction mechanisms between incident wave-field and structural resonance frequencies at increased elevations.

The phenomenon arises from multiple contributing factors, including wave reflection patterns at structural discontinuities, height-dependent modal participation during vibration cycles, and potential superposition effects from reflected wave components. Research indicates that buildings exceeding three stories in height begin to exhibit measurable elevation amplification, with effects becoming pronounced above six stories under typical blasting excitation frequencies [15]. This vertical response gradient necessitates site-specific monitoring strategies that incorporate elevation-distributed sensor arrays, rather than relying solely on ground-level instrumentation. Furthermore, the interaction between incident vibration frequencies and structural natural frequencies becomes particularly critical at elevated levels [21], where even moderate ground motion inputs may induce disproportionate dynamic responses due to reduced damping ratios and altered mass participation factors. For the dynamic response of high-rise buildings subjected to blast-induced vibrations, it shows that the vertical vibration velocity is usually greater than that in the horizontal directions [22]. As a complex dynamic system, buildings have different responses to vibrations of different frequencies on different floors, which may cause certain frequencies of vibration to resonate or be amplified on specific floors. Research has found that the amplification effect of structures on energy components of different frequencies is selective, which means that vibration energy of certain frequencies is more easily absorbed and amplified by buildings, thereby leading to a greater structural response [23]. These findings underscore the importance of three-dimensional dynamic analysis in vulnerability assessments for multi-story buildings subjected to blast-induced vibrations.

Reviewing the current research status of blast-induced vibrations and structural dynamic response, there are relatively few studies on specific rural residential buildings, and there is also a lack of systematic experimental data. Therefore, based on the blasting project site of Qianshi Mountain in the Wenzhou section of Hangzhou–Wenzhou High-Speed Railway, and taking the typical rural residential buildings located in the adjacent area of the blasting site as the research object, this research conducts the dynamic response of the rural residential building subjected to blast-induced vibrations. This research provides an important reference basis for evaluating the impact of blast-induced vibrations on rural residential buildings and provides an important safety reference basis for related blasting projects.

2. Generalization of Experimental Site

2.1. Geological Conditions of Experimental Site

In order to meet the requirements of railway engineering construction and associated ancillary projects, comprehensive blasting and excavation works were undertaken on Qianshi Mountain within the Wenzhou section of the Hangzhou–Wenzhou Railway. The blasting zone extends 460 m from east to west, with a maximum north-to-south span of 260 m. The mountain reaches a peak elevation of 90 m, and the total excavation is approximately 3.8 million cubic meters. The aerial comparison before and after the blasting clearly presents the proximity relationship between the blasting source and the village building complex, as shown in Figure 1. The blast-induced vibrations have a particularly significant dynamic impact on the adjacent rural residential buildings. In this test, the blasting construction operation adopts the method of small steps, weak loosening, deep holes combined with shallow blasting, mainly controlled blasting, and mechanical-assisted excavation operation is combined with the construction. The unique spatial relative position characteristics of the test site provide test element guarantees for this research.

Figure 1. Comparison of the current situation of unmanned aerial vehicle (UAV) aerial photography of the blasting area: (**a**) before blasting construction; (**b**) after blasting construction.

The test site is situated within a river terrace-valley system and exhibits pronounced geological stratigraphic characteristics. The surface water network, comprising seasonal streams and irrigation ditches, is densely distributed across adjacent agricultural land. The groundwater level fluctuates at depths of 0–1.0 m below ground level, demonstrating marked seasonal variations. The upper geological section is a heterogeneous sedimentary sequence, including clay, silt, silty clay, silty clay interspersed with crushed stones, silty sand, coarse gravel soil, and the quaternary residual slope accumulation of gravel-bearing silty clay layers, which were formed through complex sedimentary processes (marine, alluvial, and fluvial). The underlying bedrock is the fused tuff of the Zhuji Formation of the Late Jurassic. However, due to the thick overburden, the surface soil with low bearing capacity may amplify the ground vibration caused by blasting. In accordance with the requirements of the engineering safety oversight authority, this trial increased the quantity of blast holes, reduced the charge volume per hole, and employed delayed detonation techniques to distribute energy more rationally. These measures aim to mitigate risks associated with rock projection caused by concentrated energy release, while minimizing the impact of blast-induced vibrations on adjacent structures.

2.2. Relative Positional Relationship

During single blast trials, limitations on the quantity and resources allocated to monitoring instrumentation preclude comprehensive testing of all rural residential structures adjacent to the test site. Consequently, this study selected typical representative rural residential buildings as its primary research. Field inspections and surveys revealed that the majority of local rural residential buildings comprise reinforced concrete frame structures constructed or renovated between 1990 and 2010. A seven-story reinforced concrete frame structure, exhibiting the greatest architectural prominence and characteristic features of the region, was chosen as the principal investigation target. Situated approximately 580 m from the blasting origin, its relative spatial configuration is illustrated in Figure 2.

Figure 2. The relative positional relationship between the blasting origin and the seven-story rural residential building.

2.3. Structural Dimension Measurement

Following site inspections and occupant surveys, the selected rural residential observation structure was constructed circa 2005. It is arranged in an east–west orientation and adopts a seven-story and seven-bay reinforced concrete frame structure. With the exception of terraces on the eastern and western extremities of the uppermost floor, the structural layout remains consistent across all levels. The overall elevation of the building reaches 24 m (including roof structures), establishing it as the tallest rural residence within the near field. To ascertain geometric dimensions and structural particulars, methodologies encompassing occupant interviews and direct measurements were employed, utilizing standard instrumentation such as infrared laser rangefinders, floor thickness gauges, and steel tape measures. The on-site household investigation is shown in Figure 3.

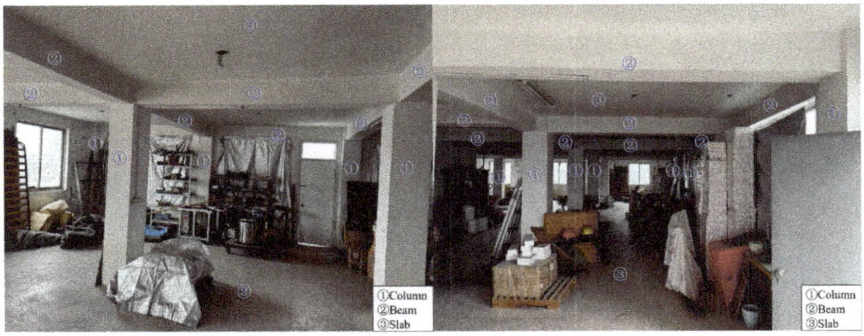

Figure 3. The on-site investigation of the rural residential building.

In terms of specific geometric dimensions, the floor height is 2.8 m, the floor slab thickness is 0.12 m, the thickness of both the interior and exterior walls is 0.3 m, the overall width of the room is 24.5 m, the depth is 14.9 m, and the width of each room is 3.5 m. Additionally, an elevator shaft and a staircase are set on the east side, with a width of 3.5 m and a depth of 2.9 m. The size of the frame columns is 400 mm × 500 mm. The frame beam size is 300 mm × 480 mm. The ground floor partially adopts filled partition walls, and there are no partition walls on the second to seventh floors. The facade layout of the rural residence building is shown in Figure 4, and its floor plan layout is shown in Figure 5.

Figure 4. The facade layout of the rural residential building.

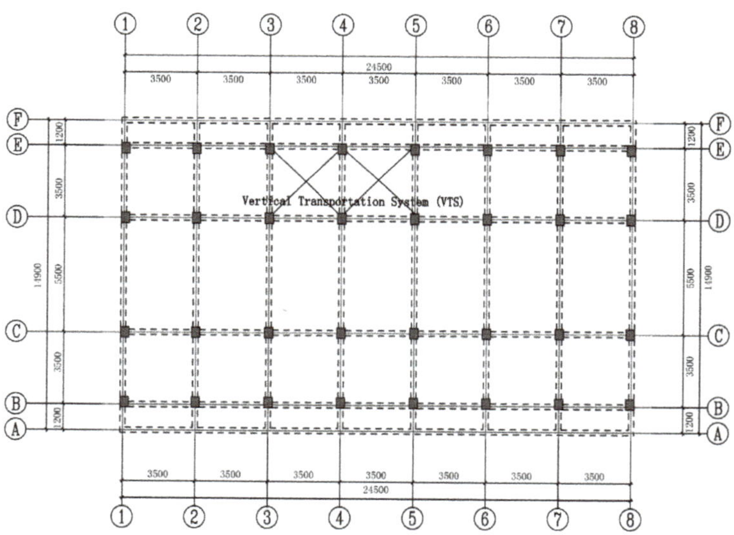

Figure 5. The floor plan layout of the rural residential building.

3. Experimental Design

For the constraints on instrumentation quantity and resource allocation, it is imperative to identify strategically critical locations for instrument deployment to ensure the relevance and reliability of measured data. Therefore, during the experimental design, by establishing a finite element model and combining it with the method of calculating modal analysis, we can fully grasp the structural modes of rural residential buildings, thereby determining the key parts of rural residential buildings, designing the observation system, and reasonably placing the instrument positions.

3.1. FEM Modeling

Considering the substantial computational demands of nonlinear analysis for reinforced concrete frame structures, optimization of modeling complexity and calculation duration is achievable while preserving a reasonably accurate representation of the inherent dynamic characteristics of the rural residential building. During the modeling process, non-structural elements such as partition walls, doors, and windows were intentionally simplified, as these components do not contribute to load-bearing capacity. Consequently, the finite element model exclusively incorporates primary structural elements, including beams, columns, and slabs, which constitute the load-bearing framework of the reinforced concrete structure.

In terms of material arrangement, the primary reinforcement configuration for the reinforced concrete frame structure employs HRB400 grade for longitudinal bars and HPB300 grade for stirrups. Cross-sectional reinforcement detailing has been streamlined for computational efficiency. Regarding steel reinforcement properties, fundamental parameters including Density, Young's modulus, and Poisson's ratio were established at 7850 kg/m^3, 206 GPa, and 0.3, respectively. Considering the mechanical behaviors of reinforcing bars and concrete, both fall within the scope of nonlinear analysis. The characteristics of reinforcement after yield can be described by the elastic stage (with the slope being the Young's modulus E) and the plastic stage (with a slope of 0.01E after yield) when the double-line model is adopted for reinforcing bars. A bilinear kinematic hardening model was adopted to characterize reinforcing bars' behavior, effectively capturing post-yield strengthening effects while maintaining reasonable simulation accuracy. Its constitutive relationship is shown in Figure 6a. The concrete composition was specified as C30 grade, with Density, Young's modulus, and Poisson's ratio parameters set at 2500 kg/m^3, 300 GPa, and 0.2, respectively. To model the cumulative damage evolution in concrete elements, an elastoplastic damage-coupled constitutive framework capable of representing irreversible plastic deformation was selected (The Concrete Damaged Plasticity model) as outlined in referenced studies [24,25]. Detailed material parameters are tabulated in Table 1. At present, many building industry research institutes have established standardized concrete constitutive relations through extensive experimental statistics and theoretical analysis. By adjusting the segmented equations (such as the ascending and descending segments) and damage parameters, it can cover concrete of different strength grades. In the elastic stage, the stress–strain relationship is defined by the elastic modulus and Poisson's ratio of the concrete. In the inelastic stage, the stress–strain numerical correspondence directly given by the design requirements is adopted. The specific relationship is shown in Figure 6b. The contact mode between the steel bars and the concrete is set as embedded, which means the steel bars are embedded into the concrete.

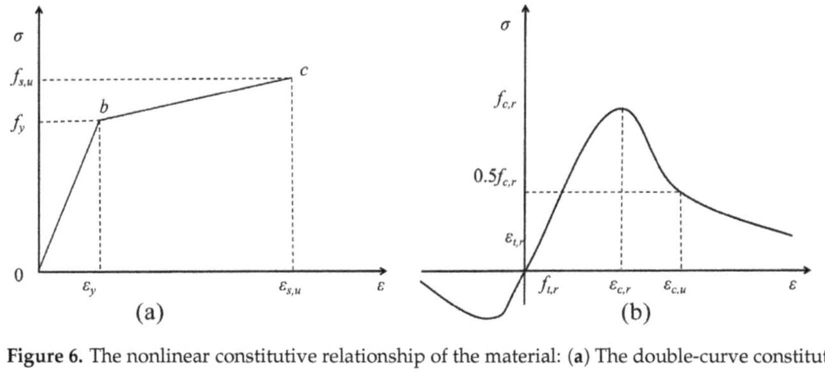

Figure 6. The nonlinear constitutive relationship of the material: (**a**) The double-curve constitutive model curve of the reinforcing bar; (**b**) the uniaxial stress–strain relationship curve of the concrete.

Table 1. The required CDP model parameters.

Dilation Angle	Eccentricity	Stress Ratio	Shape Coefficient	Viscosity Parameter
30°	0.1	1.16	0.66667	0.0005

In terms of grid division, there exists a critical balance between computational efficiency and accuracy: excessively coarse meshes compromise the fidelity of numerical results, while overly refined grids impose disproportionate computational burdens. For the rural residential building, mesh resolution must be judiciously calibrated to ensure adequate geometric representation while adhering to the element size guidelines recommended by Kuhlemeyer et al. [26] and Lysmer et al. [27]. This approach optimizes both analytical precision and computational feasibility.

In terms of boundary conditions and constraint settings, through the part of this field regarding the vibration of buildings caused by boundary events [28], the rural residential buildings have reached a state of geotechnical equilibrium following over three decades of consolidation. Consequently, the foundation boundary may be idealized as a fully fixed constraint to reflect the absence of further differential settlement. Regarding structural connectivity, the frame beams and columns, constructed as monolithic cast-in-place elements, were computationally unified through Boolean merging operations during finite element assembly, thereby eliminating interfacial discontinuities. While the floor slabs also comprise cast-in-place concrete, their sequential installation post-dating the primary frame necessitates distinct boundary treatment. Specifically, the interface between slabs and the pre-existing frame components was defined using tie constraint formulations to accommodate coordinated deformation behavior while preserving the methodological distinction between primary and secondary structural elements.

To facilitate a comprehensive comparison of natural vibration characteristics across different stories within the rural residential building and inform the instrumentation deployment strategy, distinct finite element models were formulated to analyze the global structural dynamic behavior. Two primary modeling paradigms were adopted. The first simplifies the structural system by focusing on primary load-bearing components, represented by a beam–column framework model (denoted as Model A). This configuration incorporates only the essential gravitational and lateral force-resisting elements. The second paradigm extends this framework by integrating floor diaphragms, thereby constituting a complete three-dimensional frame model (denoted as Model B). Both finite element representations were independently developed to evaluate the influence of slab-frame interaction on modal response parameters. The specific structural components included in each computational model are detailed in Table 2, and the finite element modeling is shown in Figure 7.

Table 2. The structural composition of the calculation model.

Model Number	Model Name	Structural Composition
Model A	Beam–column frame model	Column + Beam
Model B	Complete three-dimensional frame model	Column + Beam + Slab

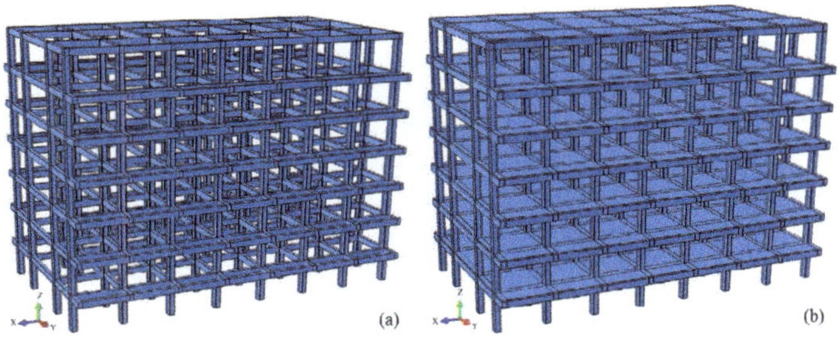

Figure 7. The rural residential building modeling diagram: (**a**) Beam–column frame model (Model A); (**b**) complete three-dimensional frame model (Model B).

3.2. Modal Analysis of Rural Residential Buildings

Modal parameters of the rural residential building encapsulate their intrinsic dynamic attributes and serve as the foundational prerequisite for comprehending the global dynamic response characteristics of structures. The theoretical underpinnings of modal analysis are firmly rooted in structural dynamics and vibration theory, with the fundamental objective revolving around the extraction of modal parameters through the resolution of the structural eigenvalue problem [29,30]. Modern finite element analyses typically employ advanced eigenvalue solution algorithms to extract modal parameters from the system. Notable methodologies include the subspace iteration method, the Lanczos algorithm, and the Automatic Multi-Level Substructuring (AMS). Among these, the Lanczos algorithm is the most commonly adopted method.

3.2.1. Modal Analysis of Beam–Column Frame Model (Model A)

The rural residential building exhibits planar symmetry, with uniform facade configurations observed across all stories except for the seventh-floor terrace. This architectural regularity not only streamlines construction procedures but also reduces the complexity of finite element discretization, thereby enhancing computational efficiency. The natural frequency spectrum for the initial 20 modes of the beam–column framework model (Model A) is presented in Table 3, while the corresponding deformation patterns for the first six vibration modes are illustrated in Figure 8. The modal characteristics provide critical insights into the dynamic coupling behavior of the structural system, enabling targeted instrumentation placement for experimental modal analysis.

Table 3. The natural frequency spectrum for the initial 20 modes of beam–column framework model (Model A).

Mode	Frequency (Hz)	Mode	Frequency (Hz)	Mode	Frequency (Hz)
1	1.7028	8	2.2771	15	3.0022
2	1.7601	9	2.3442	16	3.0890
3	1.8004	10	2.5649	17	3.1582
4	1.8612	11	2.7861	18	3.2262
5	1.9579	12	2.9200	19	3.2606
6	1.9794	13	2.9432	20	3.3808
7	2.1325	14	2.9498		

Through analysis, the frequencies of the first six vibration modes are 1.7028 Hz, 1.7601 Hz, 1.8004 Hz, 1.8612 Hz, 1.9579 Hz, and 1.9794 Hz, respectively. The characteristics of Mode 1 are overall bending in the binning direction and longitudinal beam bending, and the reaction of the top frame beam is the most obvious. The characteristics of Mode 2 are overall torsional deformation, longitudinal beam bending, and frame column bending. Among them, the corner columns of the frame columns and the top frame beams have the most obvious reactions. The characteristics of Mode 3 are the overall bending in the depth direction, the bending of the crossbeam, and the bending of the frame column. Among them, the reactions of the frame column and frame beam at the middle part between the sixth and seventh floors are the most obvious. The characteristics of Mode 4 are overall torsional deformation, bending of frame beams, and bending of frame columns. Among them, the reactions of frame columns and frame beams at both ends of the sixth and seventh floors are the most obvious. The characteristics of Mode 5 are overall torsional deformation, frame column bending, and beam bending. Among them, the reaction of the frame beam in the middle part of the depth direction of each floor is the most obvious. The characteristics of Mode 6 are the overall bending and torsion in the depth direction, as well as the bending of frame columns and beams. Among them, the reactions of frame columns and beams at the middle and end parts of the sixth and seventh floors of rural residential buildings are the most obvious.

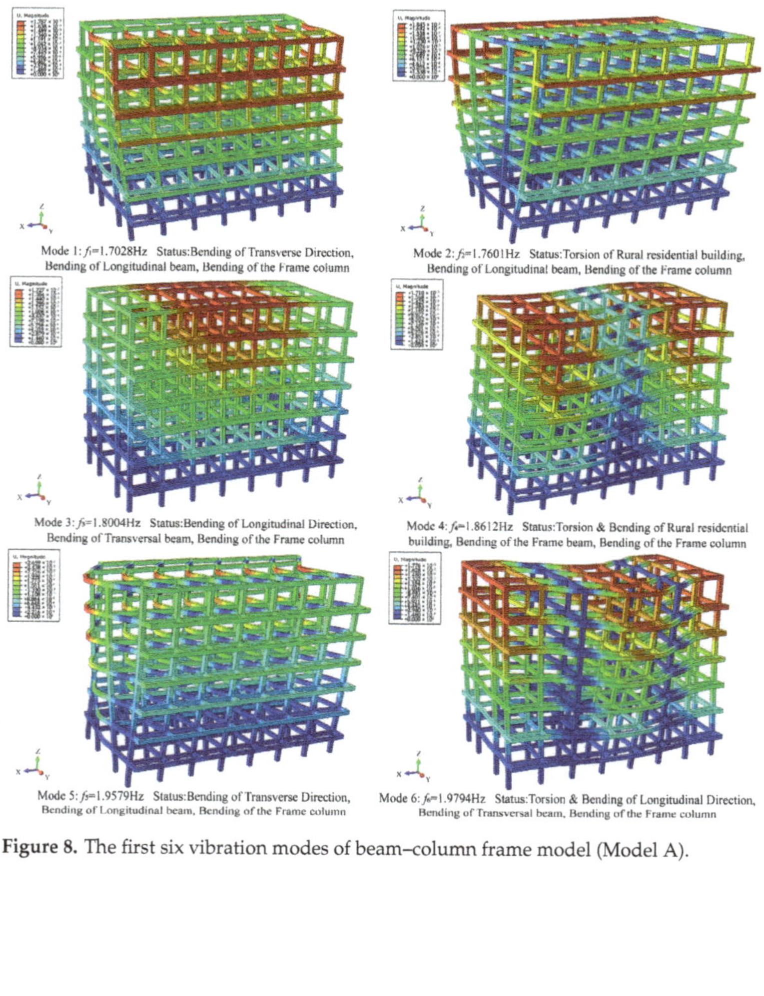

Figure 8. The first six vibration modes of beam–column frame model (Model A).

3.2.2. Modal Analysis of Complete Three-Dimensional Frame Model (Model B)

The natural frequency spectrum for the initial 20 modes of the complete three-dimensional frame model (Model B) is shown in Table 4, and its vibration patterns in the first six orders are shown in Figure 9.

Table 4. The natural frequency spectrum for the initial 20 modes of complete three-dimensional frame model (Model B).

Mode	Frequency (Hz)	Mode	Frequency (Hz)	Mode	Frequency (Hz)
1	1.8634	8	6.5160	15	6.5669
2	2.1281	9	6.5168	16	6.5678
3	2.2054	10	6.5177	17	6.5687
4	5.5448	11	6.5185	18	6.5691
5	6.4524	12	6.5190	19	6.5698
6	6.4564	13	6.5463	20	6.6544
7	6.5156	14	6.5621		

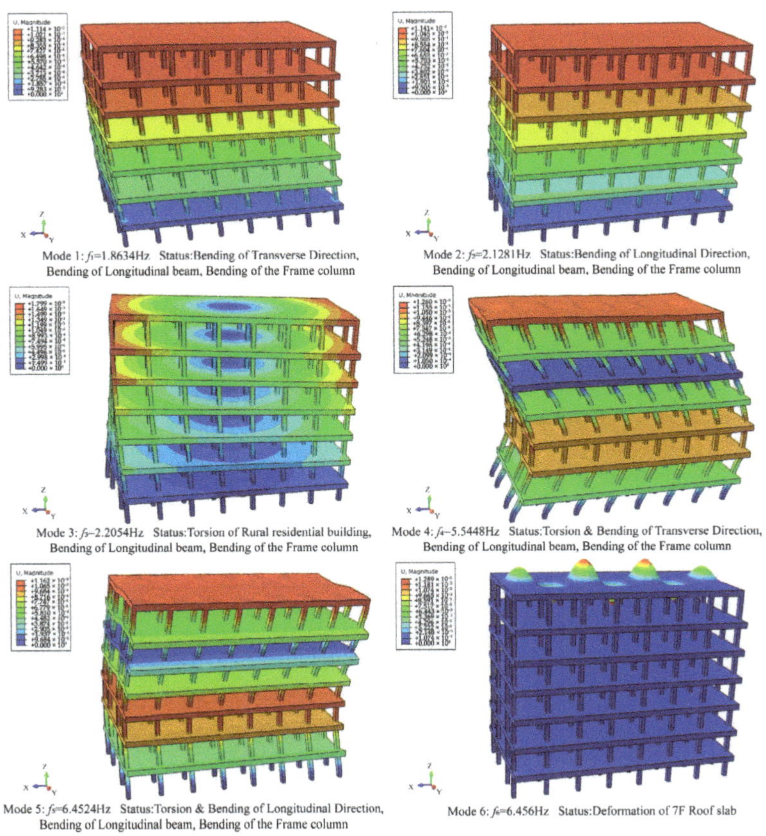

Figure 9. The first six vibration modes of complete three-dimensional frame model (Model B).

Through analysis, the frequencies of the first six vibration modes are 1.8634 Hz, 2.1281 Hz, 2.2054 Hz, 5.5448 Hz, 6.4524 Hz, and 6.4564 Hz, respectively. Among them, the characteristics of the first three vibration modes are consistent with the part of the above-mentioned beam–column structure Model (Model A), due to the addition of floor slab components, the vibration mode characteristics of each order have undergone certain changes. Specifically, the characteristics of the first-order vibration mode are the overall bending in the bay direction and the bending of the frame columns. The floor slabs of each floor respond uniformly, and the response becomes more obvious as the height of the floor increases. The characteristics of the second-order vibration mode are the overall bending in the depth direction and the bending of the frame columns. Its characteristics are similar to those of the first-order vibration mode, also manifested as the uniform response of the floor slabs on each floor. As the height of the floor increases, the response becomes more obvious. The third-order vibration mode feature is the overall torsional deformation of rural residential buildings. Similarly, the floor slabs of each floor twist along the center position of the floor slabs, and the corner edges of the floor slabs and the corner columns of the frame columns react most obviously. The characteristics of the fourth-order vibration mode are the overall bending in the barium direction of rural residential buildings and the bending of frame columns. With the increase of floor height, the reaction characteristics also show a trend of strengthening, weakening, and strengthening. The floor of the sixth floor hardly reacts, while the floor of the third and fourth floors reacts more obviously with the roof of the seventh floor. The characteristics of the fifth-order vibration mode are the overall bending and torsion of rural residential buildings along the depth direction, the bending of frame columns, and the deformation of floor slabs. With the increase of floor height, the reaction characteristics also show a trend of strengthening, weakening, and strengthening. The floor slab of the sixth floor hardly reacts, while the floor slab of the fourth floor reacts more obviously with the roof slab of the seventh floor. The characteristic of the sixth vibration mode is the deformation of the roof slab of the seventh floor, and the reaction is most obvious at the center of the floor slab.

3.3. Natural Vibration Characteristics of Rural Residential Buildings

A comparative analysis of the beam–column framework model (Model A) and the complete three-dimensional frame model (Model B) of rural residential buildings reveals the following structured insights:

(1) By comparing the frequency differences of the first 20 vibration modes, it is found that the overall vibration of the observed objects in the test mainly occurs in the low-frequency band. Among them, the vibration mode frequencies of the rural residential building are all within 7 Hz, and the superposition effect of high-order vibration modes is prominent. Such vibration mode frequency values are caused by the stiffness, overall height, and structural characteristics of the rural residential building, which are also related to the overall planar layout of the structure.

(2) By comparing the differences in modes and vibration forms of each order, it is found that the overall vibration mode characteristics of the first few orders of the structure are all manifested as binning and depth direction bending, frame column bending, frame beam bending, overall torsional deformation, floor slab deformation, especially the reactions of the top layer of the structure are relatively obvious. The main reason for the differences is also related to its structural and construction characteristics. Therefore, it is necessary to focus on the layout of instruments in the frame beams, frame columns, and the top layer of the structure.

(3) Compared with Model A, Model B takes the floor slab into account on this basis. The floor slab forms a whole, and the vibration mode frequency of the rural residence building increases as a whole. Many floor slab responses also occur in the first 20 modes. Moreover, in the characteristics of the fourth and fifth modal vibration

modes, it is observed that as the height of the floor increases, the reaction characteristics also show a trend of strengthening, weakening, and strengthening. The floor of the sixth floor hardly reacts, while the floor of the third and fourth floors reacts more obviously with the roof of the seventh floor. Therefore, it is not only necessary to install the instruments on the top floor, but also to pay special attention to the placement of instruments on the middle floors such as the third and fourth floors, as well as between adjacent floors.

3.4. Design of Experimental Observation System

In this test, the ETNA-2 digital strong-motion accelerograph produced by Kinemetrics Co., Ltd. (Pasadena, CA, USA) was employed to acquire blast-induced vibration data. This instrument is renowned for its exceptional performance and incorporates triaxial EpiSensor accelerometers, enabling simultaneous recording of ground acceleration components in the east–west (EW), north–south (NS), and vertical (UD) directions. Key attributes include its compact form factor, lightweight construction, broad dynamic range (±2 g~±10 g), robust stability, and user-friendly operation, facilitating straightforward deployment in field conditions. The technical specifications of the ETNA-2, as outlined in Table 5. The instrument's seismic-grade design and compatibility with standard installation protocols render it suitable for rugged field applications, while its triaxial configuration permits comprehensive characterization of blast-induced vibrations.

Table 5. The technical specifications of the ETNA-2 digital strong-motion accelerograph.

 ETNA-2 digital strong seismograph	***Sensor***	
	Type:	Triaxial EpiSensor force balance accelerometers, orthogonally oriented
	Full scale range:	User selectable at ±1 g, ±2 g or ±4 g
	Bandwidth:	DC to 200 Hz
	Dynamic range:	155 dB+
	Digitizer	
	Channels:	3 24-bit sensor channels for the internal sensors bandwidth-optimized 32-bit data path
	Sample rates and Acquisition modes:	100sps/continuous (ring buffer)

To ensure the relevance and efficacy of the observation, the principle of 'layered arrangement and composite measurement point layout' along the vertical axis is adopted. Priority is assigned to monitoring the bottom and top layers to capture the primary global vibration modes and localized dynamic responses. Simultaneously, practical considerations such as instrument accessibility, structural compatibility, and measurement repeatability inform the selection of deployment locations, ensuring the identification of reasonable, effective, and operationally viable key positions for sensor placement. To facilitate the later data comparison and analysis, the various deployment positions of the instruments are named by floor numbers. The corresponding floors from bottom to top are respectively named the ground, fourth, seventh, and roof levels, as shown in Figure 10. Concurrently, the instrumentation grid must be systematically aligned with the orthogonal coordinate framework relative to the building's geometric axes, wherein the X-axis corresponds to the transverse width dimension, the Y-axis to the longitudinal depth dimension, and the Z-axis to the vertical elevation axis.

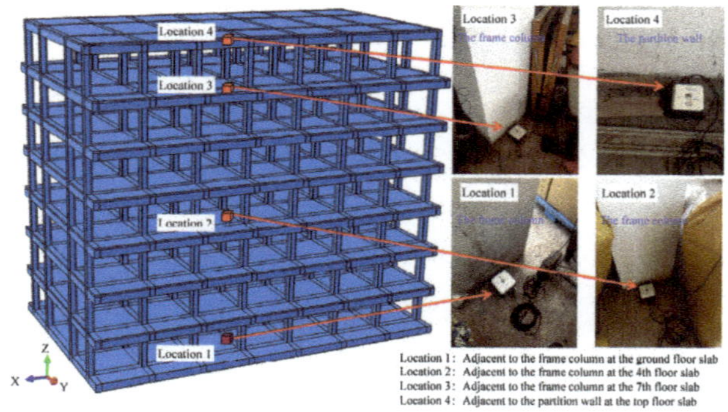

Figure 10. The installation and layout positions of the instruments of the rural residential building.

4. Data Processing and Analysis

In accordance with the designed experimental observation system, the original data signals observed still need to adopt a series of methods and measures for data processing. Combined with the technical regulations for strong vibration observation, this section systematically elaborates the data processing methods:

(a) Time–Domain Analysis: Evaluation of temporal waveforms to quantify peak particle velocities, acceleration amplitudes, and duration parameters.
(b) Frequency–Domain Analysis: Spectral decomposition via fast Fourier transform (FFT) to identify modal frequencies and energy distribution across the vibration spectrum.
(c) Structural Dynamic Characterization: Assessment of dynamic amplification factors (DAFs) and damping ratios to quantify the vibration transmission characteristics of the rural residential building.

These analyses facilitate the correlation of field measurements with theoretical predictions of structural response, thereby enabling critical evaluations of building performance under transient loading conditions.

4.1. Data Processing

4.1.1. Data Pre-Processing

Based on systematic characterization of the test site, the typical duration of the explosion ranges from 0.5 to 3 s. To optimize computational efficiency and mitigate superfluous background noise processing, a 10 s temporal window is systematically captured for each event, encompassing 2 s of pre-trigger baseline data and 8 s of post-trigger transient response. This protocol ensures comprehensive acquisition of the blast-induced vibration envelope while minimizing extraneous signal components. After the baseline correction is completed, it provides a certain reference basis for the later comparative analysis of the acceleration time history of each measurement point.

4.1.2. Schedule Conversion

Time–history transformation constitutes a critical process in blast-induced vibration signal analysis, enabling the conversion of acceleration time–domain records into equivalent velocity and displacement histories. The underlying mathematical principle involves the numerical integration of discrete time-series data through quadrature algorithms [31]. However, the theoretical continuum integration must be approximated via discrete summation techniques, for blast-induced vibration signals are inherently non-stationary and discretely sampled. The discretization process transforms the analytical integral into a tractable algebraic operation, where the Riemann sum approximation of the area under the acceleration–time curve yields the corresponding velocity history, and subsequent iterative integration generates the displacement profile. Despite the numerical implementation, the procedure remains mathematically close to the definite integral operation, thereby preserving the physical fidelity of the dynamic response reconstruction. The numerical integration of acceleration time histories to derive velocity and displacement profiles is typically accomplished through quadrature algorithms such as the Trapezoidal Rule or Simpson's Rule [32–34]. The Trapezoidal Rule, as a fundamental method, approximates the integral as a piecewise linear function, with its discrete formulation expressed as:

$$v_k = v_{k-1} + \frac{a_{k-1} + a_k}{2}\Delta t, \tag{1}$$

$$d_k = d_{k-1} + \frac{v_{k-1} + v_k}{2}\Delta t, \tag{2}$$

The approach effectively computes the area under the acceleration–time curve by summing trapezoidal segments between consecutive data points. While computationally efficient, the Trapezoidal Rule introduces cumulative phase errors proportional to the square of the sampling frequency, necessitating careful consideration of time-step discretization and high-frequency noise filtering to mitigate aliasing effects. By contrast, Simpson's Rule employs quadratic interpolation between data points, while offering improved accuracy for smooth acceleration histories at the expense of increased computational complexity. Therefore, when calculating the velocity time history, using the Trapezoidal Rule can take into account both efficiency and high-frequency noise to meet the requirements of the calculation results.

4.2. Analysis of Experimental Characteristics of Rural Residential Buildings

4.2.1. Time–Domain Analysis

After systematic pre-processing of raw observational data, including coherent signal extraction and baseline drift compensation, a corrected acceleration time–history curve is obtained. Figure 11 illustrates the X, Y, and Z directions' acceleration time histories recorded at the ground, fourth, seventh, and top floors of the rural residential building. Key observations include:

(1) The acceleration amplitude variation on the top floor is notably greater than on the ground floor, while the time–history curve exhibits tail oscillations. This phenomenon likely arises from the top floor's reduced mass and stiffness, such as lightweight roofing materials or ancillary attic structures. As blast-induced seismic waves propagate upward, inertial effects intensify seismic amplification, yielding heightened acceleration responses accompanied by persistent oscillations.

(2) The variation of the acceleration amplitude in the vertical direction (Z-direction) is significantly greater than that in the horizontal direction (X and Y directions). The reason for this phenomenon should be the stiffness of the multi-story reinforced concrete frame structure in the vertical direction is greater than that in the horizontal directions. Therefore, the natural frequency of the rural residential building in the vertical direction is relatively closer to the main shock frequency of blast-induced seismic waves, which is prone to cause the structure to generate an acceleration response higher than that in the horizontal directions.

(3) From the ground floor to the top floor of the rural residential building, the overall change of peak acceleration in three directions is not consistent. It does not gradually increase according to the height of the building. There is a process of decreasing peak acceleration in the middle part of the floors.

Presently, the prevailing norms and standards across numerous nations predominantly adhere to the joint criterion system for blast-induced vibration safety, which integrates peak particle velocity (PPV) and dominant vibration frequency parameters [35–38]. Within this safety framework, it is stipulated that when the blast-induced dominant frequency range falls within 10 Hz $< f \leq 50$ Hz, the corresponding permissible particle vibration velocity threshold ranges between 2.0 and 2.5 cm/s. To facilitate analysis of velocity variation patterns across different floors in the vertical space, this study examines the time–history velocity profiles in the X, Y, and Z directions for the rural residential building in Figure 12. It can be found:

(1) The variation in velocity amplitude on the top floor is markedly more pronounced than on other floors. A comparative analysis of peak velocities between the ground floor and the top floor reveals that the peak velocity increases from 1.57×10^{-3} cm/s to 1.99×10^{-3} cm/s in the X-direction. In the Y-direction, it rises from 1.51×10^{-3} cm/s to 3.07×10^{-3} cm/s, while in the Z-direction, it escalates from 5.63×10^{-3} cm/s to 9.24×10^{-3} cm/s. Consequently, the peak velocity increments in the X, Y, and Z directions on the top floor exhibit a consistent upward trend, attributable to the inherent structural rigidity of rural residential buildings.

(2) Analogous to the acceleration time–history patterns, the blast-induced vertical velocity amplitude (Z-direction) demonstrates a considerably greater magnitude than horizontal components (X and Y directions). Notably, the maximum peak velocity is recorded in the Z-direction at the uppermost story, reaching 9.24×10^{-3} cm/s. While this velocity remains substantially below the threshold prescribed by safety regulations, the dynamic amplification effect on the upper levels of the rural residential building is particularly pronounced, indicating significant vibration amplification at the uppermost floor despite the low absolute velocities.

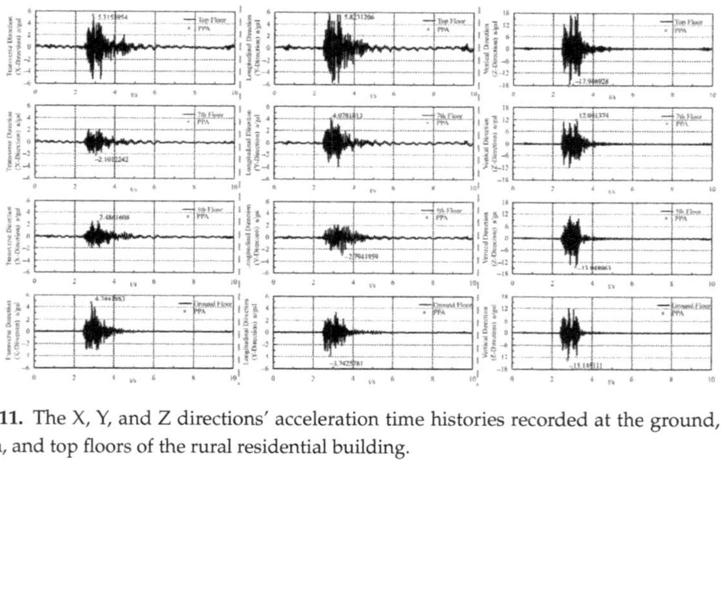

Figure 11. The X, Y, and Z directions' acceleration time histories recorded at the ground, fourth, seventh, and top floors of the rural residential building.

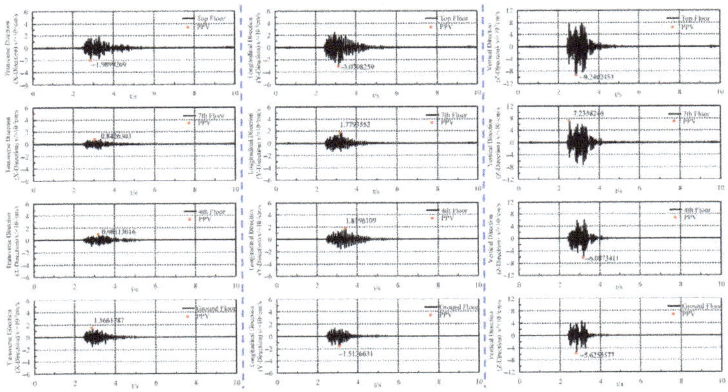

Figure 12. The X, Y, and Z directions' velocity time histories conversed at the ground, fourth, seventh, and top floors of the rural residential building.

4.2.2. Frequency–Domain Analysis

Although the amplitude and duration of the time–domain signal can be intuitively understood, the frequency–domain characteristics cannot be observed for the blast-induced seismic signal, which belongs to a mixture of interference waves containing various frequency components. The Fourier transform constitutes a cornerstone of digital signal processing, with the discrete Fourier transform (DFT) serving as the optimal tool for frequency–domain analysis [39]. Application of DFT to discretely sampled blast-induced seismic signal enables derivation of the Fourier amplitude spectrum for acceleration time-histories at each observation point. The Nyquist frequency resolvable from the 100 Hz sampling rate [40–42] corresponds to a maximum decomposable frequency component of 50 Hz. The fast Fourier transform (FFT) algorithm integrated within MATLAB is typically employed in engineering applications, which offers substantially enhanced computational efficiency relative to the discrete Fourier transform (DFT). To facilitate comparative analysis of frequency distribution patterns across vertical floors of the rural residential building, the Fourier amplitude spectra for each directional component were independently analyzed in Figure 13.

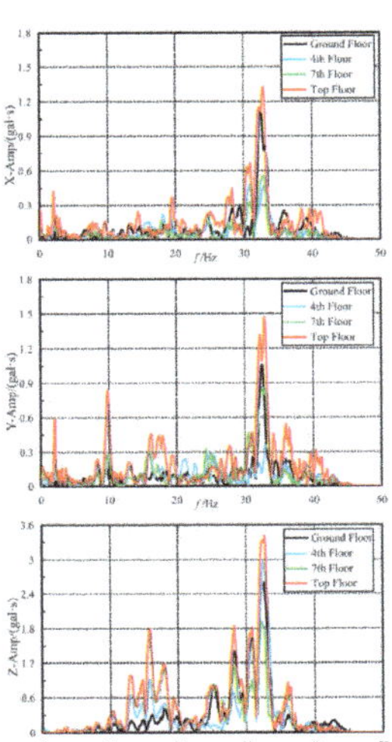

Figure 13. The X, Y, and Z directions' Fourier amplitude spectra of the rural residential building.

(1) A comparative analysis of the Fourier amplitude spectra in the X, Y, and Z directions for the ground floor reveals that the dominant frequency of the rural residential building subjected to blast-induced vibrations approximates 33 Hz, and there are also many dominant frequencies.
(2) By comparing the Fourier amplitude spectrum curves for the X, Y, and Z directions on the seventh and top floors of the rural residential building, it was observed that multiple distinct dominant frequencies emerged between 10 and 30 Hz in the Y and Z directions. The 33 Hz dominant frequency also exhibited more pronounced amplitudes across all directions. This indicates that both the seventh floor and the top floor demonstrate comparable amplification effects on specific frequency components under blast-induced vibrations.
(3) Through comparative analysis of the Fourier amplitude spectra in the X, Y, and Z directions for the rural residential building, it is evident that the frequency–domain components of the signals on both the seventh floor and the top floor exhibit greater complexity. The dynamic amplification effect in rural residential structures manifests not only through high-frequency oscillations during velocity time histories but also via broadened frequency bandwidths with the floor height increases. Specifically, the number of dominant frequencies within the low-frequency range progressively increases at higher stories, accompanied by the emergence of additional dominant frequencies with enhanced spectral responses. This phenomenon collectively demonstrates a multi-frequency band selective amplification effect across vertical stories.
(4) Combining Figures 12 and 13, compared with natural seismic vibrations, blasting-induced vibrations have the characteristics of higher frequency, faster attenuation of vibration intensity, shorter duration, and smaller source energy. The spectral characteristics of the measured blasting-induced signals are affected by the explosives (blast source), the site (propagation path), and the inherent characteristics of the structures.

4.3. Analysis of Dynamic Amplification Characteristics of Rural Residential Buildings

4.3.1. Characteristics of Acceleration Response Spectrum

The response spectrum analysis theory establishes a dynamic relationship between the structural dynamic characteristics and the properties of the ground motion [43,44]. Based on the motion equation of a single-degree-of-freedom system under ground motion and combined with the Duhamel integral [45] for solution, the response spectrum of the structural system can be obtained. If the relative displacement response spectrum, relative velocity response spectrum, and absolute acceleration response spectrum of the system are represented by S_d, S_v and S_a respectively, when the damping ratio ξ of the system is constant, the absolute acceleration response spectrum can be expressed as:

$$S_a = \left|\ddot{U}(t) + \ddot{U}_g(t)\right|_{\max} = \omega_0 \left| \int_0^t \ddot{U}_g(\tau) e^{-\xi\omega_0(t-\tau)} \sin\omega_0(t-\tau) d\tau \right|_{\max}, \quad (3)$$

The acceleration response spectra for the rural residential building systems, computed with a 5% damping ratio, were derived for various floor levels using a logarithmic coordinate system to emphasize high-frequency characteristics. The horizontal axis (period, T) spans 0.02–1.28 s on a base-2 logarithmic scale, while the vertical axis (spectral acceleration, Sa) ranges from 0.0001–0.1 on a base-10 logarithmic scale. Floor-specific spectra are presented in Figure 14 with the following key observations:

(1) The overall contour profiles of acceleration response spectrum curves for the rural residential building exhibit notable similarities across floor levels. In the X and Y

directions, spectral accelerations demonstrate an initial rise followed by gradual attenuation with increasing characteristic periods, accompanied by multiple secondary peaks indicative of complex modal interactions. In the Z-direction, a comparable trend is observed, marked by a pronounced single peak followed by a gradual decline, reflecting limited higher-mode participation compared to horizontal responses. These directional disparities arise from differences in structural dynamics: horizontal responses are influenced by lateral torsional and flexural modes, while vertical behavior is governed by axial and short-period flexural modes.

(2) Comparison of acceleration response spectra for the ground, fourth, seventh, and top floors in X, Y, and Z directions reveals that the abscissa of the primary peak consistently occurs at 0.03 s across all elevations. This corresponds to the previously established fundamental frequency of approximately 33 Hz for the site under blast-induced vibrations, with numerous spectral components aligning with dominant frequencies observed in corresponding Fourier amplitude spectra. The consistency of peak locations across floor levels underscores the site-specific vibration response characteristics governed by blast-related frequency content.

(3) Comparison of acceleration response spectra in X and Y directions across the ground, fourth, seventh, and top floors reveals consistent amplitude rankings and shape similarities. Spectral magnitudes follow a uniform progression from largest to smallest: the top floor, the seventh floor, the fourth floor, and the ground floor. No residual peaks are observed in the Z-direction spectra. Amplitudes increase progressively with floor elevation, attributable to the structural symmetry of the rural residential building and central axis instrumentation deployment, which minimizes torsional effects and emphasizes vertical accumulation of dynamic responses.

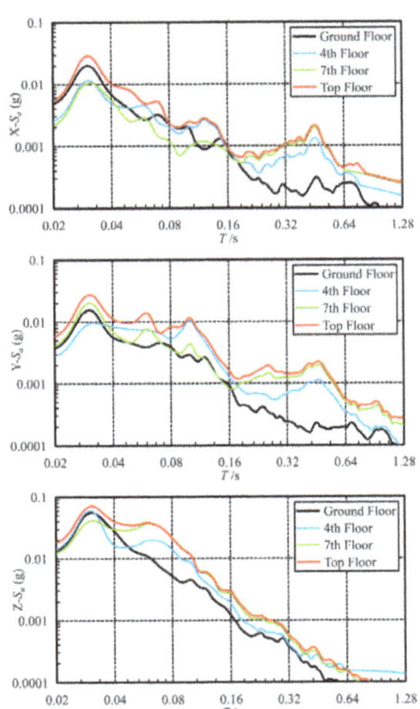

Figure 14. The X, Y, and Z directions' acceleration response spectrum of the rural residential building.

4.3.2. Characteristics of Dynamic Amplification Coefficient

The standard response spectrum, defined as the relationship between the ratio of a structural system's maximum dynamic response (acceleration, velocity, or displacement) to the peak ground motion and the system's inherent properties (natural vibration period, damping ratio), encapsulates the frequency-selective amplification of blast-induced seismic wave components by the structure. This spectrum quantifies the dynamic response characteristics of the system under blast loading conditions and is synonymously termed the dynamic amplification coefficient, categorized into acceleration, velocity, and displacement variants [46]. The acceleration dynamic amplification coefficient derived from this experimental analysis is expressed mathematically in Equation (4).

$$\beta_a = \frac{S_a}{|a|_{\max}}, \tag{4}$$

The acceleration dynamic amplification coefficients in the X, Y, and Z directions of each floor of the rural residential building were calculated, respectively, and the comparison curves are shown in Figure 15. The characteristics are summarized as follows:

(1) The contour profiles of amplification coefficient curves exhibit directional consistency. The acceleration dynamic amplification coefficients demonstrate an initial rise followed by gradual attenuation as the period increases, which is a trend mirroring the acceleration response spectrum. Notably, directional anisotropy is evident: the X-direction exhibits a unimodal profile, the Y-direction a trimodal configuration, and the Z-direction a bimodal pattern. These variations reflect differential modal participation across orthogonal axes, likely stemming from structural asymmetries, non-uniform stiffness distribution, or anisotropic foundation interactions. Despite these directional discrepancies, the overall variation pattern aligns with the acceleration response spectrum, underscoring the frequency-dependent amplification mechanism governed by the structure's dynamic properties and site-specific excitation characteristics.

(2) The comparison of acceleration dynamic amplification coefficient curves in the X-direction across the fourth, seventh, and top floors reveals consistent spectral peaks at a characteristic period of 0.03 s, corresponding to the site's dominant blast-induced frequency of 33 Hz. The pronounced amplification effect on the top floor is attributed to vertical variability in dynamic response, where upper floors exhibit reduced damping and increased participation in higher-order vibration modes. This trend mirrors the acceleration response spectrum findings, where spectral magnitudes diminish with decreasing floor elevation due to cumulative energy dissipation and foundation proximity effects.

(3) The dynamic amplification effect is most pronounced on the top floor across all directions. This phenomenon reveals selective amplification characteristics at elevated levels: The directional anisotropy is evident, with distinct response patterns in orthogonal axes, alongside spectral selectivity tied to specific characteristic periods. The top floor's dominance arises from cumulative modal participation and reduced damping, as intermediate floors may exhibit comparable or lesser amplification depending on resonance with site-specific excitation frequencies and structural natural vibration modes.

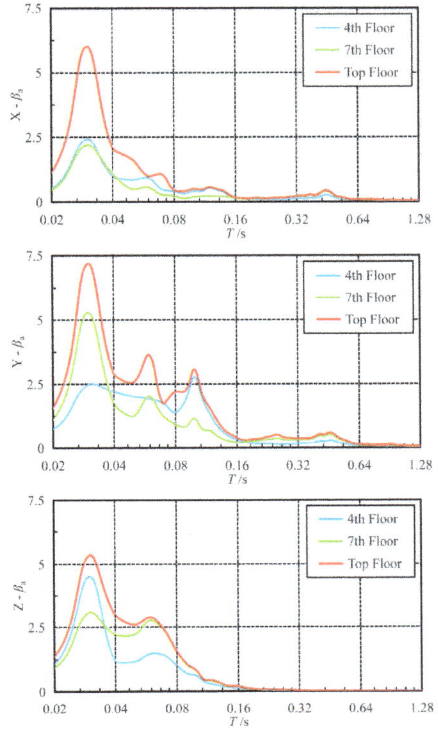

Figure 15. The X, Y, and Z directions' acceleration dynamic amplification coefficients of the rural residential building.

4.3.3. Characteristics of Spectral Ratio Curve

The acceleration dynamic amplification coefficients in the X, Y, and Z directions of each floor of the rural residential building were calculated, respectively, and the comparison curves are shown in Figure 16. The characteristics are summarized as follows:

To elucidate the vertical dynamic characteristics of the rural residential building under blasting-induced vibrations, the instrument positions at the ground floor level serve as reference points for the transfer function spectral ratios. Leveraging the acceleration response spectrum data, spectral ratio curves were derived for each floor relative to the ground floor reference in the X, Y, and Z directions. By definition, the spectral ratio at the ground floor ground position equates to a horizontal line at unity (1.0) on the vertical axis, as depicted in Figure 15. This reference line facilitates comparison across floor levels, highlighting variations in dynamic amplification with elevation. The spectral ratio characteristics of each floor are summarized as follows:

(1) The spectral ratio curves for the X-direction acceleration response spectra of the rural residential building's ground, fourth, seventh, and top floors reveal distinct peaks at a characteristic period of 0.40 s, corresponding to a frequency of 2.5 Hz. Notably, under the characteristic period of 0.12 s, peaks emerge on the top and fourth floors, whereas the seventh floor exhibits a comparatively smaller spectral ratio value.

(2) The spectral ratio curves for the Y-direction acceleration response spectra of the rural residential building's first, fourth, seventh, and top floors similarly exhibit prominent peaks at a characteristic period of 0.40 s, corresponding to a dominant frequency of 2.5 Hz. Additionally, the top floor's curves display distinct peaks under

the characteristic periods of 0.06 s and 0.10 s, though the fourth and seventh floors do not exhibit concurrent peaks at these periods.

(3) The spectral ratio curves for the Z-direction acceleration response spectra of the rural residential building's ground, fourth, seventh, and top floors exhibit notable peaks at a characteristic period of 0.07 s, corresponding to a dominant frequency of 14.3 Hz. This observation aligns with the signal spectrum analysis findings, indicating heightened sensitivity to short-period, high-frequency vertical excitations.

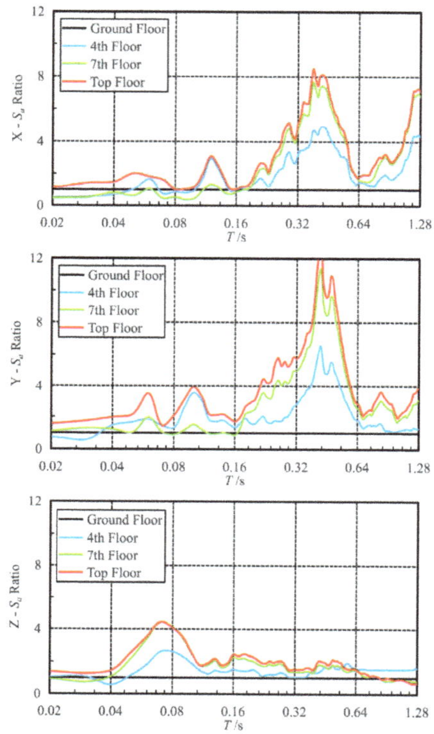

Figure 16. The X, Y, and Z directions' acceleration spectral ratio of the rural residential building.

5. Discussion

5.1. The Difference of Dynamic Response

Since acceleration effectively captures the transient structural response to blasting actions, while velocity reflects energy input and dissipation characteristics, the peak acceleration and peak velocity values at the ground floor reference points exhibit variability across different rural residential buildings and operational conditions. Consequently, the peak acceleration and peak velocity data were normalized relative to the ground floor references, yielding the peak acceleration ratio (PPAR, Equation (5)) and peak velocity ratio (PPVR, Equation (6)). The inter-floor comparisons derived from these ratios are illustrated in Figure 17.

$$PPAR = \frac{PPA_{\text{target}}}{PPA_{\text{reference}}}, \tag{5}$$

$$PPVR = \frac{PPV_{\text{target}}}{PPV_{\text{reference}}} \tag{6}$$

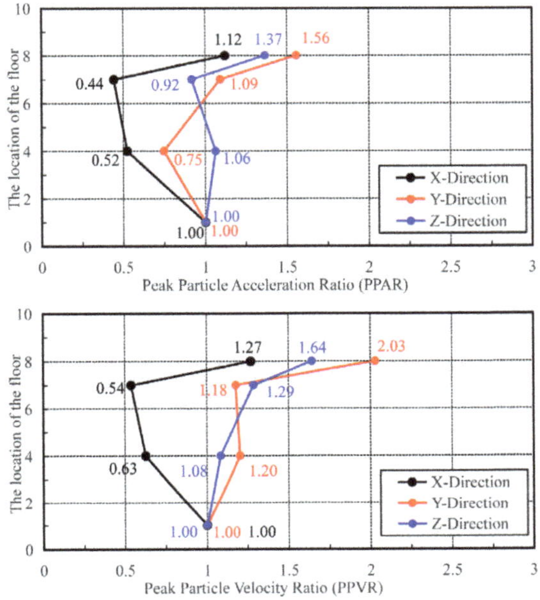

Figure 17. Normalized comparison of peak acceleration and peak velocity on each floor of the rural residential building.

The longitudinal and transverse stiffness differentials within the structure result in non-uniform amplification of peak acceleration and velocity at the uppermost story across the X, Y, and Z axes. Notably, peak acceleration amplifications attain 12%, 56%, and 37% in the respective directions, while peak velocity amplifications reach 27%, 103%, and 64% across the same axes. This phenomenon primarily arises from the structural idiosyncrasies of rural residential construction, wherein the absence of load-bearing elements such as shear walls between framework columns engenders comparatively greater horizontal stiffness (in X-Y planes), contrasted with reduced vertical stiffness (along the Z-axis), thereby modulating vibration response patterns. Furthermore, the progression of peak acceleration across story heights does not adhere to a linear trend, instead being influenced by interactions between horizontal and vertical vibration modes. Horizontal vibration signatures exhibit heightened sensitivity to structural stiffness variations and mass distribution, whereas vertical oscillations are constrained by rigid connections between floor slabs and framework components. This latter effect induces pronounced stiffness degradation at the uppermost structural level, consequently influencing acceleration response dynamics.

In the actual investigation, it was found that the fourth floor (the middle floor) is used as a storage warehouse. Around the placement of the instruments, some goods and materials (mainly small keychains, metal hangers, and other small hardware handicrafts) may have formed "additional mass dampers" [47–51]. The loading situation of the goods is shown in Figure 18. These items absorbed horizontal medium-to-high-frequency vibrations, thereby dissipating corresponding energy at the fourth floor level and reducing peak acceleration values. Concurrently, cargo loading altered the local mass distribution of the structure, suppressing modal participation rates in intermediate zones. This phenomenon underscores the considerable influence of structural characteristics and internal load distribution within rural residential buildings on their vibration responses. The incorporation of measures such as 'additional mass dampers' offers a viable means of enhancing structural vibration performance, thereby reducing damage to rural residential buildings.

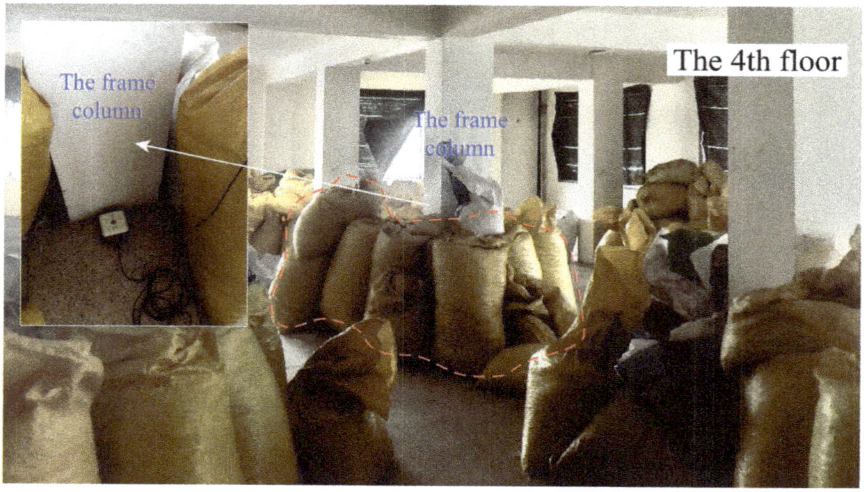

Figure 18. The on-site storage investigation on the fourth floor of the rural residential building.

5.2. Optimize Monitoring Point Layout Criteria

At present, the selection and placement of monitoring points for most rural residential buildings frequently rely on historical practices rather than evidence-based criteria, with industry standards lacking explicit guidelines regarding critical vibration velocity monitoring zones in conventional civil structures. Through dynamic response analysis of rural residential buildings, this study reveals that peak particle velocity (PPV) exhibits marked nonlinear vertical amplification, with structural summit measurements contributing disproportionately to damage assessments of rural dwellings. The following optimization suggestions are proposed for rural construction blast-induced vibration monitoring point criteria:

(1) Primary Monitoring Points: Set at ground-bearing wall-column junctions in typical rural structures to capture basal vibration data inputs.

(2) Supplementary Monitoring Points: Positioned at the four corners and ridge line midpoints of rooftops in conventional rural buildings to detect PPV amplification arising from the whip effect.

(3) Selective Monitoring Points: Targeted at weak structural zones based on building significance, natural frequencies, and foundation conditions. These may include mid-floor stiffness transition areas, fenestration openings, or other vulnerable zones. Where feasible, wireless sensor networks are recommended for multi-story synchronous monitoring.

The horizontal amplification effect at the summit of the rural residential building manifests as pronounced low-frequency enhancement, indicative of the structural selectivity toward long-period, low-frequency components within blast-induced vibration signals. Concurrently, to counter velocity amplification at elevated positions, tuned liquid dampers (TLDs) are advocated for rooftop installations, such as attic spaces, to enhance low-frequency energy dissipation [52–56]. Based on the findings of this experiment, it can be surmised that variations in the layout characteristics, floor heights, and structural stiffness of rural residential buildings significantly influence the dynamic amplification effect under vibration conditions at blasting sites. The influence of the building structure's slabs also deserves further in-depth analysis. Furthermore, these factors are closely linked to the conditions of the blasting site and the surrounding environment. The influencing factors in this domain necessitate further exploration and validation.

6. Conclusions

Based on the dynamic response test of the Mountain Qianshi blasting project in the Wenzhou segment of the Hangzhou–Wenzhou High-Speed Railway, the dynamic response characteristics of rural residential buildings subjected to blast-induced vibrations were analyzed. Through a combination of modal analysis, strong-motion instrumentation, and advanced signal processing techniques, the study systematically examined vibrational characteristics and dynamic amplification patterns across different story levels, offering technical insights into structural performance. The main conclusions are as follows:

(1) Time–domain dynamic characteristic analysis revealed that the acceleration amplitude variation at the top floor of the structure was markedly greater than that on the ground floor, with vertical acceleration and velocity amplitudes exhibiting significantly greater variability compared to the horizontal direction. This phenomenon arises because the vertical stiffness of the multi-story reinforced concrete frame structure exceeds the horizontal stiffness, contributing to the observed disparities in dynamic response.

(2) Frequency–domain distribution analysis indicated that rural residential buildings exhibit a site-specific optimal frequency of approximately 33 Hz under blasting loads, with multiple advantageous frequencies observed. Notably, the frequency–domain components of signals recorded on the seventh and the top floor levels demonstrated greater complexity. As floor elevation increased, predominant frequencies exhibited a tendency to shift toward lower frequency bands. Blast-induced vibration energy manifested low-frequency concentration and high-frequency attenuation, resulting in a multi-band, multi-directional selective amplification effect across the structural height.

(3) Analysis of characteristic curves, including floor acceleration response spectra, dynamic amplification coefficients, and spectral ratios, indicated that the dynamic amplification effect is most pronounced on the top floor of the structure. However, this amplification did not exhibit a direct proportionality with story elevation. Certain intermediate floors, such as the fourth story in the Y-direction, also demonstrated significant amplification under specific periodic conditions. The amplification phenomenon observed at elevated stories exhibited directional selectivity and period-dependent variability, suggesting a complex interaction between structural resonance characteristics and blast-induced excitation frequencies.

(4) Analysis of the dynamic response revealed that the "additional mass damper" effect induced by accidental loading facilitated the dissipation of medium and high-frequency vibrational energy. This energy attenuation manifested as reduced local acceleration and velocity amplitudes, demonstrating pronounced nonlinear variation across story heights. Such behavior elucidated the regulatory mechanism through which irregular construction loading patterns in rural residences influence the redistribution of blast-induced vibration energy, particularly by leveraging inadvertent mass-damping effects to modulate structural response characteristics.

(5) In light of the experimental findings, the analysis of the disparities in floor dynamic responses was conducted. Subsequently, optimization criteria for the layout of monitoring points were proposed, offering valuable insights for further research and engineering applications. Additionally, practical guidance was provided to

enhance the seismic resilience and safety of rural residential buildings exposed to blast-induced vibrations.

Author Contributions: Conceptualization, J.P. and D.Z.; methodology, J.P., D.Z. and Z.Z.; software, Z.Z., C.P. and Y.H.; validation, D.Z., S.W. and Z.Z.; formal analysis, J.P., J.H. and L.Z.; investigation, Z.Z.; resources, Z.Z. and S.W.; data curation, J.P., D.Z., Z.Z., J.H., L.Z., C.P., Y.H. and S.W.; writing—Original draft preparation, J.P., D.Z. and Z.Z.; writing—review and editing, D.Z. and Z.Z.; visualization, D.Z.; supervision, J.P.; project administration, J.P. and D.Z.; funding acquisition, J.P. and D.Z. All authors have read and agreed to the published version of the manuscript.

Funding: This research was funded by the National Key Research and Development Program of China (Grant No.2023YFC3008604), the National Natural Science Foundation of China (Grant No.42174062), the Emergency Management Research and Development Project of Zhejiang Emergency Management Department (Grant No.2024YJ019), and the Key Scientific Research Project of Zhejiang College of Security Technology (Grant No.AF2023Z03), which are gratefully acknowledged.

Data Availability Statement: The original contributions presented in this study are included in the article. Further inquiries can be directed to the corresponding author.

Acknowledgments: We would like to express our gratitude to China Railway 24th Bureau Group Co., Ltd. and the Emergency Management Department of Zhejiang Province for their support in the elements of blasting engineering tests.

Conflicts of Interest: Author Sishun Wang was employed by the company China Railway 24th Bureau Group Co., Ltd. The remaining authors declare that the research was conducted in the absence of any commercial or financial relationships that could be construed as a potential conflict of interest.

References

1. Dogan, O.; Anil, Ö.; Akbas, S.O.; Kantar, E.; Tuğrul Erdem, R. Evaluation of blast-induced ground vibration effects in a new residential zone. *Soil Dyn. Earthq. Eng.* **2013**, *50*, 168–181. [CrossRef]
2. Yan, Z.X.; Bao, T.; Wang, H.Y. Research on Blast-Induced Seismic Action. *Adv. Mater. Res.* **2010**, *163–167*, 4415–4421. [CrossRef]
3. Ercins, S. Vibration prediction with a method based on the absorption property of Blast-induced seismic waves: A case Study. *Open Geosci.* **2024**, *16*, 20220633. [CrossRef]
4. Yang, J.; Wu, Z.; Sun, W.; Yao, C.; Wang, Q. Numerical simulation on radiation and energy of Blast-induced seismic waves in deep rock Masses. *J. Cent. South. Univ.* **2022**, *29*, 645–662. [CrossRef]
5. Yu, C.; Wang, Z.; Han, W. A prediction model for Amplitude-frequency characteristics of blast-induced seismic Waves. *Geophysics* **2018**, *83*, T159–T173. [CrossRef]
6. Yu, C.; Wang, Z.; Han, W. A prediction model for frequency spectrum of blast-induced seismic wave in viscoelastic Medium. *Geophys. Prospect.* **2017**, *66*, 87–98. [CrossRef]
7. Yang, J.; Sun, J.; Jia, Y.; Yao, Y. Energy Generation and Attenuation of Blast-Induced Seismic Waves under In Situ Stress Conditions. *Appl. Sci.* **2022**, *12*, 9146. [CrossRef]
8. Gong, J.; Wang, Q.; Nie, B.; Ge, Z. Experimental study on ground vibration caused by the blast loading of an explosion vessel. *Int. Struct. Stab. Dyn.* **2022**, *22*, 2250003. [CrossRef]
9. Singh, P.K.; Roy, M.P. Damage to surface structures due to blast vibration. *Int. Rock. Mech. Min. Sci.* **2010**, *47*, 949–961. [CrossRef]
10. Lawal, A.I.; Kwon, S.; Hammed, O.S.; Idris, M.A. Blast-induced ground vibration prediction in granite quarries: An application of gene expression programming, ANFIS, and sine cosine algorithm optimized ANN. *Int. Min. Sci. Technol.* **2021**, *31*, 265–277. [CrossRef]
11. Fattahi, H.; Hasanipanah, M. Prediction of blast-induced ground vibration in a mine using relevance vector regression optimized by metaheuristic algorithms. *Nat. Resour. Res.* **2021**, *30*, 1849–1863. [CrossRef]
12. Norén-Cosgriff, K.M.; Ramstad, N.; Neby, A.; Madshus, C. Building damage due to vibration from rock blasting. *Soil Dyn. Earthq. Eng.* **2020**, *138*, 106331. [CrossRef]
13. Fan, H.; Yu, D.; Zhao, D.; Lai, J.; Xie, Y. Experimental and numerical investigation of blasting-induced ground vibration from tunnel undercrossing a village. *Stavební Obz.-Civ. Eng.* **2017**, *26*, 404–417.

14. Chen, C.; Gan, D.Q.; Zhang, Y.B. Safety assessment method based on response spectrum analysis of building structures to blasting vibration. *Appl. Mech. Mater.* **2013**, *351–352*, 1669–1672. [CrossRef]
15. Akinbinu, V.A. Assessment of blast induced vibrations based on regulatory criteria (a case study of limestone mines, nigeria). *Int. J. Min. And. Miner. Eng.* **2018**, *9*, 81. [CrossRef]
16. Ozcelik, M. Back analysis of ground vibrations which cause cracks in buildings in residential areas karakuyu (dinar, afyonkarahisar, turkey). *Nat. Hazards* **2018**, *92*, 497–509. [CrossRef]
17. Hudaverdi, T.; Akyildiz, O. Prediction and evaluation of blast-induced ground vibrations for structural damage and human response. *Arab. Geosci.* **2021**, *14*, 378. [CrossRef]
18. Faramarzi, F.; Ebrahimi Farsangi, M.A.; Mansouri, H. Simultaneous investigation of blast induced ground vibration and airblast effects on safety level of structures and human in surface blasting. *Int. Min. Sci. Technol.* **2014**, *24*, 663–669. [CrossRef]
19. Yao, Q.; Yang, X.; Li, H. Development of absorbed blasting vibration energy index for the evaluation of human comfort in multistorey buildings. *Shock. Vib.* **2017**, *1*, 9567657. [CrossRef]
20. Ghosh, S.; Behera, C.; Mishra, M.K. Prediction of blast-induced ground vibration using multivariate statistical analysis in the opencast chromite mines of the indian state of odisha. *Sādhanā* **2024**, *49*, 141. [CrossRef]
21. Yu, Z.; Shi, X.; Zhou, J.; Gou, Y.; Huo, X.; Zhang, J.; Armaghani, D.J. A new multikernel relevance vector machine based on the HPSOGWO algorithm for predicting and controlling blast-induced ground vibration. *Eng. Comput.* **2022**, *38*, 1905–1920. [CrossRef]
22. Yang, L.J.; Liu, D.W.; Guo, Z.L.; Li, J. Engineering mechanics in whipping effect of high-rise building. *Appl. Mech. Mater.* **2014**, *540*, 173–176. [CrossRef]
23. Mohanty, M.; Sarkar, R.; Das, S.K. Effect of blast induced vibration on coal mine overburden dump slope through discrete element method. *Structures* **2023**, *56*, 105013. [CrossRef]
24. Lee, J.; Fenves, G.L. Plastic-damage model for cyclic loading of concrete structures. *J. Eng. Mech.* **1998**, *124*, 892–900. [CrossRef]
25. Lee, J.; Fenves, G.L. A plastic-damage concrete model for earthquake analysis of dams. *Earthq. Eng. Struct. Dyn.* **1998**, *27*, 937–956. [CrossRef]
26. Kuhlemeyer, R.L.; Lysmer, J. Finite element method accuracy for wave propagation problems. *J. Soil Mech. Found. Div.* **1973**, *99*, 421–427. [CrossRef]
27. Lysmer, J.; Kuhlemeyer, R.L. Finite dynamic model for infinite media. *J. Eng. Mech. Div.* **1969**, *95*, 859–877. [CrossRef]
28. Ruggieri, S.; Bruno, G.; Attolico, A.; Uva, G. Assessing the dredging vibrational effects on surrounding structures: The case of port nourishment in Bari. *J. Build. Eng.* **2024**, *96*, 110385. [CrossRef]
29. Majkut, L. Free and forced vibrations of timoshenko beams described by single difference equation. *J. Theor. Appl. Mech.* **2009**, *47*, 193–210.
30. Carrera, E.; Pagani, A. Free vibration analysis of civil engineering structures by component-wise models. *J. Sound. Vib.* **2014**, *333*, 4597–4620. [CrossRef]
31. Yu, C.; Wu, J.; Li, H.; Ma, Y.; Wang, C. Study on the equivalence transformation between blasting vibration velocity and acceleration. *Sensors* **2024**, *24*, 1727. [CrossRef] [PubMed]
32. Rahman, M.S.; Pervin, M.M.; Azad, M.A.K. Comparison of the methods for numerical integration. *Jahangirnagar Univ. Sci.* **2019**, *42*, 63–78.
33. Fornberg, B. Improving the accuracy of the trapezoidal rule. *SIAM Rev.* **2021**, *63*, 167–180. [CrossRef]
34. Sazeli, Z.A.; Ali, N.A.A.; Talib, M.A.A. Comparison of the trapezoidal rule and simpson's rule. *Multidiscip. Appl. Res. Innov.* **2024**, *5*, 118–123.
35. Lu, Z.; Li, K.; Zhou, Y. Comparative studies on structures with a tuned mass damper and a particle damper. *J. Aerosp. Eng.* **2018**, *31*, 4018090. [CrossRef]
36. Hudaverdi, T. Application of multivariate analysis for prediction of blast-induced ground vibrations. *Soil Dyn. Earthq. Eng.* **2012**, *43*, 300–308. [CrossRef]
37. Shan, R.; Zhao, Y.; Wang, H.; Liu, Z.; Qin, H. Blasting vibration response and safety control of mountain tunnel. *Bull. Eng. Geol. Environ.* **2023**, *82*, 166. [CrossRef]
38. Prassetyo, S.H.; Simangunsong, G.M.; Wattimena, R.K.; Rai, M.A.; Sidiq, T.P.; Widodo; Aditya, F.; Fahrudinsyah, M.H.; Putra, D.; Prabowo, J.D.; et al. Evaluation of blast-induced ground vibration in urban area for theakarta-bandung HSR tunnel. *Geotech. Geol. Eng.* **2024**, *42*, 7255–7285. [CrossRef]
39. Wang, Z. Fast algorithms for the discrete W transform and for the discrete fourier transform. *IEEE Trans. Acoust. Speech Signal Process.* **1984**, *32*, 803–816. [CrossRef]
40. Robinson, E.; Clark, D. Sampling and the nyquist frequency. *Lead. Edge* **1991**, *10*, 51–53. [CrossRef]
41. McLean, R.F.; Alsop, S.H.; Fleming, S. Nyquist—Overcoming the limitations. *J. Sound. Vib.* **2005**, *280*, 1–20. [CrossRef]

42. Gohshi, S. A new signal processing method for video: Reproduce the frequency spectrum exceeding the nyquist frequency. In Proceedings of the 3rd Multimedia Systems Conference, Chapel Hill, NC, USA, 22–24 February 2012; pp. 47–52.
43. Douglas, J. Earthquake ground motion estimation using strong-motion records: A review of equations for the estimation of peak ground acceleration and response spectral ordinates. *Earth-Sci. Rev.* **2003**, *61*, 43–104. [CrossRef]
44. Akkar, S.; Özen, Ö. Effect of peak ground velocity on deformation demands for SDOF systems. *Earthq. Eng. Struct. Dyn.* **2005**, *34*, 1551–1571. [CrossRef]
45. Moser, W.; Antes, H.; Beer, G. A duhamel integral based approach to one-dimensional wave propagation analysis in layered media. *Comput. Mech.* **2004**, *35*, 115–126. [CrossRef]
46. Savin, E. Dynamic amplification factor and response spectrum for the evaluation of vibrations of beams under successive moving loads. *J. Sound. Vib.* **2001**, *248*, 267–288. [CrossRef]
47. Moon, K.S. Integrated damping systems for tall buildings: Tuned mass damper/double skin facade damping interaction system. *Struct. Des. Tall Spec. Build.* **2016**, *25*, 232–244. [CrossRef]
48. Guenidi, Z.; Abdeddaim, M.; Ounis, A.; Shrimali, M.K.; Datta, T.K. Control of adjacent buildings using shared tuned mass damper. *Procedia Eng.* **2017**, *199*, 1568–1573. [CrossRef]
49. Shi, W.; Wang, L.; Lu, Z. Study on self-adjustable tuned mass damper with variable mass. *Struct. Control. Health Monit.* **2018**, *25*, e2114. [CrossRef]
50. Tophøj, L.; Grathwol, N.; Hansen, S.O. Effective mass of tuned mass dampers. *Vibration* **2018**, *1*, 192–206. [CrossRef]
51. Jiang, Q.; Rong, M.; Wei, W.; Chen, T. A quantitative seismic topographic effect prediction method based upon BP neural network algorithm and FEM simulation. *J. Earth Sci.* **2024**, *35*, 1355–1366. [CrossRef]
52. Tait, M.; Isyumov, N.; El Damatty, A.A. Performance of tuned liquid dampers. *J. Eng. Mech.* **2008**, *134*, 417–427. [CrossRef]
53. Zhang, Z.; Staino, A.; Basu, B.; Nielsen, S.R.K. Performance evaluation of full-scale tuned liquid dampers (TLDs) for vibration control of large wind turbines using real-time hybrid testing. *Eng. Struct.* **2016**, *126*, 417–431. [CrossRef]
54. Konar, T.; Ghosh, A.D. Flow damping devices in tuned liquid damper for structural vibration control: A review. *Arch. Comput. Methods Eng.* **2021**, *28*, 2195–2207. [CrossRef]
55. Lee, B.G.; Kim, J.Y.; Jung, J.S.; Lee, K.S. Seismic protection provided by a new diamond-shaped bracing system with a horizontally layered friction damper. *J. Build. Eng.* **2024**, *92*, 109709. [CrossRef]
56. Tang, Z.; Sheng, J.; Dong, Y. Effects of tuned liquid dampers on the nonlinear seismic responses of high-rise structures using real-time hybrid simulations. *J. Build. Eng.* **2023**, *70*, 106333. [CrossRef]

buildings

MDPI

Article

Finite Element Modeling and Performance Evaluation of a Novel 3D Isolation Bearing

Jianjun Li [1], Lvhong Sun [1], Yanchao Wu [1], Yun Chen [2], Dengzhou Quan [3], Tuo Lei [3,*] and Sansheng Dong [3]

[1] CCCC First Highway Northwest Engineering Co., Ltd., Xi'an 710075, China; 17582922636@163.com (J.L.)
[2] School of Civil Engineering and Architecture, Hainan University, Haikou 570228, China
[3] School of Civil Engineering, Chang'an University, Xi'an 710061, China
* Correspondence: leituo616@163.com

Abstract: A numerical investigation is conducted to examine the mechanical properties of a novel three-dimensional (3D) isolation bearing. This device is primarily composed of a lead rubber bearing (LRB), disc springs, and U-shaped dampers. A finite element model is developed and validated against the previous experimental results. Subsequently, comprehensive analyses are performed to evaluate the influence of vertical loadings, shear strains, and the number of U-shaped dampers on the horizontal behavior, as well as the effects of displacement amplitudes and the number of dampers on the vertical performance. Under horizontal loading conditions, the bearing demonstrates reliable energy dissipation capabilities. However, the small lead core design limits its energy dissipation capacity. Compared with the bearing without U-shaped dampers, the bearing's energy dissipation capacity increases by 628%, 1300%, and 2581% when employing 1, 2, and 4 dampers on each side, respectively. Regarding vertical performance, the innovative disc spring group design effectively reduces the tensile displacement of the LRB under tension, thereby enhancing the overall tensile capacity of the bearing. Furthermore, in comparison to their contribution to horizontal energy dissipation, the U-shaped dampers play a relatively minor role in vertical energy dissipation.

Keywords: three-dimensional (3D) isolation bearing; mechanical property; disc spring; U-shaped damper; finite element modeling

1. Introduction

Currently, various horizontal seismic isolation technologies have reached a relatively mature stage and are widely applied in engineering structures [1]. However, due to the multi-dimensional characteristics of earthquake ground motions, the vertical responses of the structures, including long-span structures, cantilever structures, and high-rise buildings, might be much higher than the horizontal responses, especially when these structures are located at the epicenter of earthquakes or near fault zones [2,3]. To ensure the safety of structures, the research and innovation of three-dimensional (3D) isolation technologies and related devices are still desperately needed.

The prototype of 3D isolation technologies can be traced back to the 1970s, when Seigenthaler [4] used thick natural rubber blocks as isolation bearings in post-disaster reconstruction. Over the past half century and more, 3D isolation technologies and devices have made great progress, and some have been applied in practice. To isolate both horizontal and vertical ground motions, the existing 3D isolation bearings can be classified into two categories from the perspective of structural configurations: independent-type and combined-type. The independent 3D isolation bearings couple horizontal and vertical responses into single components, such as thick rubber bearings (TRBs) [5,6], helical springs [7], and air springs [8]. Although the construction of such bearings is relatively simple, their analysis and design is complicated due to the mutual influence of horizontal and vertical mechanical properties [9]. At present, this type of bearing, especially thick rubber bearings, is mainly applied in nuclear power facilities [10]. The combined 3D bearings can be decomposed into horizontal isolation devices and vertical isolation devices, which are used to isolate horizontal and vertical motions, respectively. In this way, scholars have creatively developed a variety of 3D isolation bearings.

By integrating a horizontal rubber bearing and disc springs in series, Fujita and Kato [11], Xiong [12], and Zhao et al. [13] previously conducted 3D isolation bearing tests. To further improve the tension-resistant capacity and overturn resistance of 3D bearings, wire rope [14], an anti-extraction device [15], a steel plate damper [16], and SMA strands [17] were also added to this type of bearing. Kashiwazaki et al. [18] and Mo et al. [9,19] proposed a 3D isolation bearing composed of rubber bearing and air spring and completed a shaking table test. Liu et al. [20] and Xu et al. [21] proposed a 3D isolation device by connecting several inclined laminated lead rubber bearings (LRBs) in series with a horizontal LRB and completed a series of static tests and a shaking table test. By connecting a horizontal LRB with multiple hydraulic cylinders in series, Chen et al. [22] developed a new 3D isolation device. Similarly, Liang et al. [23] developed a new 3D isolation bearing by connecting a horizontal LRB with a vertical ring spring bearing (RSB) arranged vertically in series. Based on multiple inclined LRBs, Wei et al. [24] developed an inclined sliding 3D seismic isolation device in which sliders and limiters are also included. The finite element simulation results show that the device has stable hysteretic performance and good energy dissipation capacity. Based on a laminated natural rubber bearing improved with a pre-compressed disc spring, Luo et al. [25] proposed a 3D isolation bearing by adding spring bearings in series. Scholars [26,27] also sought to develop new 3D isolation bearings, which are made of LRBs and TRBs in series. The standard LRBs isolate the horizontal component of earthquakes, and the TRBs can be used for vertical isolation.

Based on friction pendulum bearings, Han et al. [28] developed a 3D isolation device, using a friction pendulum system to isolate horizontal motions and an air spring to isolate vertical motions, respectively. Cao et al. [29] developed a 3D isolation bearing with disc springs as the vertical vibration isolation component and a single friction pendulum as the horizontal vibration isolation component. The bearing not only achieved vertical isolation of building structures under environmental excitation but horizontal isolation under seismic excitation. Cao et al. [30] also presented a 3D isolation bearing to mitigate the horizontal seismic action caused by earthquakes and the vertical vibration generated by train operation. The 3D bearing consists of two modules: one is a thick-layer rubber bearing for vertical vibration isolation and the other is a friction pendulum for horizontal vibration isolation. Kitayama et al. [31] proposed a 3D isolation system which is composed of triple friction pendulum isolators acting as a horizontal isolation module and a coil spring-damper system acting as a vertical isolation module. They further conducted a series of probabilistic seismic analyses for non-isolated, horizontally isolated, and 3D-isolated electrical transformers. Connecting a quasi-zero stiffness system with a horizontal friction pendulum, Zhou et al. [32] developed a 3D bearing and applied it to a high-rise building model.

Other combined-type 3D isolation bearings have also been proposed in recent studies. For instance, to address the insufficient energy dissipation capacity and to reduce the horizontal constraint imposed by disc springs, Yu et al. [33] developed a 3D isolation device consisting primarily of disc springs, U-shaped dampers, and a friction pendulum system. Furthermore, Yu et al. [34] presented a 3D isolator capable of achieving self-centering behavior under horizontal loadings and exhibiting quasi-zero stiffness under vertical loadings. The self-centering capability is provided by shape memory alloy U-shaped dampers (SMA-UDs), and the quasi-zero stiffness (QZS) property is realized by the negative stiffness of a disc spring (DS) isolator and the positive stiffness of SMA-UDs. To enhance vertical seismic isolation performance and improve the seismic resilience of isolation bearings, Sha et al. [35] proposed a 3D isolation bearing integrating a disc-spring system (DSS), a high-damping rubber bearing (HDR), and a horizontal spring system (HSS). In this design, the DSS is engineered with low vertical stiffness to optimize vertical isolation, whereas the HSS contributes a self-centering force that enhances the seismic resilience of the HDR. This study presents a comprehensive theoretical analysis of both horizontal and vertical stiffness components. The vertical performance of the 3D isolation bearing is primarily attributed to the DSS, while the horizontal performance arises from the combined contributions of the HSS and HDR.

Among existing 3D isolation bearings, the primary horizontal isolation devices are LRBs and friction pendulum bearings, which are widely adopted due to their mature technology and ease of manufacture. The vertical isolation devices mainly consist of disc springs, air springs, ring springs, and helical springs. To further enhance the performance of these systems, the integration of additional functional components has become increasingly common. However, from a practical application perspective, the manufacturing complexity and cost of 3D isolation bearings are critical factors that must be carefully considered.

Inspired by the advantages of LRBs and disc springs, this paper proposes a novel 3D isolation bearing comprising an LRB serving as the horizontal isolation module and two groups of disc springs acting as the vertical isolation module. Additionally, U-shaped steel dampers are incorporated to improve the energy dissipation and safety of the bearing. Compared to existing 3D isolation bearings, the proposed design offers several notable features: (1) The vertical disc springs are divided into two groups: one for compression and the other for tension, which enhances the stability of the disc springs and facilitates independent control of the compression and tension stiffness; (2) The inclusion of U-shaped dampers allows for a reduction in the lead core diameter of the bearing while simultaneously enhancing its energy dissipation capacity; (3) The incorporation of wire ropes serves as a displacement-limiting mechanism, thereby improving the overall safety of the bearing at extreme displacements.

To evaluate the mechanical performance of the proposed bearing, a numerical investigation is conducted based on previous experimental data. First, finite element models are established and validated using the prior horizontal test results by the authors. Subsequently, a detailed parametric analysis is carried out to examine the horizontal and vertical mechanical behaviors of the bearing under various conditions, considering key parameters such as the vertical load magnitude, the shear or vertical displacement levels, and the number of U-shaped dampers employed.

2. Configuration and Working Mechanism of Bearing

2.1. Configuration

The structural configuration of the 3D isolation bearing is illustrated in Figure 1. The bearing primarily consists of disc spring group 1, disc spring group 2, a lower LRB arranged in series, and U-shaped dampers, along with upper and lower sleeves made of steel plates and other auxiliary connecting components. Disc spring group 1 comprises five sets of disc springs, with each set containing two disc springs connected in parallel and four in series. Similarly, disc spring group 2 also includes five sets, where each set consists of four disc springs in parallel and four in series. Disc spring groups 1 and 2 are symmetrically mounted on the upper and lower sides of the middle steel plate, respectively. The upper end of the U-shaped damper is connected to the middle steel plate, while its lower end is fastened via bolts to the lower connection plate of the LRB. The upper sleeve is welded to the top plate and the middle steel plate, whereas the lower sleeve is welded to the middle steel plate and merely contacts the upper connection plate of the LRB. Five supporting shafts, comprising one centrally positioned shaft and four symmetrically distributed around it, pass through holes in the middle steel plate, with their lower ends welded to the upper connection plate of the LRB.

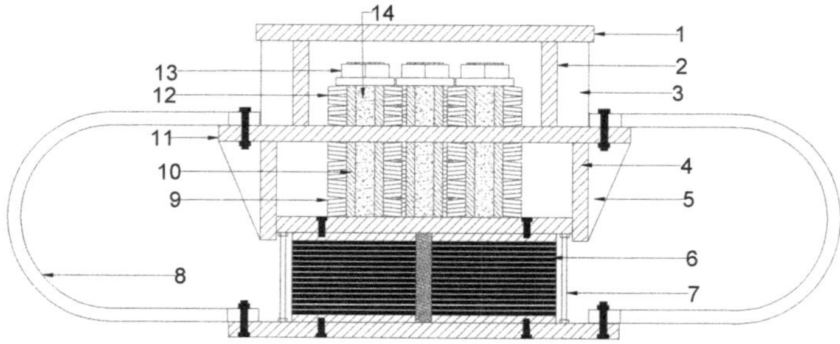

Figure 1. Structural configuration of 3D bearing. 1–Upper plate; 2–Upper sleeve; 3–Stiffeners for upper sleeve; 4–Lower sleeve; 5–Stiffeners for lower sleeve; 6–LRB; 7–Wire rope; 8–U-shaped damper; 9–Disc spring group 2; 10–Supporting shaft; 11–Middle plate; 12–Disc spring group 1; 13–Hexagon nut; 14– High-strength cement mortar.

2.2. Working Mechanism

The working mechanism of the bearing can be analyzed from five key aspects: load-bearing capacity, horizontal isolation behavior, vertical isolation behavior, energy dissipation, and post-earthquake reparability. Regarding load-bearing capacity, disc spring group 2 provides sufficient vertical stiffness and strength to support the dead load of the superstructure. Under horizontal loading conditions, the LRB undergoes significant shear deformation, thereby elongating the structural natural period and dissipating seismic energy. In terms of vertical loading, when the bearing experiences compressive forces, disc spring group 2 is compressed while disc spring group 1 remains inactive. Conversely, under tensile loads, disc spring group 1 is compressed while disc spring group 2 does not work. This design enables the bearing to function as a specialized device for enhancing tensile performance. Consequently, the compressive deformations of disc spring groups 1 and 2 help mitigate vertical seismic responses. Additionally, the U-shaped dampers not only contribute supplementary damping but also moderately increase the initial stiffness of the bearing. Furthermore, upon yielding, the U-shaped dampers allow for adjustable horizontal and vertical stiffness characteristics and can be conveniently replaced after an earthquake.

3. Introduction to Specimen Design

3.1. Design of LRB

In this study, the dead load of the superstructure is set to 1000 kN. The basic parameters of LRB are presented in Table 1. The selected rubber has a Shore hardness of 40 HA. The lead core of LRB is made of lead with a purity exceeding 99.9%, a density of 11.3×10^3 kg/m^3, an elastic modulus of 1.7×10^4 MPa, a Poisson's ratio of 0.42, and a yield stress of 8.5 MPa. The internal steel layers, upper and lower sealing plates, and connecting plates are all made of Q355B steel, with a density of 7.8×10^3 kg/m^3, an elastic modulus of 2.1×10^5 MPa, and a Poisson's ratio of 0.3.

Table 1. Main parameters of LRB.

Diameter (mm)	Lead Core Diameter (mm)	Thickness of Rubber Layer (mm)	Internal Plate Thickness (mm)	Rubber Layer Numbers	Rubber Shear Modulus (MPa)
500	30	7	3	14	0.4

3.2. Disc Spring Groups 1 and 2

As previously mentioned, disc spring groups 1 and 2 are both assembled using individual disc springs. The selected disc springs belong to standard series A [36,37], made from high-strength 50 CrVA steel, with an elastic modulus of 2.06×10^5 MPa, a Poisson's ratio of 0.3, and a yield strength of 1400 MPa. The main parameters of the individual disc spring are summarized in Table 2.

Table 2. Main parameters of single disc spring.

Outside Diameter (mm)	Inside Diameter (mm)	Thickness (mm)	Ultimate Displacement (mm)	Free Altitude (mm)	Bearing Capacity (kN)
140	72	8	3.2	11.2	85.3

To restrict horizontal displacements of the disc spring groups, five supporting shafts are employed. These supporting shafts utilize composite tubes constructed from 45# steel, each equipped with a matching hexagonal flange nut. The outer diameter of the selected steel tube corresponds to series 2 [36,37], with a dimension of 70 mm and a wall thickness of 17 mm. All steel tubes are filled with micro-expansion high-strength cement mortar.

3.3. U-Shaped Dampers

The U-shaped steel damper is manufactured from Q235B steel, characterized by an elastic modulus of 2.06×10^5 MPa, a Poisson's ratio of 0.3, and a density of 7.8×10^3 kg/m^3. The primary parameters of the U-shaped steel damper are listed in Table 3. The damper is connected to the middle or lower plate of the bearing via two M14 bolts. Additionally, since the middle and bottom plates feature evenly distributed holes along all four sides, it is convenient to symmetrically install 1, 2, or 4 U-shaped dampers on each side of the bearing in both horizontal directions.

Table 3. Main parameters of U-shaped steel damper.

Radius of Arc Segment (mm)	Breadth (mm)	Thickness (mm)	Straight Section Length (mm)
175	40	25	220

3.4. Sleeves and Other Plates

As illustrated in Figure 1, the upper sleeve and stiffening ribs, the lower sleeve and stiffening ribs, as well as the upper and middle plates are all constructed from Q355B steel. The wall thickness of the sleeves is 30 mm, while the stiffening ribs have a thickness of 20 mm. Detailed dimensional specifications can be found in reference [38].

4. Finite Element Modeling

4.1. Material Constitutive Model

To develop the finite element model of the 3D bearing, five types of material models, including rubber, lead, steel plate, U-shaped dampers, and disc springs, are needed.

Due to the large Poisson's ratio and entangled molecular chains, rubber materials can undergo large deformations without significant volume changes. They are approximately incompressible materials and have the characteristic of being regarded as isotropic bodies. Their mechanical properties can be characterized by their elastic strain energy. To simulate the hyper-elastic characteristics of the rubber material of LRB, the Mooney–Rivlin strain energy potential function is adopted. This function can be expressed as [39]:

$$W = C_{10}(I_1 - 3) + C_{01}(I_2 - 3) + \frac{(J-1)^2}{D_1} \quad (1)$$

where W is the strain energy per unit volume; C_{10}, C_{01}, and D_1 are temperature-dependent material parameters; I_1 and I_2 are the first and second deviatoric strain invariants, respectively; J denotes the elastic volume ratio. The material parameters are calculated to be C_{10} = 0.192 MPa, C_{01} = 0.048 MPa, and D_1 = 0.002 mm^2/N. Assuming rubber is incompressible, so J is taken as 1.

A bilinear elastic–plastic model is employed to simulate the behavior of the lead and U-shaped damper materials, whereas an elastic model is used for the other steel components. The corresponding material parameters are summarized in Table 4.

Table 4. Material parameters of steel plate, lead, and U-shaped steel damper.

Material	Elastic Modulus (GPa)	Poisson's Ratio	Yield Stress (MPa)	Hardening Modulus (MPa)
Steel plate	206.00	0.30	355	-
Lead	17.00	0.42	8.50	17.00
U-shaped damper	206.00	0.30	298.88	2060

To simulate the vertical disc spring groups composed of stacked disc springs, one approach involves constructing a detailed finite element model of each individual disc spring. However, this method requires modeling the geometric dimensions, material properties, and interactions of each disc spring, resulting in high computational costs and convergence difficulties. Alternatively, a simplified spring element can be used. To reduce computational complexity, a simplified spring-damper element can be used to substitute each set of disc springs. Under normal conditions, the stiffness of each disc spring set can be considered constant and is calculated using the following equations [36]:

$$K_s = \frac{4E}{1-\mu^2}\frac{t^3}{K_1 D^2}\left[\left(\frac{h_0}{t}\right)^2 - 3\frac{h_0}{t}\frac{f}{t} + \frac{3}{2}\left(\frac{f}{t}\right)^2 + 1\right] \quad (2)$$

$$K_1 = \frac{1}{\pi}\left(\frac{C-1}{C}\right)^2 / \left(\frac{C+1}{C-1} - \frac{2}{\ln C}\right) \quad (3)$$

where E represents the elastic modulus of the disc spring; μ denotes Poisson's ratio; t and h_0 are the thickness and solid height of each disc spring, respectively; f denotes the deformation of each disc spring; C is the ratio of the outer diameter D to the inner diameter d; K_1 is a correlation coefficient. For a disc spring set with m slices in parallel and n in series, its stiffness is equal to m times K_s divided by n.

Furthermore, the nonlinear damping effect of the disc spring group can be approximated using a constant damping coefficient assigned to each spring set [40]. Based on Equations (2) and (3), the stiffness of the simplified spring element for each set in disc spring group 1 is determined to be 18.19 kN/mm, with a damping coefficient of 1.28 kN/(mm/s). Considering disc spring group 2, the stiffness and damping coefficients of the simplified spring element of each set are 36.39 kN/mm and 30 kN/(mm/s), respectively.

4.2. Element Type and Meshing

To establish a reliable finite element model, two key factors must be carefully considered in addition to employing accurate material constitutive models: first, the appropriate simplification of model components and the interaction mechanics between them; second, achieving an optimal balance between element mesh densities and computational efficiency [41,42]. The finite element model of the bearing is developed using ABAQUS (version 6.12) [39]. Based on the structural configuration and experimental observations of the bearings, the deformation of the bearings is mainly concentrated in the LRB, disc springs, and U-shaped dampers. Consequently, this study excluded non-critical components, namely, the supporting shafts, the upper and lower sleeves, as well as the corresponding stiffening ribs, from the finite element model to enhance computational efficiency and convergence performance. The functional roles of these components will be elaborated in the subsequent section.

Given that rubber exhibits typical nonlinear behavior and is nearly incompressible, the C3D8RH solid element is selected to define the inner rubber layer. The cover rubber of the LRB is neglected. In addition, the C3D8R solid element is employed to discretize the upper and lower plates, middle plate, closure plates, inner steel layers, upper connection plate of LRB, lead core, and U-shaped dampers. From the working mechanism of the bearing described earlier, it can be inferred that when the bearing is subjected to vertical compression, the disc spring group 2 is compressed, while the disc spring group 1 remains inactive. Conversely, under vertical tension, the disc spring set 1 is compressed, whereas the disc spring group 2 does not function. To accurately simulate the disc spring groups in this paper, the nonlinear spring/dashpot element, Spring A provided by ABAQUS software is employed. A total of ten such elements are implemented. Specifically, the upper five elements are utilized to replace the disc springs of group 1, and the lower five are employed to substitute the disc springs of group 2.

The detailed finite element model of the novel 3D bearing is illustrated in Figure 2a–e. It is worth mentioning that the mesh density of the closure plates, each rubber layer, and each internal steel layer of the LRB within the plane is the same, with 5, 3, and 1 layers defined along the thickness direction, respectively. To accurately capture the large plastic deformation of the lead core and obtain its stress and strain distributions, the mesh size was controlled within 2 mm, as shown in Figure 2d. Figure 2e presents the complete finite element model of the bearing, with only one U-damper connected to each side.

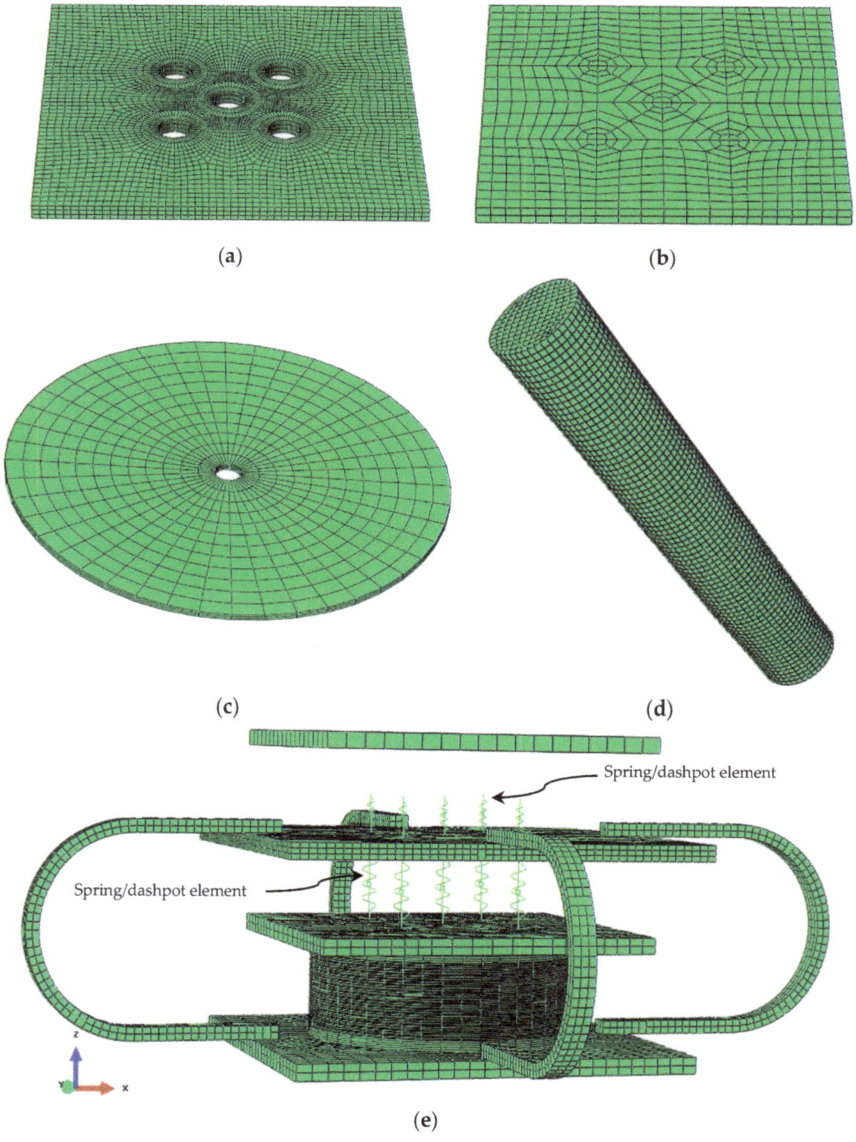

Figure 2. Finite element model of bearing. (**a**) Middle plate; (**b**) Upper connection plate of LRB; (**c**) Rubber layer of LRB; (**d**) Lead core of LRB; (**e**) Complete FE model.

4.3. Definition of Interaction and Boundary Conditions

To ensure proper force transmission and deformation compatibility among the components, the following interaction definitions are established:

Firstly, kinematic coupling constraints are enforced separately between reference point 1 (RP-1) and the top surface of the upper plate, as well as between RP-1 and the side surfaces of the middle plate. In both cases, all six degrees of freedom, namely, U1, U2, U3, UR1, UR2, and UR3, are constrained. This approach effectively couples the deformations of the upper and middle plates, thereby eliminating the need to explicitly model the upper sleeve and related stiffeners. In addition, one more kinematic coupling constraint is implemented between RP-1 and the upper connection plate of the LRB. In this case, only the U3 degree of freedom (vertical direction) is left unconstrained, while the remaining five (U1, U2, UR1, UR2, and UR3) are constrained. This ensures that the upper connection plate of the LRB moves horizontally in conjunction with the upper and middle plates while allowing free vertical movement. As a result, the explicit modeling of the lower sleeve and related stiffeners was avoided. Details of these three coupling constraints are presented in Figure 3a–c.

Secondly, five MPC beam constraints were established on the upper connection plate of the LRB to simulate the disc spring shafts, as shown in Figure 3d. Ten reference points were defined at the center positions of the holes on the upper and lower surfaces of the middle plate to assign the spring elements. Among these, the upper five reference points serve as the lower nodes of the upper five spring elements, and the lower five reference points act as the upper nodes of the lower five spring elements. Each reference point is coupled with the annular region within a 70 mm radius around the corresponding hole on the upper and lower surfaces of the middle plate.

To establish the 23 coupling constraints mentioned above, a total of 21 reference points are employed. In the model of the LRB, the inner rubber layers and the inner steel layers, the upper and lower connection steel plates, the closure plates, and the lead core are all connected to each other by tie constraints. Finally, the upper end of the U-shaped dampers is connected to the middle steel plate, and the lower end is connected to the lower connection plate of the LRB. The connection method of both also uses tie constraints. For the boundary conditions of the bearing, all translational degrees of freedom of the lower connection plate are constrained, that is, U1 = U2 = U3 = 0.

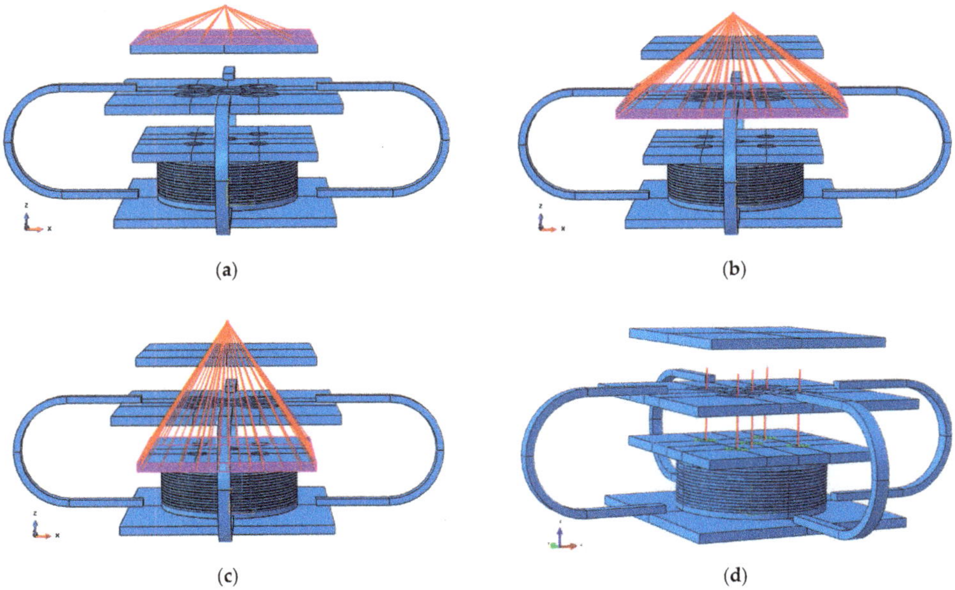

Figure 3. Coupling constraints of bearing. (**a**) Coupling constraint for upper plate; (**b**) Coupling constraint for middle plate; (**c**) Coupling constraint for upper connection plate of LRB; (**d**) MPC beam coupling constraints.

5. Validation of FE Model

The experimental setup and a typical loading photograph are presented in Figure 4. To validate the accuracy of the finite element (FE) model, Figure 5a–c depict the comparison between the numerical results and the experimental outcomes regarding the horizontal hysteretic performance of the bearing under vertical loads of 1000 kN, 1200 kN, and 1500 kN, respectively. In each loading test, the maximum horizontal displacement is set to 98 mm for three cycles [43], with a loading frequency of 0.01 Hz. The test data shown in Figure 5 correspond to the third cycle of the compression–shear tests. Further analysis of Figure 5a reveals that the equivalent stiffness and peak force at 98 mm derived from the simulation are 7.8% and 13.55% higher than the experimental values, respectively. In Figure 5b, these differences are 2.9% and 10.92%, respectively. Similarly, in Figure 5c, the simulated equivalent stiffness and peak force exceed the experimental results by 2.9% and 8.2%, respectively.

Additionally, Figure 5d illustrates the hysteresis behavior of the bearing at various horizontal shear strain levels: 50%, 75%, 100% (i.e., 98 mm), 125%, and 150%, under a constant vertical load of 1000 kN. The comparison indicates that the numerical results are in good agreement with the experimental data. Therefore, it can be concluded that the finite element model accurately predicts the horizontal hysteretic behavior of the bearing.

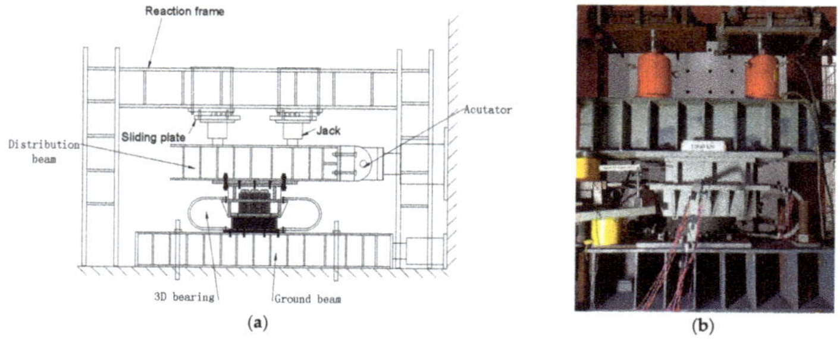

Figure 4. Experimental setup and typical photograph: (**a**) Experimental setup; (**b**) Typical compression—shear test at 100% strain level under vertical load of 1000 kN.

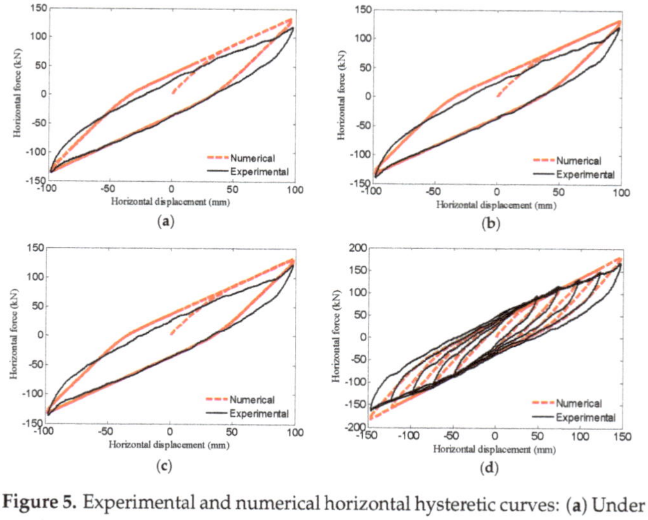

Figure 5. Experimental and numerical horizontal hysteretic curves: (**a**) Under vertical load of 1000 kN; (**b**) Under vertical load of 1200 kN; (**c**) Under vertical load of 1500 kN; (**d**) At five shear strain levels: 50%, 75%, 100%, 125%, 150% under vertical load of 1000 kN.

6. Horizontal Performance Analysis

To comprehensively evaluate the influence of vertical loads, shear strains, and the number of U-shaped dampers on the horizontal isolation behavior of the 3D bearing, numerical simulations involving 12 loading cases are conducted, as summarized in Table 5.

Table 5. Parametric analysis in horizontal direction.

Parameter Classification	Load Case Number	Constant Vertical Force (kN)	Horizontal Displacement (mm)	Equivalent Horizontal Stiffness (kN/mm)	Equivalent Horizontal Damping ratio (%)	Horizontal Natural Period (s)
Vertical load	1	1000	[−98, +98]	1.38	12.28	1.71
	2	1200	[−98, +98]	1.37	12.34	1.88
	3	1500	[−98, +98]	1.36	12.53	2.11
Shear strain	4	1000	[−47, +47]	1.71	12.06	1.53
	5	1000	[−74, +74]	1.47	13.27	1.66
	6	1000	[−98, +98]	1.38	12.28	1.71
	7	1000	[−123, +123]	1.28	11.41	1.77
	8	1000	[−147, +147]	1.24	10.38	1.80
0 damper	9	1000	[−98, +98]	0.99	2.35	2.02
1 damper	10	1000	[−98, +98]	1.38	12.28	1.71
2 dampers	11	1000	[−98, +98]	1.70	18.62	1.54
4 dampers	12	1000	[−98, +98]	2.42	25.05	1.29

Based on the numerical simulation results, the equivalent stiffness, equivalent damping ratio, and natural period of the 3D bearing in the horizontal direction can be further obtained. The formulas are as follows:

$$K_{\mathrm{eq,h}} = \frac{Q_1 - Q_2}{X + |-X|} \tag{4}$$

$$\zeta_{\mathrm{eq,h}} = \frac{W_{\mathrm{h}}}{2\pi K_h X^2} \tag{5}$$

$$T_{\mathrm{h}} = 2\pi\sqrt{\frac{F_{\mathrm{v}}}{g K_{\mathrm{h}}}} \tag{6}$$

where $K_{\mathrm{eq,h}}$, $\zeta_{\mathrm{eq,h}}$ and T_{h} denote the equivalent stiffness, equivalent damping ratio, and natural period of the 3D bearing in the horizontal direction, respectively; Q_1 and Q_2 represent the positive and negative peak shear forces; X and -X are the positive and negative shear displacements corresponding to Q_1 and Q_2, respectively; W_{h} denotes the area of the hysteresis loop in the horizontal load–displacement curve; F_{v} represents the applied vertical load.

6.1. Different Vertical Loads

To investigate the horizontal hysteresis behaviors of the 3D bearing under varying load capacities, vertical loads of 1000 kN, 1200 kN, and 1500 kN are individually applied to the bearing. A sinusoidal excitation with a frequency of 0.01 Hz is horizontally applied in a displacement-controlled manner.

Figure 6a presents the horizontal hysteretic curves of the bearing under different vertical loads. It can be observed that the variations in vertical loading result in minimal differences in the horizontal load–displacement characteristics, indicating that the bearing's horizontal hysteresis performance is not significantly affected by changes in the vertical load. This behavior is primarily due to the near-incompressibility of rubber and the reciprocating plastic shear deformation of the lead core. As the vertical load increases, the equivalent horizontal stiffness of the bearing exhibits a decreasing trend, as illustrated in Figure 6b. Regarding the horizontal equivalent damping ratio of LRBs, previous studies have indicated that when the pressure is below 10 MPa, the damping ratio increases linearly with pressure, while for pressures up to 15 MPa, the increase becomes nonlinear [44,45]. As shown in Figure 6c, the equivalent damping ratio increases with compressive stress, although the variation remains relatively small. This observation aligns with findings reported for LRBs. The natural period is closely related to the equivalent stiffness and is inversely proportional to the square root of the equivalent stiffness when the vertical load is constant. Consequently, its variation trend is evident, as demonstrated in Figure 6d.

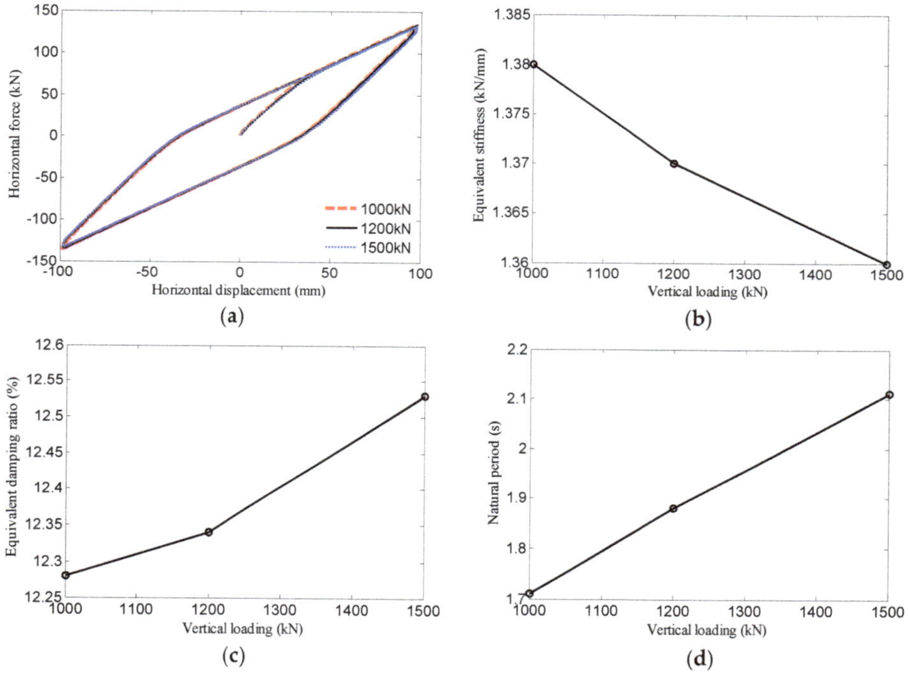

Figure 6. Horizontal performance of bearing under different vertical loads. (**a**) Hysteretic curves; (**b**) Equivalent stiffness; (**c**) Equivalent damping ratio; (**d**) Natural period.

6.2. Different Horizontal Shear Strains

To further investigate the horizontal hysteresis behaviors of the 3D bearing under different horizontal shear strains, the displacement-controlled loading method is utilized to simulate the behavior of the bearing under combined compressive and shear loading conditions. First, a vertical load of 1000 kN is applied, and then sinusoidal excitations are imposed incrementally. The selected peak shear strains are 50%, 75%, 100%, 125%, and 150%, respectively.

As shown in Figure 7a, under moderate shear strain conditions, the load–displacement curve of the bearing exhibits a fusiform shape without pinching or noticeable degradation in strength and stiffness. This phenomenon can be attributed to the contribution of the U-shaped dampers. Figure 7b illustrates that with increase in the shear strain, the equivalent horizontal stiffness of the bearing shows a decreasing trend. The reason is that the rubber molecular chains gradually untangle and align directionally, leading to softening, and the lead core constantly loses its ability to contribute stiffness. Figure 7c presents the variation law that as the shear strain changes from 50% to 150%, the equivalent damping ratio of the bearing changes from 12.06% to the maximum value of 13.27% when the shear strain is 75%, and then continuously decreases to 10.38%. This trend can be explained such that as the shear strain increases, the rubber, lead core, and U-shaped dampers transition from the elastic stage to the plastic stage and subsequently enter the hardening stage. This process initially leads to an increase in energy dissipation capacity, followed by a subsequent decrease.

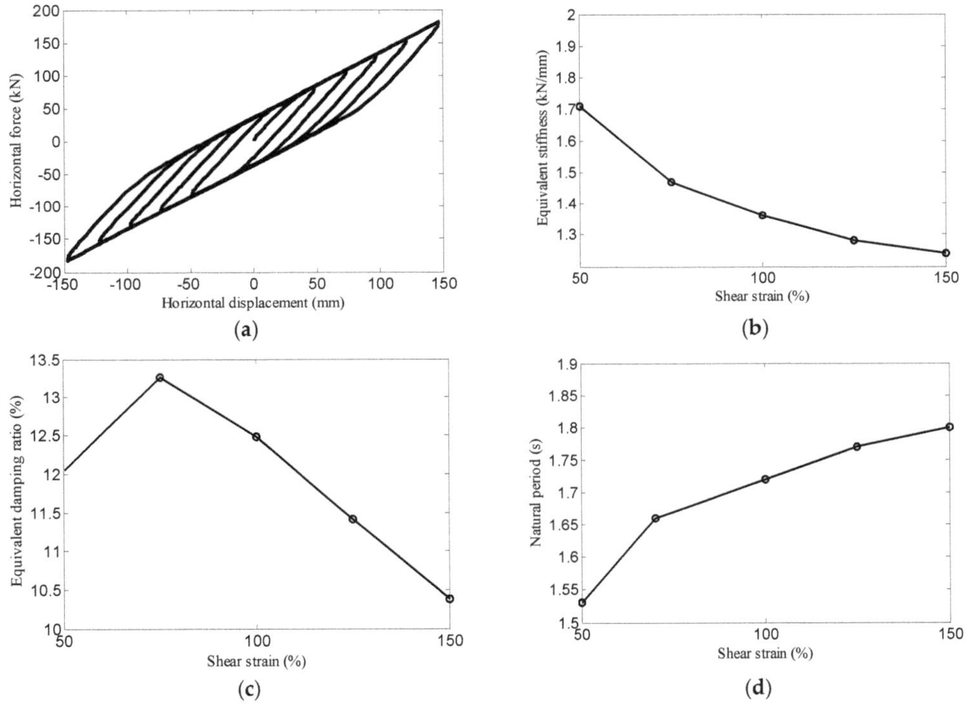

Figure 7. Horizontal performance of bearing under different shear strains. (**a**) Hysteretic curve; (**b**) Equivalent stiffness; (**c**) Equivalent damping ratio; (**d**) Natural period.

6.3. Different Numbers of U-Shaped Dampers

To further study the influence of the number of U-shaped dampers on the horizontal hysteresis performance of the bearings, compression–shear performance simulations are conducted for the bearings with different numbers of U-shaped dampers (see Figure 8), respectively. The loading is controlled by displacement, in sinusoidal form, with the peak shear strain taken as 100%. The vertical load of 1000 kN remains unchanged during the loading processes.

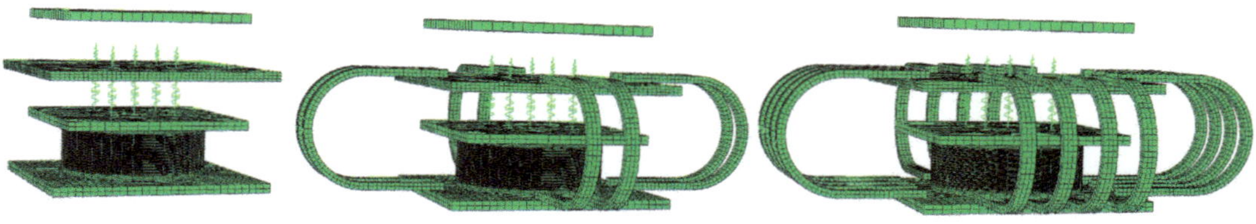

(**a**) (**b**) (**c**)

Figure 8. Finite element models with varying numbers of dampers on each side. (**a**) 0 damper. (**b**) 2 dampers. (**c**) 4 dampers.

Figure 9 presents the horizontal behaviors of the 3D bearings with varying numbers of U-shaped dampers. It can be observed that as the number of U-shaped dampers increases, the energy dissipation capacity, equivalent horizontal stiffness, and equivalent damping ratio of the bearing all increase. Therefore, the U-shaped damper significantly enhances the horizontal isolation performance of the bearing discussed in this study. Notably, considering that the lead core diameter (30 mm) used in this bearing is considerably smaller than that of conventional lead rubber bearings (LRBs), the energy dissipation capacity of the bearing without U-shaped dampers is markedly limited in the horizontal direction, as demonstrated in Figure 9a. It is obvious that the addition of U-shaped dampers can partially substitute for the lead core. In this way, not only is the energy dissipation capacity of the bearing effectively improved, but also an environmentally sustainable design objective is achieved.

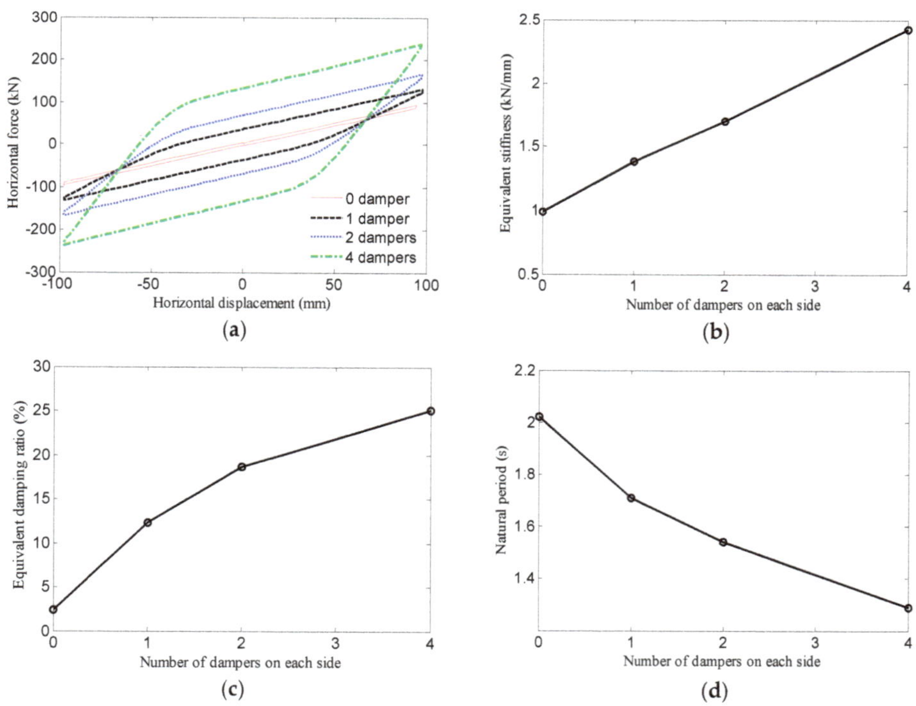

Figure 9. Horizontal Performance of bearing with different numbers of U-shaped dampers. (**a**) Hysteretic curves; (**b**) Equivalent stiffness; (**c**) Equivalent damping ratio; (**d**) Natural period.

7. Vertical Performance Analysis

To explore the vertical isolation performance of the novel device, numerical simulations under 8 loading cases are carried out, as shown in Table 6.

Table 6. Parametric analysis in the vertical direction.

Parameter Classification	Load Case Number	Vertical Displacement (mm)	Equivalent Vertical Stiffness (kN/mm)	Equivalent Vertical Damping Ratio (%)	Vertical Vibration Period (s)
Vertical displacement	1	[−1, +1]	232.68	9.59	0.132
	2	[−2, +2]	220.80	11.53	0.146
	3	[−4, +4]	229.34	10.36	0.137
	4	[−6, +6]	237.66	9.34	0.130
0 damper	5	[−6, +6]	227.36	9.08	0.133
1 damper	6	[−6, +6]	237.66	9.34	0.130
2 dampers	7	[−6, +6]	240.49	10.55	0.129
4 dampers	8	[−6, +6]	252.72	9.65	0.126

Similarly, we can also obtain the equivalent stiffness, equivalent damping ratio, and natural vibration period of the bearings in the vertical direction according to the following formulas,

$$K_{\mathrm{eq,v}} = \frac{P_1 - P_2}{Y + |-Y|} \tag{7}$$

$$\zeta_{\mathrm{eq,v}} = \frac{W_{\mathrm{v}}}{2\pi K_v Y^2} \tag{8}$$

$$T_{\mathrm{v}} = 2\pi\sqrt{\frac{F_{\mathrm{v}}}{gK_{\mathrm{v}}}} \tag{9}$$

where $K_{\mathrm{eq,v}}$, $\zeta_{\mathrm{eq,v}}$, and T_{v} are the equivalent vertical stiffness, equivalent damping ratio, and vertical natural period of the bearing; P_1 and P_2 are the maximum and minimum vertical forces; Y and -Y are the positive and negative vertical displacements corresponding to P_1 and P_2, respectively; W_{v} denotes the area enclosed by the vertical load–displacement hysteresis loop; F_{v} refers to the vertical loading force.

7.1. Different Vertical Displacements

Figure 10a displays the vertical hysteretic curve of the bearing, which exhibits a pronounced asymmetry. This characteristic arises since when the bearing is under tension, only spring group 1 is compressed, and when the bearing is under compression only spring group 2 is compressed. Moreover, spring group 1 and spring group 2 have different compressive stiffness. As shown in Figure 10b, with increasing amplitude of the tension–compression cycles, the equivalent stiffness of the bearing initially decreases and then increases. This phenomenon can be attributed to the transition of the rubber and lead core from the elastic phase to the yielding phase, combined with the hardening effect exhibited by the U-shaped dampers under large displacements. For similar reasons, the energy dissipation capacity in the vertical direction first increases and then decreases, as illustrated in Figure 10c.

Figure 11 presents the load–displacement hysteresis curves of spring groups 1 and 2 during the vertical tension–compression cyclic loading process of the bearing. The following observations can be made: (1) The mechanical behavior of the spring groups during the tension–compression cycles aligns with the intended design expectations. (2) Throughout the tension–compression cycles, the displacements of the spring groups remain lower than the overall vertical displacement of the bearing. For instance, when the bearing undergoes a tensile displacement of 6 mm, the displacements of the central and outer springs within group 1 are only 1.33 mm and 1.57 mm, respectively. This indicates that the vertical displacement of the LRB component ranges between 4.43 mm and 4.67 mm, thereby contributing to an enhancement in the tensile performance of the bearing. (3) Owing to the deformation of the middle plate, the reaction force generated in the central spring is always less than that generated in the outer springs.

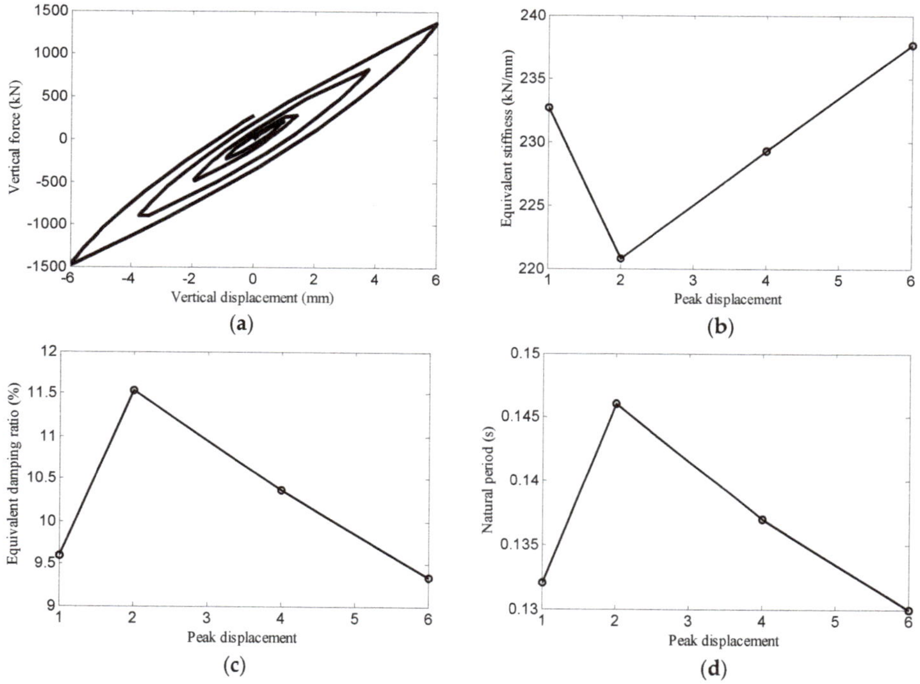

Figure 10. Vertical performance of bearing under different displacement amplitudes. (**a**) Hysteretic curve; (**b**) Equivalent stiffness; (**c**) Equivalent damping ratio; (**d**) Natural period.

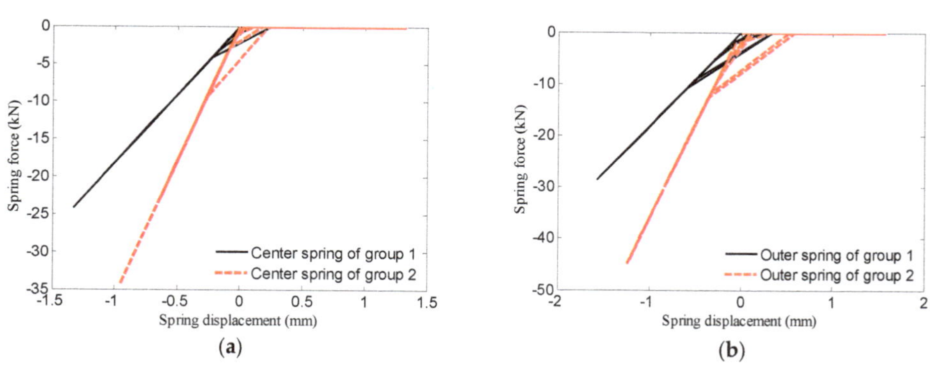

Figure 11. Vertical hysteretic curves of springs. (**a**) Central springs; (**b**) Outer springs.

Figure 12 shows the stress distributions of the inner rubber layers at different positive peak displacements of the tension–compression cyclic loading process of the bearing. It can be seen that the stress distribution has obvious symmetry, that is, the stress is the greatest at the center and gradually decreases along the radial direction. To precisely control the tensile stress on the rubber layer, careful design is needed.

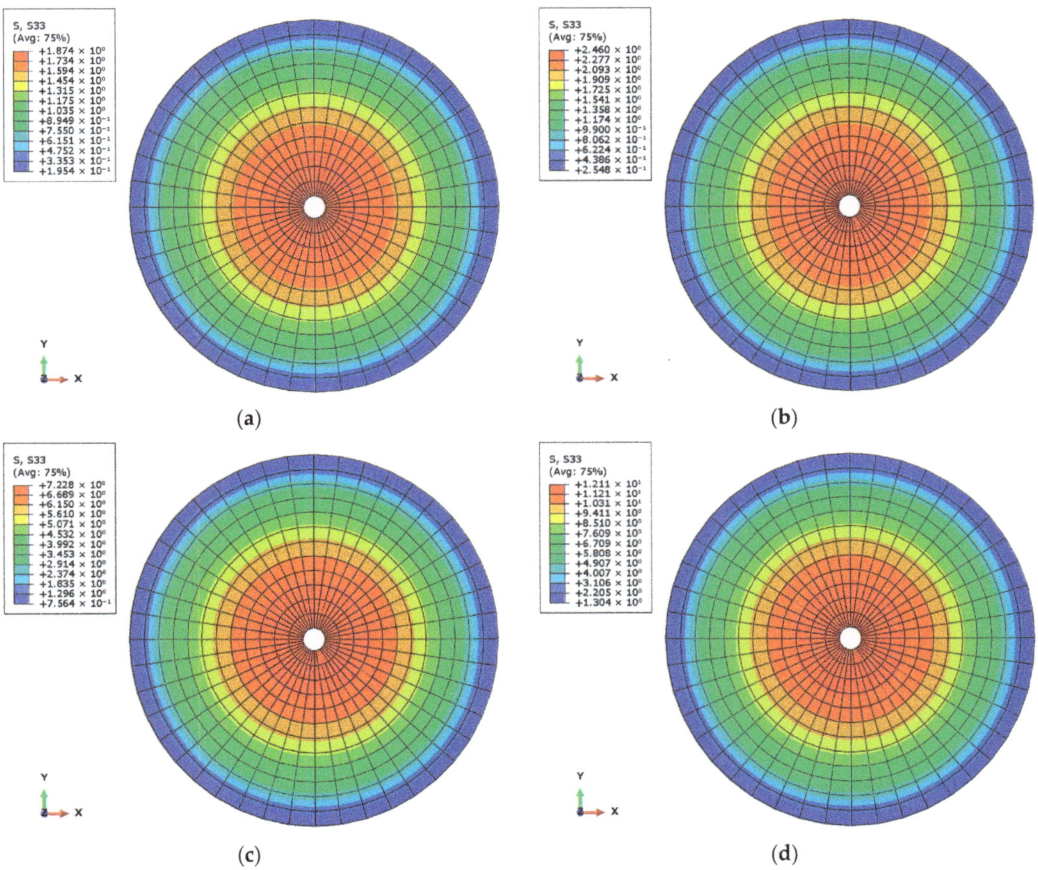

Figure 12. Stress distributions of rubber layers under different positive displacement amplitudes. (**a**) +1 mm; (**b**) +2 mm; (**c**) +4 mm; (**d**) +6 mm.

7.2. Different U-Shaped Dampers

Figure 13 illustrates the vertical isolation performance of bearings equipped with varying numbers of U-shaped dampers. As shown in Figure 13a, the energy dissipation capacity does not significantly change with an increasing number of U-shaped dampers, indicating that these dampers have a limited effect on enhancing the vertical isolation performance of the bearings. In Figure 13b, it can be observed that the equivalent vertical stiffness of the bearings continues to increase as the number of dampers increases, which is primarily due to the initial stiffness provided by the U-shaped dampers. According to Figure 13c, the impact of the U-shaped dampers on the equivalent damping ratio of the bearings is relatively complex. Initially, increasing the number of dampers enhances the energy dissipation mechanism of the system, specifically through the plastic deformation of the U-shaped dampers. However, when the number of dampers becomes excessive, they remain predominantly in the elastic state, thereby reducing the overall energy dissipation capacity of the bearings.

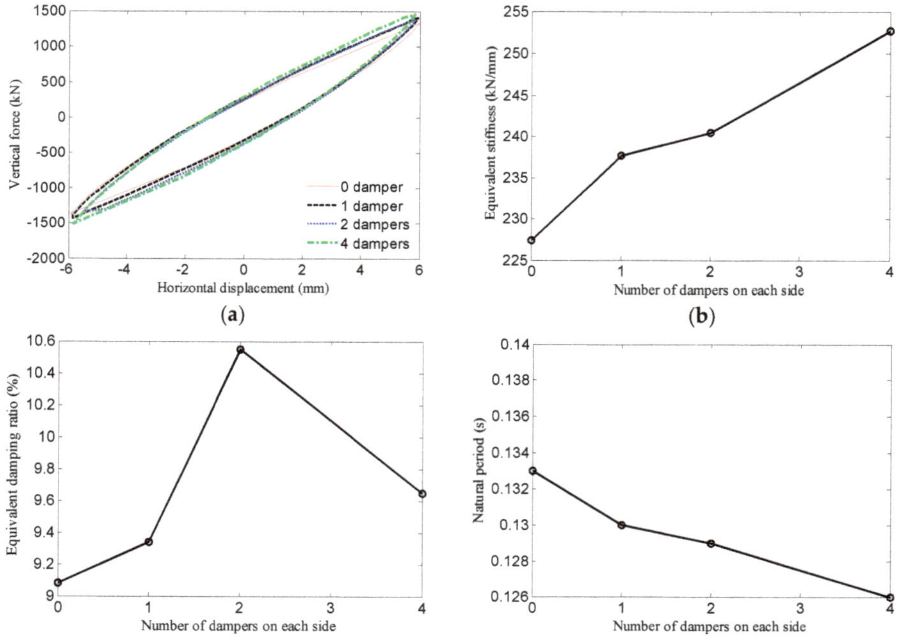

Figure 13. Vertical performance of bearing with different numbers of U-shaped dampers. (**a**) Hysteretic curves; (**b**) Equivalent stiffness; (**c**) Equivalent damping ratio; (**d**) Natural period.

8. Conclusions

This study proposes a novel 3D isolation bearing and develops its finite element model, followed by a series of simulation analyses. The main contributions are summarized as follows:

(1) A finite element model of the 3D isolation bearing is established. Given the structural complexity of the bearing, certain components, such as the stiffened sleeves, disc springs, and support shafts, are simplified during the modeling process. The validity of the model is confirmed through comparison with existing experimental data. The results indicate that the finite element model can accurately predict the mechanical behavior of the bearing.

(2) In terms of load-bearing capacity, the numerical analysis demonstrates that the proposed bearing provides stable load-carrying performance. The distribution of vertical loads between the LRB and U-shaped dampers can be flexibly adjusted by varying the number of dampers.

(3) Under compressive-shear loading conditions, the bearing exhibits reliable energy dissipation characteristics. Although the small lead core design imposes limitations on energy dissipation, increasing the number of U-shaped dampers significantly enhances this capability. Specifically, compared with the bearing without U-shaped dampers, the energy dissipation capacity of the bearing increases by 628%, 1300%, and 2581% when employing 1, 2, and 4 dampers on each side, respectively.

(4) With respect to vertical performance, the numerical results show that when the bearing is subjected to a tensile displacement of 6 mm, the displacement of the LRB remains within the range of 4.43 to 4.67 mm. This indicates that the disc spring group design effectively mitigates the tensile deformation of the LRB, thereby improving

its tensile resistance. Additionally, the U-shaped dampers contribute minimally to vertical energy dissipation compared to their impact on the vertical stiffness and load-bearing capacity.

(5) This study primarily employs finite element analysis to conduct a preliminary investigation into the mechanical properties of a new 3D bearing. To facilitate further research and practical application, future studies should include vertical performance testing, shaking table experiments, and theoretical analysis, which are urgently needed.

Author Contributions: Conceptualization, J.L. and L.S.; Methodology, Y.W., Y.C. and D.Q.; Software, J.L. and T.L.; Writing—original draft, T.L.; Writing—review and editing, J.L., Y.C., D.Q. and S.D. All authors have read and agreed to the published version of the manuscript.

Funding: This research received the support of the Natural Foundation of Shaanxi Province, China (No. 2025 JC-YBMS-535; No. 2014 JM7278).

Data Availability Statement: Data are contained within the article.

Conflicts of Interest: Authors Jianjun Li, Lvhong Sun, and Yanchao Wu were employed by the company CCCC First Highway Northwest Engineering Co., Ltd. The remaining authors declare that the research was conducted in the absence of any commercial or financial relationships that could be construed as a potential conflict of interest.

References

1. Pan, P.; Zeng, Y.; Cao, Y.R.; Ai, H.H.; Wang, H.S. State-of-the-art of research on the building structure isolation technologies. *Eng. Mech.* **2024**, *41*, 39–54. (In Chinese)
2. Collier, C.J.; Elnashai, A.S. A procedure for combining vertical and horizontal seismic action effects. *J. Earth Eng.* **2001**, *5*, 521–539. [CrossRef]
3. Li, N.; Liu, H.G.; Liu, P.; Li, Z.X.; Xie, L.L. Statistical analysis of vertical ground motion characteristics in near-fault regions. *Chin. Civ. Eng. J.* **2020**, *53*, 120–128. (In Chinese)
4. Seigenthaler, R. Earthquake-proof building supporting structure with shock absorbing damping elements. *Schweiz. Bauztg.* **1970**, *20*, 211–219.
5. Aiken, I.D.; Kelly, J.M.; Tajirian, F.F. *Mechanics of Low Shape Factor Elastomeric Seismic Isolation Bearings*; Report No. UCB/EERC-89/13; University of California: Berkeley, CA, USA, 1989.
6. Lu, Z.; Liang, Q.Y.; Zhou, Y.; Luo, W.L.; Li, J.R.; Jiang, K. Novel thick layer damping rubber bearing (TLDRB) with reduced vertical stiffness: Laboratory tests and mechanical models. *J. Build. Struct.* **2024**, *94*, 109839. [CrossRef]
7. Huffmann, G. Full base isolation for earthquake protection by helical springs and viscodampers. *Nucl. Eng. Des.* **1985**, *84*, 331–338. [CrossRef]
8. Fujita, S.; Minagawa, K.; Tanaka, G.; Shimosaka, H. Intelligent seismic isolation system using air bearings and earthquake early warning. *Soil Dyn. Earthq. Eng.* **2011**, *31*, 223–230. [CrossRef]
9. Mo, Z.; Lai, B.L.; Shu, G.P.; Yang, T.Y.; Ventura, C.E.; Richard Liew, J.Y. Enhancing seismic resilience in modular steel building through three-dimensional isolation. *Eng. Struct.* **2025**, *323*, 119269. [CrossRef]
10. Zhou, Z.G.; Wong, J.; Mahin, S. Potentiality of using vertical and tree-dimensional isolation systems in nuclear structures. *Nucl. Eng. Technol.* **2016**, *48*, 1237–1251. [CrossRef]
11. Fujita, S.; Kato, E.; Kashiwazaki, A.; Shimoda, I.; Sasaki, K. Shake table tests on three-dimensional vibration isolation system comprising rubber bearing and oil spring. In Proceedings of the 11th World Conference on Earthquake Engineering, Acapulco, Mexico, 23–28 June 1996; pp. 23–28.
12. Xiong, S.S. Theoretical and Experimental Research of Three-Dimensional Seismic Base Isolation System. Ph.D. Thesis, Huazhong University of Science and Technology, Wuhan, China, 2004. (In Chinese).
13. Zhao, Y.M.; Su, J.Y.; Lu, M. Experimental study on three-dimensional base-isolated model with 3DIB. *Adv. Mater. Res.* **2011**, *255–260*, 2325–2329. [CrossRef]
14. Yan, X.Y.; Zhang, Y.S.; Wang, H.D.; Wei, L.S. Experimental study on high-rise structure with three-dimensional base isolation and overturn resistance devices. *J. Build. Struct.* **2009**, *30*, 4–11. (In Chinese)
15. Xue, S.D.; Chang, L.H.; Li, X.Y.; Liang, S.Z. Research on mechanical characteristics of anti-extraction three-dimensional seismic isolation bearing. *Build. Struct.* **2022**, *52*, 81–87. (In Chinese)

16. Jia, J.F.; Ou, J.P.; Liu, M.; Zhang, Z. Mechanical performance tests of a novel three-dimensional isolation bearing. *J. Civ. Archit. Environ. Eng.* **2012**, *34*, 29–34, 53. (In Chinese)
17. Liu, H.Q. SMA Strands-Laminated Rubber Compound Bearing and Its Seismic Isolation Behavior. Ph.D. Thesis, Tianjin University, Tianjin, China, 2006. (In Chinese)
18. Kashiwazaki, A.; Tanaka, M.; Tokuda, N. Shaking test of seismic isolation floor system by using 3-dimensional isolator. In Proceedings of the Ninth World Conference on Earthquake Engineering, Tokyo and Kyoto, Japan, 2–9 August 1988; pp. 845–850.
19. Shu, G.P.; Lai, B.L.; Ventura, C.E.; Yang, T.Y. Experimental behavior, numerical modeling and parameters evaluation of three-dimensional seismic isolation device. *J. Constr. Steel Res.* **2024**, *220*, 108817. [CrossRef]
20. Liu, W.G.; Xu, H.; He, W.F.; Yang, Q.R. Static test and seismic dynamic response of an innovative 3D seismic isolation system. *J. Eng. Struct.* **2018**, *144*, 04018212. [CrossRef]
21. Xu, H.; He, W.F.; Zhang, L.L.; Liu, W.G. Shaking table test of an novel three-dimensional seismic isolation system with inclined rubber bearings. *Eng. Struct.* **2023**, *293*, 116609. [CrossRef]
22. Chen, Z.; Ding, Y.; Shi, Y. A vertical isolation device with variable stiffness for long-span spatial structures. *Soil Dyn. Earthq. Eng.* **2019**, *123*, 543–558. [CrossRef]
23. Liang, Q.H.; Luo, W.L.; Zhou, Y.; Ke, X.B.; Li, J.R. Seismic performance of a novel three-dimensional isolation bearing. *J. Build. Struct.* **2022**, *57*, 104818. [CrossRef]
24. Wei, Y.Y.; Wang, X.P.; Wu, H.B.; Liu, B.; Zhu, N.H. A new sliding three-dimensional seismic isolation bearing mechanical model and its seismic performance. *Structures* **2025**, *72*, 108216. [CrossRef]
25. Luo, W.L.; Liang, Q.H.; Zhou, Y.; Lu, Z.C.; Li, J.R.; He, Z.M. A novel three-dimensional isolation bearing for buildings subject to metro- and earthquake-induced vibration: Laboratory test and application. *J. Build. Eng.* **2024**, *86*, 108798. [CrossRef]
26. Gao, Y. Design and Research of a Three-Dimensional Seismic Isolation Bearing. Master's Thesis, Taiyuan University of Technology, Taiyuan, China, 2019. (In Chinese)
27. Ge, G.Y. Study on the Mechanical Properties and Isolation Performance of Three-Dimensional Isolation Support. Master's Thesis, Xihua University, Chengdu, China, 2006. (In Chinese)
28. Han, Q.; Jing, M.; Lu, Y.; Liu, M. Mechanical behaviors of air spring-FPS three-dimensional isolation bearing and isolation performance analysis. *Soil Dyn. Earthq. Eng.* **2021**, *149*, 106872. [CrossRef]
29. Cao, Y.; Pan, P.; Sun, J.; Wang, H. Mechanical properties and isolation effect of disc spring-single friction pendulum 3D vibration isolation device. *J. Build. Struct.* **2022**, *7*, 44–53.
30. Cao, Y.R.; Pan, P.; Wang, H.S.; Sun, J.B.; Xiao, G.Q.; Zuo, Z.F. Development of an innovative three-dimensional vibration isolation bearing. *Eng. Struct.* **2023**, *295*, 116890. [CrossRef]
31. Kitayama, S.; Lee, D.H.; Constantinou, M.C.; Kempner, L., Jr. Probabilistic seismic assessment of seismically isolated electrical transformers considering vertical isolation and vertical ground motion. *Eng. Struct.* **2017**, *152*, 888–900. [CrossRef]
32. Zhou, Y.; Chen, P.; Mosqueda, G. Numerical studies of Three-Dimensional isolated structures with vertical Quasi-Zero stiffness property. *J. Earthq. Eng.* **2022**, *26*, 3601–3622. [CrossRef]
33. Yu, T.H.; Zhang, C.; Huang, Z.Q.; Huang, W.Y.; Wang, S.Y.; Zhong, G.Q.; Qu, D.T. Experimental and numerical studies of a novel three-dimensional isolation device incorporating disc springs with U-shaped dampers. *Soil Dyn. Earthq. Eng.* **2023**, *174*, 108164. [CrossRef]
34. Yu, T.H.; Huang, Z.Q.; Zhang, C.; Huang, W.Y.; Bao, W.; Liu, Y.Y. Numerical investigation of three-dimensional isolator and mitigation for single-layer lattice shell structure. *Structures* **2024**, *62*, 106235. [CrossRef]
35. Sha, M.; Guan, M.S.; Lin, M.J.; Tan, X.Y.; Chen, X.S. Experimental study on the performance of innovative self-centering three-dimensional isolation bearing. *Structures* **2024**, *69*, 107587. [CrossRef]
36. *GB/T 1972.1-2023*; Disc Spring—Part 1: Calculation. China National Standards: Beijing, China, 2023. (In Chinese)
37. *GB/T 1972.2-2023*; Disc Spring—Part 2: Technical Specifications. China National Standards: Beijing, China, 2023. (In Chinese)
38. Mu, J.J. Design and Mechanical Property Research of a New Three-Dimensional Seismic Isolation Bearing. Master's Thesis, Hainan University, Haikou, China, 2024. (In Chinese)
39. ABAQUS. *ABAQUS Standard User's Manual*; Version 6.12; Dassault Systems Simulia Corporation: Providence, RI, USA, 2012.
40. Xing, J.H.; Huang, H.; Zhang, J.Y.; Yang, Q.S. Mechnical properties of disc springs. *J. Vib. Shock* **2015**, *34*, 167–172. (In Chinese)
41. Bagheri, M.; Jamkhaneh, M.E.; Samali, B. Effect of Seismic Soil-Pile-Structure Interaction on Mid- and High-Rise Steel Buildings Resting on a Group of Pile Foundations. *Int. J. Geomech.* **2018**, *18*, 04018103. [CrossRef]
42. Asgari, A.; Ranjbar, F.; Bagheri, M. Seismic resilience of pile groups to lateral spreading in liquefiable soils: 3D parallel finite element modeling. *Structures* **2025**, *74*, 108578. [CrossRef]
43. *GB/T 20688.1-2007*; Rubber Bear—Part 1: Seismic-Protection Isolators Test Methods. China National Standards: Beijing, China, 2007. (In Chinese)

44. Benzoni, G.; Casarotti, C. Effects of vertical load, strain rate and cycling on the response of Lead-Rubber seismic isolators. *J. Earthq. Eng.* **2009**, *13*, 293–312. [CrossRef]
45. Kircher, C. Chapter 12: Seismically Isolated Structures. In *FEMA P-751, 2009 NEHRP Recommeded Provisions: Design Examples;* BSSC: Washington, DC, USA, 2012; pp. 1–63.

buildings

MDPI

Article

Seismic Performance of Tall-Pier Girder Bridge with Novel Transverse Steel Dampers Under Near-Fault Ground Motions

Ziang Pan [1], Qiming Qi [1,*], Ruifeng Yu [1], Huaping Yang [1], Changjiang Shao [2] and Haomeng Cui [2]

1 School of Architecture and Civil Engineering, Chengdu University, Chengdu 610106, China; 212023085901011@cdu.edu.cn (Z.P.); 212023085901025@cdu.edu.cn (R.Y.); yanghuaping@cdu.edu.cn (H.Y.)
2 School of Civil Engineering, Southwest Jiaotong University, Chengdu 610031, China; shao_chj@swjtu.edu.cn (C.S.); haomeng_cui@my.swjtu.edu.cn (H.C.)
* Correspondence: qiqiming@cdu.edu.cn

Abstract: This study develops a novel transverse steel damper (TSD) to enhance the seismic performance of tall-pier girder bridges, featuring superior lateral strength and energy dissipation capacity. The TSD's design and arrangement are presented, with its hysteretic behavior simulated in ABAQUS. Key parameters (yield strength: 3000 kN; initial gap: 100 mm; post-yield stiffness ratio: 15%) are optimized through seismic analysis under near-fault ground motions, incorporating pulse characteristic investigations. The optimized TSD effectively reduces bearing displacements and results in smaller pier top displacements and internal forces compared to the bridge with fixed bearings. Due to the higher-order mode effects, there is no direct correlation between top displacements and bottom internal forces. As pier height decreases, the S-shaped shear force and bending moment envelopes gradually become linear, reflecting the reduced influence of these modes. Medium- to long-period pulse-like motions amplify seismic responses due to resonance (pulse period $\approx$ fundamental period) or susceptibility to large low-frequency spectral values. Higher-order mode effects on bending moments and shear forces intensify under prominent high-frequency components. However, the main velocity pulse typically masks the influence of high-order modes by the overwhelming seismic responses due to large spectral values at medium to long periods.

Keywords: tall-pier girder bridge; seismic performance; near-fault ground motions; transverse steel dampers

1. Introduction

In recent years, numerous highway bridges have been constructed across Western China, with many piers exceeding 40 m and some even surpassing 100 m [1]. Due to their ability to span wide canyons and steep terrain, tall-pier girder bridges have been widely used in mountainous regions [2]. However, these bridges are often near active faults and face severe challenges from near-fault pulse-like ground motions [3]. Earthquakes such as the 2008 Wenchuan M8.0 [4], 2010 Yushu M7.1 [5], 2013 Lushan M7.0 [6], and 2017 Jiuzhaigou [7] events have revealed serious threats to the seismic safety of tall-pier bridges.

To improve seismic performance of bridges under earthquakes, many strategies have been employed. For instance, isolation bearings, such as high-damping rubber bearing (HDRB) [8,9], lead–rubber bearing (LRB) [10–12], and friction pendulum bearing (FPB) [13–15] have been widely used to reduce seismic responses of bridges. The isolation principle of these bearings is to elongate the fundamental period of the structure while absorbing earthquake energy [16]. Earthquake-induced inertial forces transmitted to the substructure from the superstructure can be reduced, while displacement responses are often enlarged. Fluid viscous damper (FVD) is an alternative device popularly used in long-span bridges [17–19], which upgrades seismic performance by dissipating earthquake energy and limiting bearing displacement. Usually, these isolation bearings and FVDs are effective for improving longitudinal seismic performance of bridges, but their suitability for reducing transverse seismic responses is insufficient. Specifically, elongating the transverse fundamental period of tall pier bridges is problematic due to the risk of falling girders. In addition, there is often little space to arrange FVDs in the transverse direction and allow them to function fully in tall pier bridges.

Currently, the combination of isolation bearings and shear keys is the main solution for the transverse seismic design of bridges. Reinforced concrete shear keys (RCSKs) are widely used in abutments and pier caps, which can serve as structural fuses to limit the inertial force transmitted to the substructure [20,21]. However, the strength of RCSKs degrades quickly after concrete cracking and rebar yielding, and they are not easy to repair once damaged. To address this issue, various steel shear keys have been developed. Zhou et al. [22] investigated the low cycle fatigue performance of a metallic hysteretic damper. Yue et al. [23] proposed a novel shear key and developed an analytical model based on experimental mechanical behavior. Huang et al. [24] introduced an arc-shaped shear key capable of effectively limiting girder displacement and enhancing energy dissipation.

These shear keys have been applied in various bridges. Bi et al. [25] compared the seismic responses of simply supported beam bridges with and without shear keys through numerical simulations. Xiang et al. [26] suggested setting the shear key strength at 50% to 60% of the pier yield strength for optimal control. Özşahin et al. [27] studied the appropriate gap between shear keys and superstructures in simply supported beam bridges, finding that reasonable gap control helps suppress excessive deck rotation and damage. Abbasi et al. [28] explored the effects of shear keys on seismic responses in different frame bridges, which can control overall bridge behavior under moderate earthquakes. Wu [29] analyzed shear key force distribution in skewed bridges, recommending increased gaps to reduce stress and proposing a simplified design method for low seismic zones. Shen et al. [30] validated the effectiveness of transverse steel dampers in controlling transverse seismic responses in suspension bridges under near-fault motions.

The aforementioned studies provide a useful reference for transverse seismic design of bridges. However, though existing steel dampers exhibit strength enhancement after yielding, they may not provide sufficient yield strength, plastic dissipated energy, and anti-torsion ability, since they are often composed of several parallel steel plates [22–24]. Moreover, existing publications concerning transverse shear keys mainly focus on cable-stayed bridges [22] and suspension bridges [30], as well as girder bridges with mid and short piers [26–29]. To date, little research has been conducted on seismic improvement using shear keys for tall-pier girder bridges. Compared to mid and short piers, inertial forces generated by mass distributed along tall piers are considerable and can exceed those induced by the superstructure [31]. Tall-pier girder bridges possess longer fundamental structural periods, potentially augmenting dynamic responses under pulse-like motions containing long-period velocity pulses [32–34]. Their seismic performance is notably influenced by higher-order modes [35], often resulting in mid-height plastic hinges within piers [36]. Therefore, it is important to investigate the seismic performance of tall-pier girder bridges equipped with shear keys under near-fault pulse-like ground motions.

In this study, a novel transverse steel damper (TSD) is proposed to improve the seismic performance of tall-pier girder bridges. First, the detailed design and implementation scheme of the TSD within the bridge system are described. The hysteretic behavior of the TSD is simulated using ABAQUS 2024, and a simplified analytical model is established. Second, key TSD parameters are optimized based on the seismic responses of the bridge under near-fault ground motions. Third, the influence of pulse periods and velocity pulses on bridge seismic responses is explored. Finally, significant conclusions are summarized.

2. Analytical Model and Seismic Input

2.1. Overview of the Bridge

A multi-span tall-pier composite girder bridge is selected for this investigation, as shown in Figure 1a. The substructure consists of tall piers with variable hollow cross-sections, while the superstructure is a double-box prestressed steel-concrete composite girder. In this bridge, pier No. 2 (87.4 m) is classified as the tall pier, pier No. 4 (61.0 m) as the mid pier, and pier No. 1 (46.3 m) as the short pier. C30 concrete is used for the piles and pile caps, whereas C40 concrete is specified for the piers and pier caps.

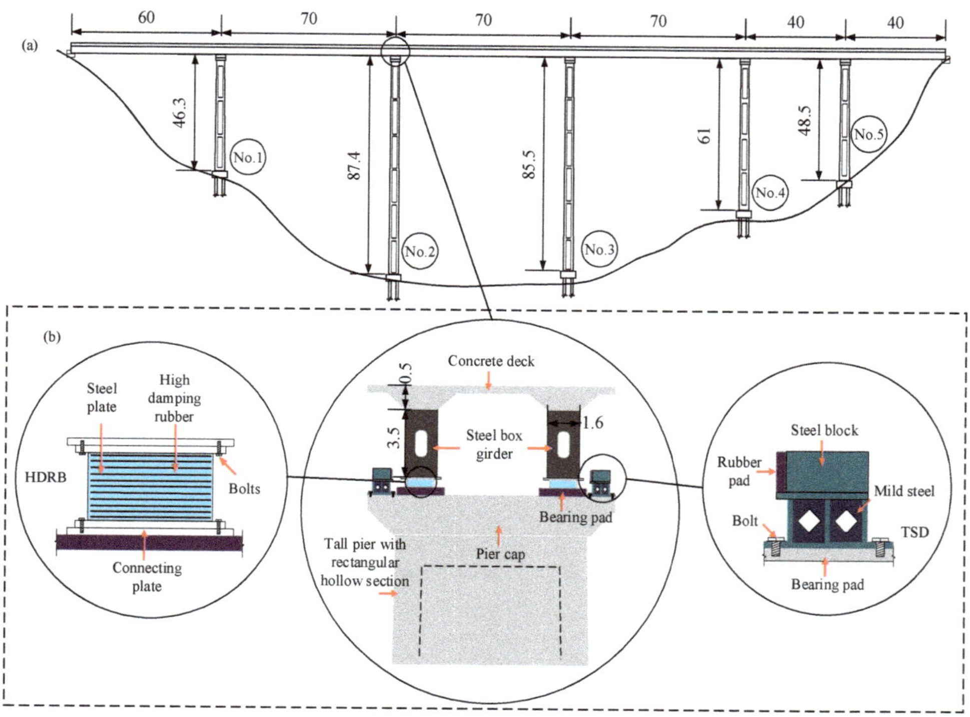

Figure 1. Basic information about the tall-pier girder bridge: (**a**) Layout of the bridge; (**b**) Arrangement of HDRBs and TSDs.

High damping rubber bearings (HDRBs) are installed between the concrete bearing pads and the bottom of the steel box girder. The detailed design parameters are presented in Table 1, according to the vertical load and the HDRB standard [37]. To restrict superstructure displacement responses under strong near-fault earthquakes and enhance the bridge's seismic performance, a novel transverse steel damper (TSD) is proposed. As illustrated in Figure 1b, the TSDs are positioned on both sides of each pier cap.

Table 1. Design parameters of the HDRB.

Bearing	Vertical Load Capacity (kN)	Compressive Stiffness (kN/mm)	Equivalent Damping Ratio (%)	Dimension (mm)		
				Transverse	Longitudinal	Height
HDRB	13,864	2713	15	1070	1070	414

2.2. Hysteretic Behavior of the TSD

The TSD is designed to meet the following objectives: (1) provide good displacement adaptability without compromising energy dissipation capacity, thereby accommodating the bridge's daily deformation requirements; (2) establish a clear and reliable horizontal load transfer path to ensure energy dissipation efficiency and safety under seismic loading; (3) facilitate straightforward modeling in finite element platforms for numerical analysis. Additionally, traditional X-shaped steel plate dampers are prone to torsional failure under in-plane torsional responses potentially induced by non-uniform motions.

To satisfy these requirements, a novel TSD is developed in this study (Figure 2). The device incorporates a hollow steel plate at its top to enhance overall yield strength and resistance. The core energy-dissipating component consists of three rectangular steel plates and six additional 8 mm slotted steel plates to enhance energy dissipation capacity. Based on this configuration, a detailed finite element model of the TSD is established using the ABAQUS platform and analyzed under cyclic loading. As shown in Figure 3a, the equivalent stress cloud indicates stress concentration primarily within the slotted steel plates, verifying the mechanical reliability of the TSD. The force–displacement hysteretic curve of the TSD is shown in Figure 3b, exhibiting full and symmetric loops that indicate excellent energy dissipation capacity. Consistent with the existing literature, the hysteretic behavior of the TSD can be characterized by a widely adopted bilinear model.

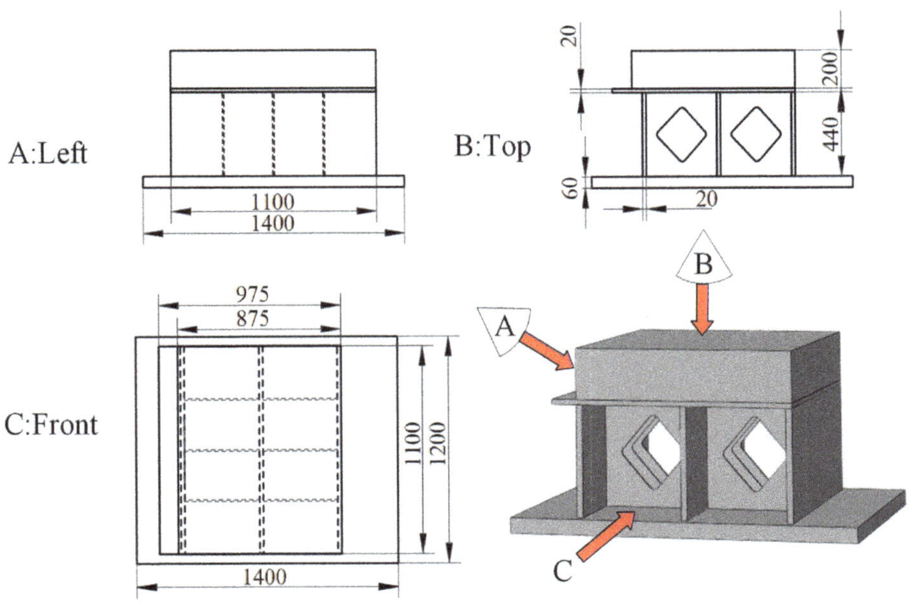

Figure 2. The configuration of the novel TSD.

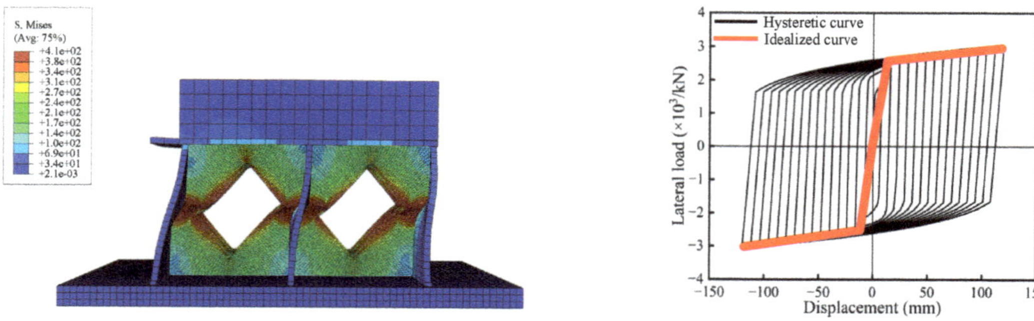

(**a**) (**b**)

Figure 3. Numerical simulation: (**a**) stress cloud of the TSD at 120 mm (unit: MPa); (**b**) hysteretic curves of the TSD.

To determine the optimal TSD parameters, a sensitivity analysis of seismic responses in the tall-pier girder bridge will be conducted in the next section. The specific parameters investigated are detailed in Table 2. Cases 1–18 examine the effect of yield strengths from 1000 kN to 6000 kN, with the initial gaps at 80 mm, 100 mm, and 120 mm and the post-yield stiffness ratio at 5%. Cases 19–36 evaluate the effect of initial gaps from 40 mm to 140 mm, with the yield strengths at 2000 kN, 3000 kN, and 4000 kN and the post-yield stiffness ratio at 5%. Cases 37–46 evaluate the effect of post-yield stiffness ratios from 5% to 25%, with the yield strengths at 3000 kN and the initial gaps at 80 mm and 120 mm. Furthermore, bridge models without TSD (the no-TSD model) and with fixed bearings (the fixed HDRB model) are also provided to check the effectiveness of the TSD.

Table 2. Parameters for the optimization of the TSD.

Case	Parameter	Yield Strength (kN)	Initial Gap (mm)	Post-Yield Stiffness Ratio (%)
1	Different yield strength	1000	80	5
2		2000	80	5
3		3000	80	5
4		4000	80	5
5		5000	80	5
6		6000	80	5
7		1000	100	5
8		2000	100	5
9		3000	100	5
10		4000	100	5
11		5000	100	5
12		6000	100	5
13		1000	120	5
14		2000	120	5
15		3000	120	5
16		4000	120	5
17		5000	120	5
18		6000	120	5

Table 2. *Cont.*

Case	Parameter	Yield Strength (kN)	Initial Gap (mm)	Post-Yield Stiffness Ratio (%)
19	Different initial gap	2000	40	5
20		2000	60	5
21		2000	80	5
22		2000	100	5
23		2000	120	5
24		2000	140	5
25		3000	40	5
26		3000	60	5
27		3000	80	5
28		3000	100	5
29		3000	120	5
30		3000	140	5
31		4000	40	5
32		4000	60	5
33		4000	80	5
34		4000	100	5
35		4000	120	5
36		4000	140	5
37	Different post-yield stiffness ratio	3000	80	5
38		3000	80	10
39		3000	80	15
40		3000	80	20
41		3000	80	25
42		3000	100	5
43		3000	100	10
44		3000	100	15
45		3000	100	20
46		3000	100	25

2.3. *Finite Element Analytical Model*

A three-dimensional (3D) finite element model of the tall-pier girder bridge is established in SAP2000, as shown in Figure 4a. The superstructure and substructure are modeled using 3D elastic frame elements. The girder's secondary dead load is applied as equivalent distribution mass. HDRBs and TSDs are modeled using a bilinear hysteretic rule (Figures 3b and 4b). The mechanical parameters of HDRBs are listed in Table 1. Pile–soil interaction is accounted for via the m-method (Figure 4c) [38], with equivalent spring stiffness calculated and calibrated using the distributed spring model. The values of m for tall piers are based on the bridge's geotechnical report and Specifications for Design of Foundation of Highway Bridges and Culverts [39].

2.4. *Selection of Ground Motions*

To investigate the influence of pulse characteristics on the transverse seismic performance of tall-pier girder bridges under near-fault earthquakes, a series of ground motions are selected from the NGA-West2 database. The primary information on the selected ground motions is shown in Table 3. The pulse-like record RSN 181 from the Imperial Valley-06 station is used for TSD parameter optimization.

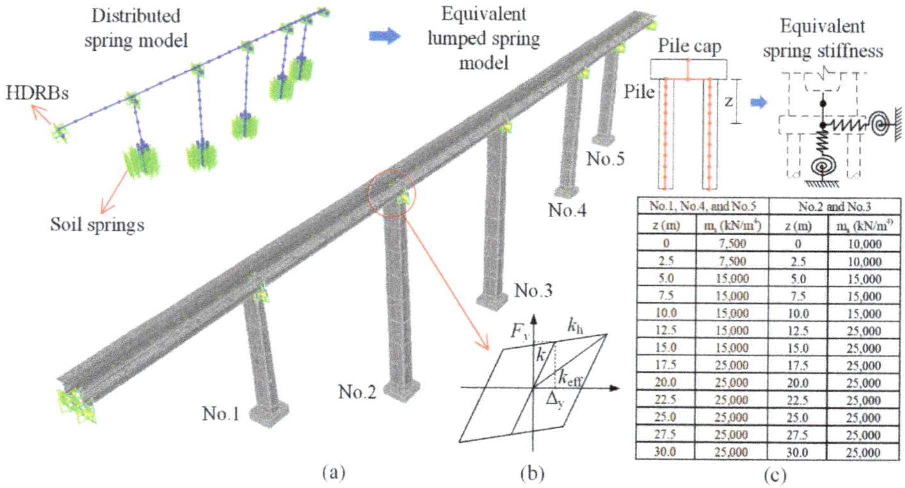

Figure 4. Analytical model: (**a**) FEM model; (**b**) bilinear hysteretic rule; (**c**) pile–soil interaction.

Table 3. The selected ground motions.

RSN	Earthquake	M_w	R_p (km)	PGA (m/s^2)	PGV (m/s)	T_g (s)	T_p (s)	Note
181	Imperial Valley-06	6.53	1.35	4.341	1.216	0.822	3.773	For the TSD optimization
2466	Chi-Chi Taiwan-03	6.2	3.52	2.005	0.3	0.822	1.057	Short-period (T_{pm} = 1.12 s)
2495	Chi-Chi Taiwan-03	6.2	22.37	4.565	0.698	1.072	1.379	
2627	Chi-Chi Taiwan-03	6.2	14.66	5.168	0.614	0.654	0.924	
2457	Chi-Chi Taiwan-03	6.2	19.65	1.838	0.327	0.919	3.185	Medium-period (T_{pm} = 2.959 s)
2734	Chi-Chi Taiwan-04	6.2	6.2	1.838	0.327	0.435	2.436	
3317	Chi-Chi Taiwan-06	6.3	35.97	1.264	0.342	1.426	3.255	
1244	Chi-Chi Taiwan	7.62	9.94	3.783	1.088	1.033	5.341	Long-period (T_{pm} = 5.208 s)
1475	Chi-Chi_ Taiwan	7.62	56.12	1.062	0.457	0.876	5.285	
1510	Chi-Chi_ Taiwan	7.62	0.89	3.045	1.048	1.033	4.998	
1513	Chi-Chi_ Taiwan	7.62	10.97	5.8067	0.7054	0.5422	-	Non-pulse
2370	Chi-Chi_ Taiwan-02	5.9	45.89	0.2989	0.0284	0.7067	-	
2381	Chi-Chi Taiwan-02	5.9	42.77	0.4352	0.0574	0.8263	-	

To specifically examine pulse period effects on seismic responses, four groups of near-fault records from the Chi-Chi Taiwan station are collected. Note that the bridge's fundamental transverse period is 3.06 s, and records RSN 2457, 2734, and 3317 (mean pulse period T_{pm} = 2.959 s) are chosen to represent medium-period motions. Using this group as a baseline, short-period motions (T_{pm} = 1.12 s) and long-period motions (T_{pm} = 5.208 s) are determined. In addition, three non-pulse records (RSN 1513, 2370, and 2381) are included for comparison. The acceleration response spectra of these records, scaled to a peak ground acceleration (PGA) of 1.0 g, along with their mean spectra, are shown in Figure 5. Reflecting the seismic intensity at the bridge site, a PGA of 0.3 g is used for both TSD optimization and pulse characteristic investigations.

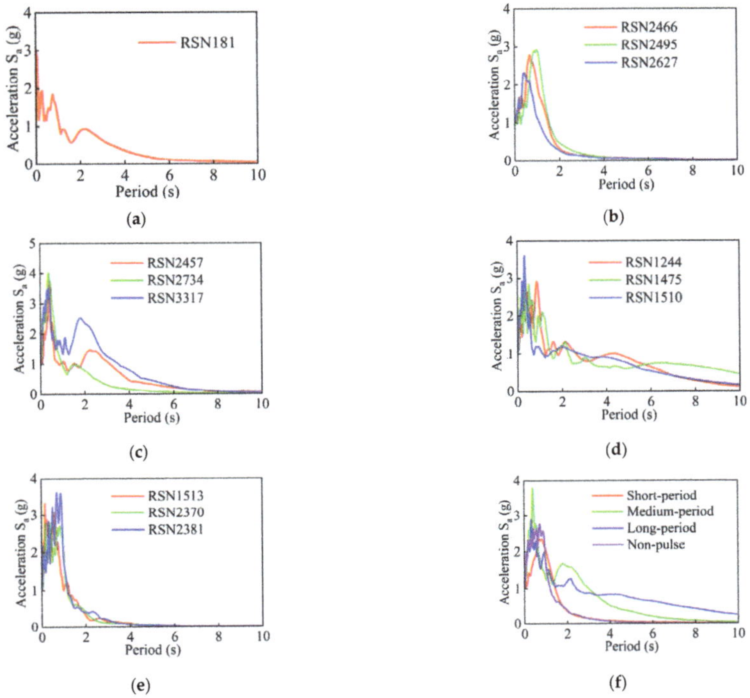

Figure 5. Acceleration response spectra: (**a**) record for TSD optimization; (**b**) short-period records; (**c**) medium-period records; (**d**) long-period records; (**e**) non-pulse records; (**f**) mean spectra.

3. Optimization of the TSD

3.1. Different Yield Strength Variations

Under strong earthquakes, excessive displacements may exceed bearing capacity limits, increasing girder unseating and structural failure risks. The TSD effectively controls maximum transverse bearing displacements, though its efficiency depends critically on the yield strength. Insufficient yield strength provides inadequate displacement resistance, while excessively high values amplify pier seismic responses. Parametric analyses are performed to determine the optimal TSD yield strength. Displacement responses of the tall pier with different yield strengths are shown in Figure 6. The force–displacement hysteretic curves of the TSD are shown in Figure 7. The corresponding bending moment and shear force at the pier bottom are shown in Figure 8.

As Figure 6a indicates, TSD displacement decreases with increasing yield strength, implying that a higher yield strength enhances the TSD's lateral resistance. Figure 6b shows that HDRB transverse displacement gradually decreases with increasing yield strength, with significant reductions compared to the no-TSD model. Figure 6c reveals that pier top displacement increases as yield strength increases and is larger than in the no-TSD model but smaller than in the fixed HDRB model. The opposite trend between Figure 6b,c indicates that enhancement of yield strength can effectively limit horizontal girder displacement but augment pier top displacement since more inertial forces are transmitted to the substructure. Moreover, HDRB displacement with an 80 mm initial gap is smallest when the yield strength ranges from 2000 kN to 6000 kN, yet produces the largest top displacement. Though a 120 mm initial gap results in the smallest pier top displacement, it yields the largest HDRB displacement when the yield strength exceeds

3000 kN. This is mainly because the strength enhancement of the TSD occurs earlier at a smaller initial gap and dissipates less earthquake energy, as shown in Figure 6c. Figure 7 indicates that higher yield strength reduces TSD energy dissipation capability. Nearly linear hysteretic behaviors are observed when the yield strengths are 4000 kN and 6000 kN.

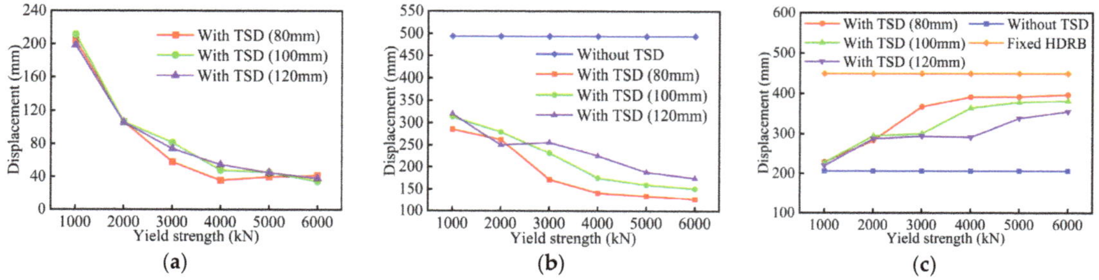

Figure 6. Displacement responses with different yield strengths: (**a**) TSD; (**b**) HDRB; (**c**) pier top.

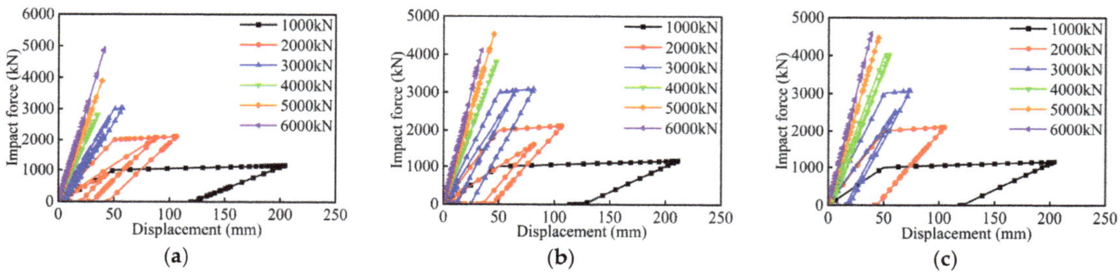

Figure 7. Hysteretic curves of the TSD with different yield strengths: (**a**) initial gap 80 mm; (**b**) initial gap 100 mm; (**c**) initial gap 120 mm.

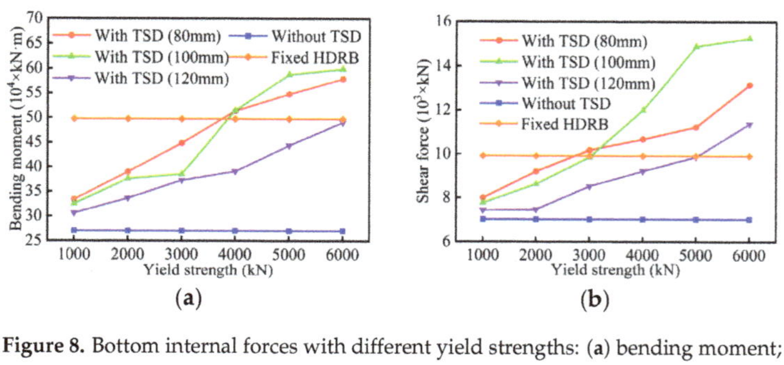

Figure 8. Bottom internal forces with different yield strengths: (**a**) bending moment; (**b**) shear force.

Figure 8a demonstrates a significantly increasing bending moment at the pier bottom with higher yield strength. While TSD-equipped bridges exhibit a larger moment than the no-TSD model, the bending moment with initial gaps of 80 mm and 100 mm will be larger than that in the fixed HDRB model when yield strength is over 4000 kN. Figure 8b shows increasing shear forces at the pier bottom with a higher yield strength. It is noteworthy that the shear force exceeds that of the fixed HDRB model when yield strength reaches certain thresholds. Compared to the fixed HDRB model, the TSD with large yield strength may produce larger bottom internal forces, even though pier top displacements are smaller.

3.2. Different Initial Gap Variations

Under strong seismic excitations, bearing responses are significantly influenced by the TSD's initial gap. The bearing–TSD interaction alters internal force transmission paths and displacement responses during seismic events. The maximum displacement responses of the tall pier with different initial gaps are shown in Figure 9. The force–displacement hysteretic curves of the TSD are shown in Figure 10. The corresponding bending moment and shear force at the pier bottom are shown in Figure 11.

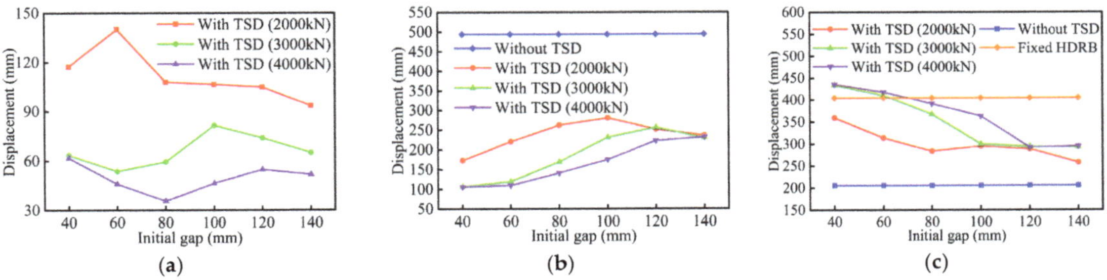

Figure 9. Displacement responses with different initial gaps: (**a**) TSD; (**b**) HDRB; (**c**) pier top.

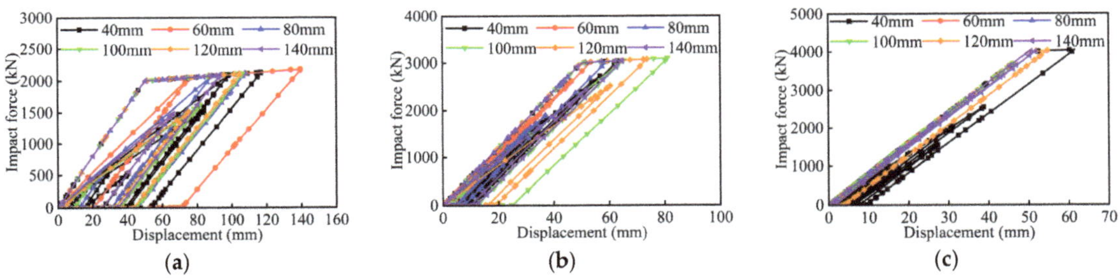

Figure 10. Hysteretic curves of the TSD with different initial gaps: (**a**) yield strength 2000 kN; (**b**) yield strength 3000 kN; (**c**) yield strength 4000 kN.

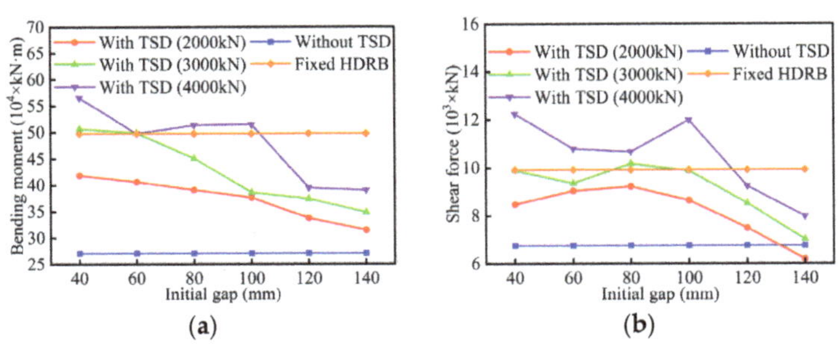

Figure 11. Bottom internal forces with different initial gaps: (**a**) bending moment; (**b**) shear force.

Figure 9a shows a complex trend of TSD displacements with varying initial gaps under different yield strengths. Figure 9b indicates that bearing displacements all increase with larger initial gaps up to 100 mm and are substantially smaller than in the no-TSD model. Overall, displacements of the TSD and HDRB decrease as yield strength increases from 2000 kN to 4000 kN. Figure 9c shows that pier top displacement gradually decreases as the initial gap increases, achieving a stable trend earlier with smaller yield strength.

These displacements are larger than those in the no-TSD model due to the greater inertial force transmitted to the pier cap. Overall, pier top displacements fall below those in the fixed HDRB model when the initial gap exceeds 80 mm for yield strengths of 3000 kN and 4000 kN. For a yield strength of 2000 kN, the pier top displacement stabilizes after 80 mm, with all displacements below those in the fixed HDRB model. Figure 10 shows fuller hysteretic loops at a smaller yield strength, indicating better energy dissipation capacity.

Figure 11a shows that for TSD models with yield strengths of 2000 kN and 3000 kN, bending moment gradually decreases with increasing initial gap, and greater yield strength leads to a larger bending moment. Figure 11b shows that with increasing initial gap, bottom shear force fluctuates, which may be due to higher-order mode effects. In addition, a higher yield strength leads to larger shear forces, which are larger than those in the no-TSD model. However, when yield strength is 4000 kN, both bending moment and shear forces are nearly always larger than in the fixed HDRB model for initial gaps below 100 mm, which is different from the variation observed for pier top displacement (Figure 9c). Both pier top displacement and bottom internal forces approach the levels in the no-TSD model when the initial gap increases. However, HDRB displacements do not approach the levels in the no-TSD model as the initial gap increases. This is mainly because the seismic responses of piers (especially shear force) are mainly contributed by their own mass instead of the steel girder mass. As the initial gap increases, inertial forces of the steel girder transmitted to the substructure are reduced, bringing responses close to those of the no-TSD model. Nonetheless, the existence of the TSD will limit girder displacement and dissipate earthquake energy, leading to an obvious reduction in HDRB displacement compared to the no-TSD model.

3.3. *Different Post-Yield Stiffness Ratio Variations*

The maximum displacement responses of a tall pier with different post-yield stiffness ratios are shown in Figure 12. The force–displacement hysteretic curves of the TSD are shown in Figure 13. The corresponding bending moment and shear force at the pier bottom are shown in Figure 14.

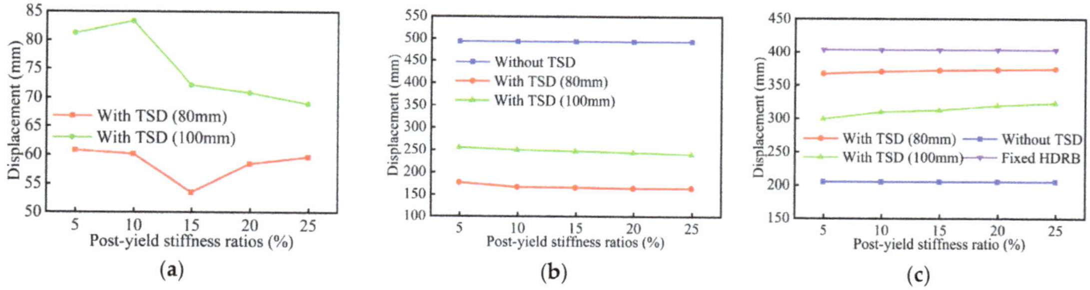

Figure 12. Displacement responses with different post-yield stiffness ratios: (**a**) TSD; (**b**) HDRB; (**c**) pier top.

Figure 12a shows that TSD displacements under a 100 mm initial gap generally decrease when post-yield stiffness ratios are over 10%, as increased stiffness provides greater post-yield strength that limits further deformation. A pronounced drop in TSD displacement occurs at a 15% post-yield stiffness ratio. Figure 12b indicates slightly decreasing HDRB transverse displacements with higher stiffness ratios, remaining substantially smaller than in the no-TSD model. In addition, models with larger initial gaps exhibit greater HDRB transverse displacement responses. Compared to yield strength and initial gap, the post-yield stiffness ratio exerts a weaker influence on bearing displacement control, as it primarily affects restoring force rather than energy dissipation capacity. While higher stiffness ratios usually enhance post-earthquake re-centering and reduce residual displacement, their contribution to energy dissipation is often limited, as shown in Figure 13. Figure 12c shows pier top displacements in models with TSD consistently exceeding those in the no-TSD model, but remaining below levels in the fixed HDRB model. Furthermore, a slight increase in pier top displacement occurs with higher post-yield stiffness ratios, attributable to enhanced impact forces and reduced energy dissipation.

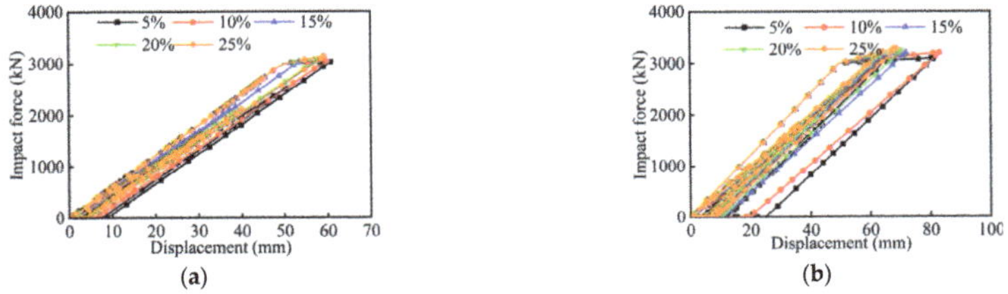

Figure 13. Hysteretic curves of the TSD with different post-yield stiffness ratios: (**a**) initial gap 80 mm; (**b**) initial gap 100 mm.

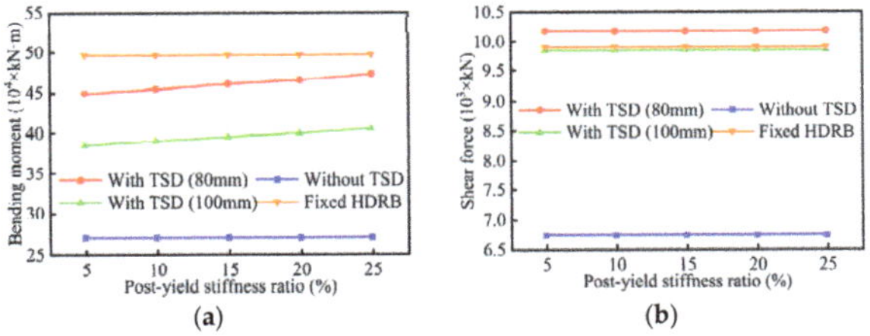

Figure 14. Bottom internal forces with different post-yield stiffness ratios: (**a**) bending moment; (**b**) shear force.

Figure 14 shows that the no-TSD model has the smallest bottom bending moments and shear forces. Bridges with TSDs exhibit smaller bending moments than the fixed HDRB model. However, for an initial gap of 80 mm, shear forces exceed those in the fixed HDRB model, which contradicts the trend observed for pier top displacements (Figure 12c). This occurs because shear forces are more susceptible to higher-order mode effects.

3.4. Seismic Performance Comparisons

Considering bearing displacement, top displacement, and bottom bending moment responses, the optimal parameters for the TSD are defined as a yield strength of 3000 kN, an initial gap of 100 mm, and a post-yield stiffness ratio of 15%. This combination provides a balanced energy dissipation capacity and effective control of internal forces. Note that there is no direct corresponding relationship between pier top displacements and bottom internal forces (especially shear force), which is quite different from the behaviors in mid- and short-piers [25–27]. To further illustrate the seismic performance of the tall-pier girder bridge, displacement and internal force envelopes for the optimized TSD model, no-TSD model, and fixed HDRB model are shown in Figure 15. In addition, simulations using reinforced concrete shear keys (RCSKs) are also presented, where the four-line skeleton and hysteretic rules of RCSKs can be found in [40].

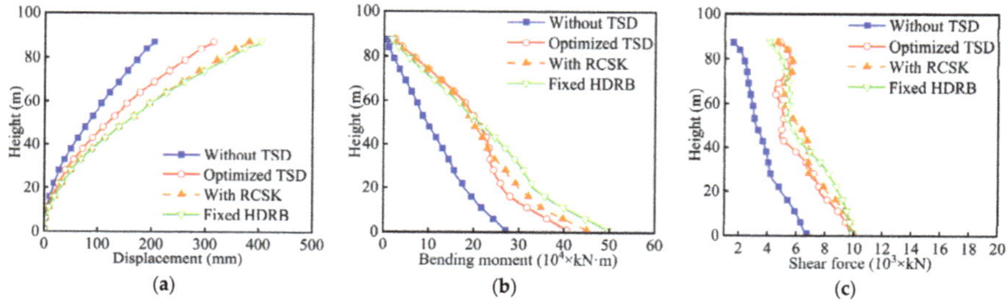

Figure 15. Envelope curves: (**a**) pier displacement; (**b**) pier bending moment; (**c**) pier shear force.

Figure 15a shows that the displacement envelope of the optimized TSD model is smaller than that of the fixed HDRB model and RCSK model but larger than that of the no-TSD model. The RCSK model exhibits a larger displacement envelope than the TSD model and slightly smaller than that of the fixed HDRB model. Figure 15b shows that the bending moment envelope of the TSD model exhibits an S-shape, with the bending moment within the lower 50 m of pier height reduced compared to the fixed HDRB model. The bending moment envelope of the TSD model is lower than that of the RCSK model within the lower 35 m. The significant bending moment in the pier's mid-height region may also induce a secondary plastic hinge away from the bottom [41,42]. Figure 15c demonstrates more complex shear force envelope shapes. Although the bottom shear forces are nearly identical for all models except the no-TSD model, shear forces along the pier body are obviously reduced when using the optimized TSD. This indicates that higher-order modes disrupt the linear correlation between top displacement and bottom internal forces in tall piers.

The time history curves of the top displacement, HDRB displacement, and bottom bending moment are shown in Figure 16. As seen, the pier top displacement and pier bottom bending moment are greatly reduced with the optimized TSD. The RCSK model can reduce pier top displacement and bottom bending moment, but it produces larger HDRB displacement than the optimized TSD model, since its strength degrades after the peak point. The axial force-bending moment (PM) curves for the key bottom section under different conditions are shown in Figure 17. The optimized TSD model remains entirely within the equivalent yield envelope throughout, ensuring elastic behavior at the pier bottom. However, obvious damage occurs at the pier bottom for the fixed HDRB model, since its PM curves break through the sectional equivalent yield envelope. Therefore, the proposed TSD parameters (yield strength: 3000 kN, initial gap: 100 mm, post-yield stiffness ratio: 15%) are validated as suitable for this bridge.

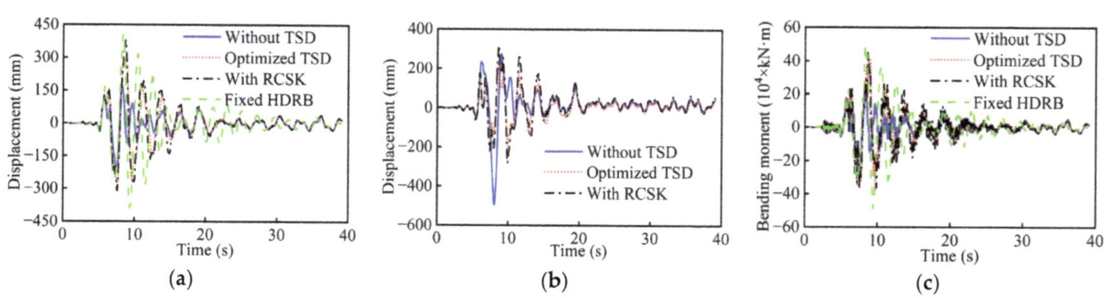

Figure 16. Time history curves: (**a**) top displacement; (**b**) HDRB displacement; (**c**) bottom bending moment.

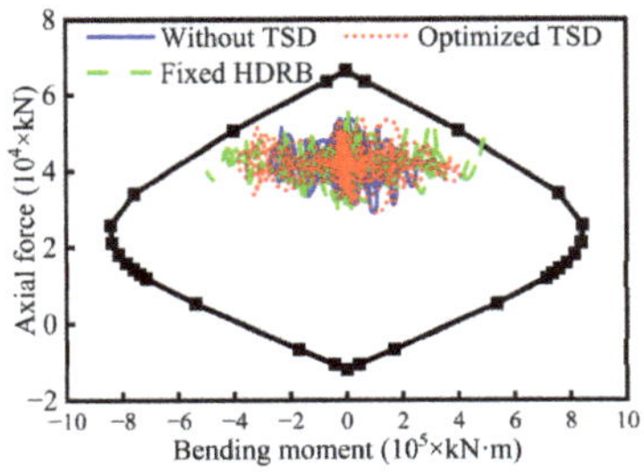

Figure 17. P-M curves of the bottom section.

4. Influence of Pulse Characteristics

For tall-pier girder bridges, the influence of higher-order modes is significant and cannot be neglected. Furthermore, the effect of near-fault pulse-like motion characteristics on the seismic responses of tall piers remains to be explored. Therefore, this section employs 12 near-fault pulse-like motions listed in Table 3 to investigate the effects of the pulse period and velocity pulse characteristics on bridge seismic responses.

4.1. *Effect of Pulse Period*

The displacement envelopes of piers are shown in Figure 18. Displacement magnitude generally increases with pier height. As shown in Figure 18b, the maximum displacement responses occur under medium-period pulse-like motions, primarily due to resonance arising from the close match between the bridge's fundamental structural period (3.06 s) and the pulse period (Figure 5c). Significant displacements also arise under long-period pulse-like motions, reflecting the slender tall-pier bridge's sensitivity to high spectral accelerations in the low-frequency range of motions (Figure 5d). Notably, displacement responses under short-period and non-pulse motions are similar, which is primarily explained by the nearly overlapping mean spectra beyond 1.8 s (Figure 5f). The spectral amplification for short-period pulse-like motions occurs between 1.2 s and 1.8 s, which is far away from the fundamental structural period of the tall-pier girder bridge (3.06 s).

The bending moment envelopes of piers are shown in Figure 19. Figure 19a,d reveal that the tall pier exhibits an S-shaped moment distribution under short-period and non-pulse excitations, indicating significant higher-order mode influence. Additionally, the large moment at the mid-height suggests a potential second plastic hinge region that is distinct from the conventional plastic hinge at the pier bottom. As pier height decreases, the mid pier shows only a slight S-shaped moment distribution. The contribution of higher-order modes diminishes substantially with reduced pier height. For the short pier, bending moment envelopes become nearly linear, demonstrating a significant reduction in higher-mode effects.

The fundamental structural period plays a crucial role in bending moment responses. Tall-pier girder bridges typically exhibit longer structural periods, shifting resonance away from motions with dominant high-frequency components. Consequently, bending moment responses under short-period motions and non-pulse motions are reduced. Conversely, significant bending moment amplification occurs in tall piers subjected to medium- to long-period excitations. This amplification stems primarily from tall piers' lower stiffness and complex higher-order mode shapes, where the fundamental structural period aligns more closely with pulse periods, inducing pronounced resonant responses. Further analyses reveal that medium-period motions induce higher bending moments than long-period motions due to closer matching between the fundamental structural period and pulse period, intensifying moments at the pier bottom. Therefore, resonance effects from period matching critically determine the magnitude and distribution of seismic responses in tall-pier girder bridges under near-fault motions.

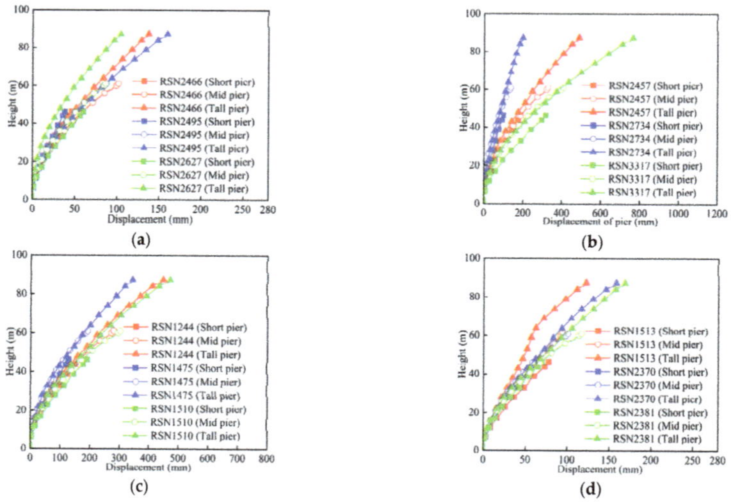

Figure 18. Displacement envelopes: (**a**) short-period motions; (**b**) medium-period motions; (**c**) long-period motions; (**d**) non-pulse motions.

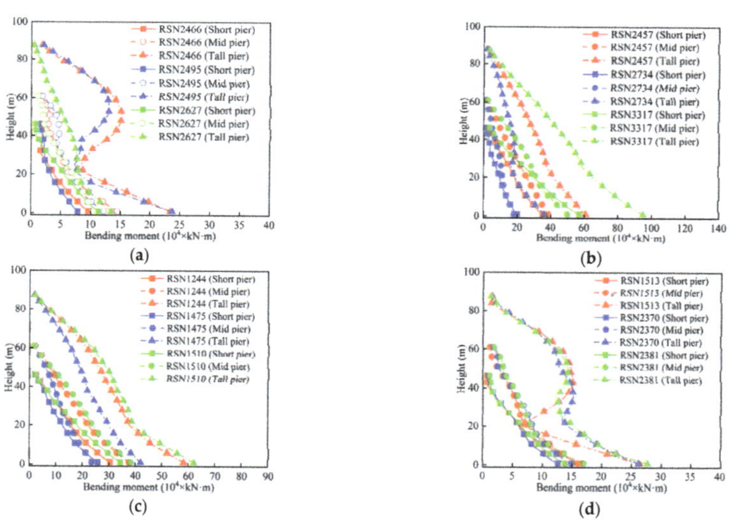

Figure 19. Bending moment envelopes: (**a**) short-period motions; (**b**) medium-period motions; (**c**) long-period motions; (**d**) non-pulse motions.

The shear force envelopes of piers are shown in Figure 20. Unlike displacement and bending moment responses, tall piers exhibit distinct S-shaped distributions under all four input motion types, indicating significant higher-order mode effects. This occurs because shear force is primarily induced by inertial forces from the partial mass, while bending moment is also influenced by the distance to the supports. For tall piers, shear forces decrease from the base to approximately 50 m height due to accumulated inertial forces from the middle to lower regions. Above 50 m, shear force distributions become less consistent, likely affected by complex pier–superstructure interaction. Medium piers show weakened S-shaped distributions, reflecting diminished higher-mode influence. For piers below 50 m, shear force envelopes progressively become linear along the pier, indicating that the dynamic responses are predominantly governed by the fundamental structural mode.

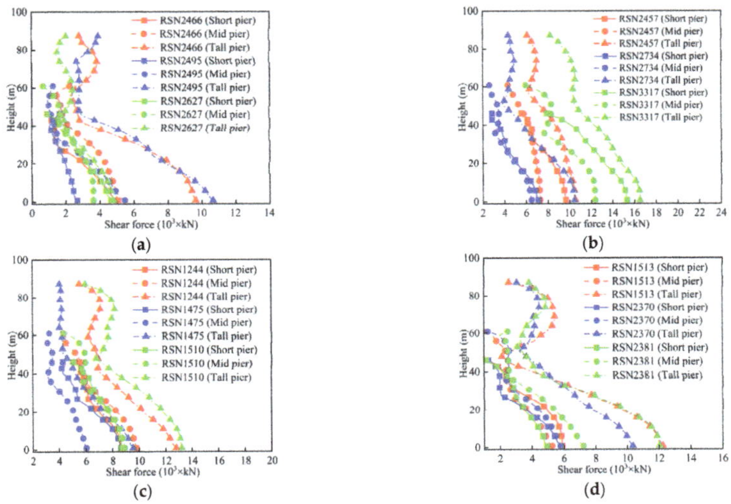

Figure 20. Shear force envelopes: (**a**) short-period motions; (**b**) medium-period motions; (**c**) long-period motions; (**d**) non-pulse motions.

To clarify the interaction mechanism between TSD and higher-order mode effects, the seismic response envelopes without the TSD under short-period motions are shown in Figure 21. Comparing Figures 19a and 21b, it can be drawn that the TSD will exacerbate higher-order mode effects, as evidenced by the S-shaped envelopes. This can be attributed to the fact that the TSD will indirectly increase the horizontal stiffness of the tall pier by limiting the pier top displacement when collision occurs. Meanwhile, the energy dissipation capacity of the TSD may reduce the seismic responses from higher-order modes, resulting in smaller pier top displacement and bottom bending moment. Given that shear forces mainly depend on local mass, the influence of the TSD and higher-order modes on shear force is more complex, which will be investigated further through additional simulations in future work.

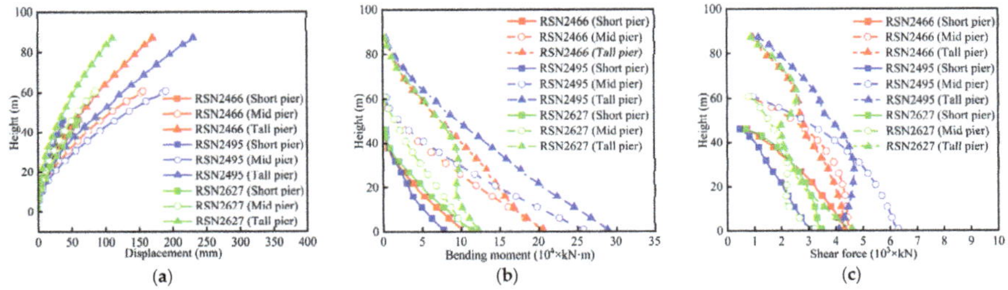

Figure 21. Seismic response envelopes without the TSD: (**a**) displacement; (**b**) bending moment; (**c**) shear force.

4.2. Effect of Velocity Pulse

The wavelet analysis framework [43] and the method proposed by Baker et al. [44] are used to separate the main velocity pulse from the original pulse-like motions, as shown in Table 3. Each motion is decomposed into two components: the main velocity pulse and the residual velocity component. The time history curves of the original waves and the separated components are shown in Figure 22. The spectral decomposition of the medium-period motions is shown in Figure 23, indicating that the main velocity pulse mainly affects the spectral values around the pulse periods. Similarly, the main velocity pulse of short-period motions and long-period motions primarily affects spectral values in the high-frequency range and low-frequency range, respectively. This decomposition assesses the dominant influence of the main velocity pulse on seismic responses of the tall-pier girder bridge.

Displacement, bending moment, and shear force envelopes of the tall pier under original and residual motions are compared in Figure 24. Removal of the main velocity pulse significantly reduces displacement and bending moment responses, with reductions exceeding 50% in some cases. This confirms the critical role of the main velocity pulse in triggering resonant responses in tall-pier bridges, highlighting the strong coupling between the fundamental structural period and pulse period. Comparisons in Figure 24e,f show evident shear force reductions under medium- and long-period motions after pulse removal. For short-period motions, however, shear distributions display a marked increase at the pier bottom in some cases, despite moderate reductions at the pier top. Furthermore, bending moment and shear force envelopes exhibit a more obvious S-shape, indicating that residual high-frequency components excite higher-order modes more obviously after removal of the main velocity pulse. However, higher-order mode effects induced by high-frequency components are masked by dominant seismic responses from large spectral values in medium to long periods.

Furthermore, TSD's sensitivity to the residual component is analyzed using record RSN2734, and the seismic responses for different TSD parameters are shown in Figures 25 and 26. The variation trends of pier top displacement and bottom bending moment under the residual component are similar to those from the original wave, though their values are much smaller due to the removal of the pulse component. Both pier top displacement and bottom bending moment under the residual component increase with increasing yield strength, while decreasing with increasing initial gap. In addition, the influence of the post-yield stiffness ratio on pier top displacement and bottom bending moment under the residual component is also not obvious.

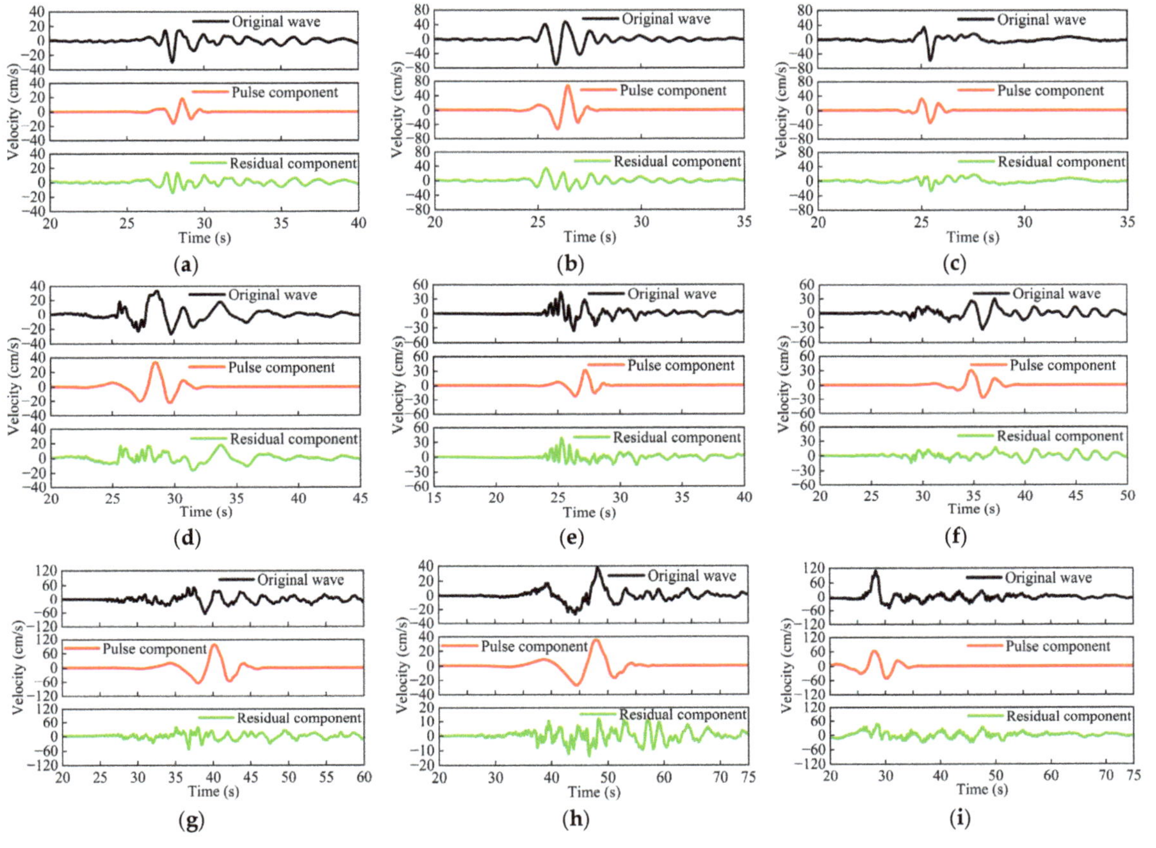

Figure 22. The decomposition of the motions: (**a**) RSN2466; (**b**) RSN2495; (**c**) RSN2627; (**d**) RSN2457; (**e**) RSN2734; (**f**) RSN3317; (**g**) RSN1244; (**h**) RSN1475; (**i**) RSN1510.

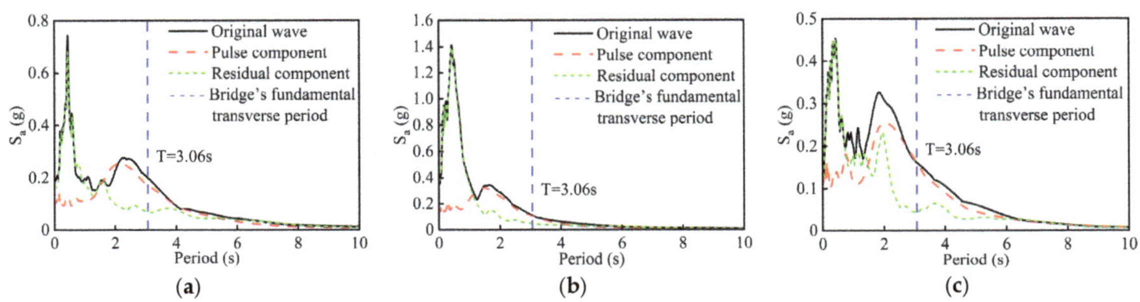

Figure 23. The spectral decomposition of the motions: (**a**) RSN2457; (**b**) RSN2734; (**c**) RSN3317.

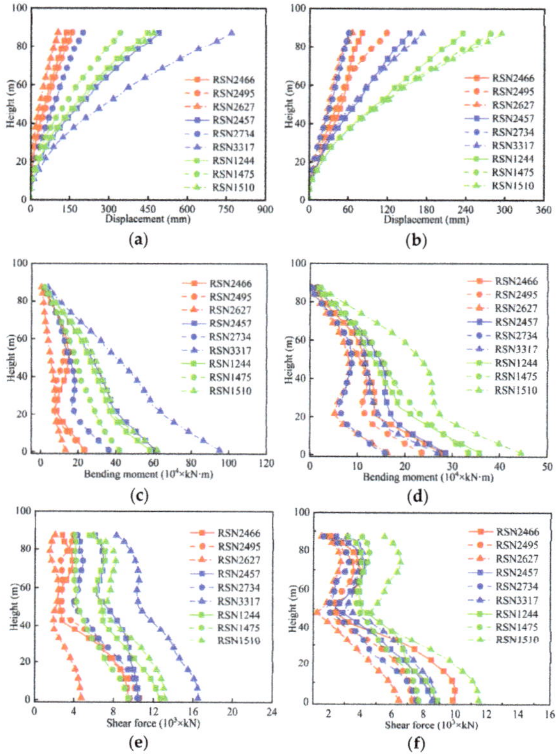

Figure 24. Seismic responses: (**a**) displacement (original); (**b**) displacement (residual); (**c**) bending moment (original); (**d**) bending moment (residual); (**e**) shear force (original); (**f**) shear force (residual).

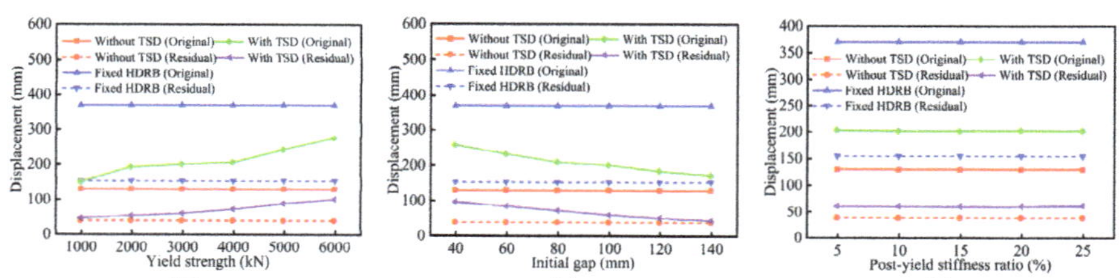

Figure 25. Pier top displacement: (**a**) yield strength; (**b**) initial gap; (**c**) post-yield stiffness ratio.

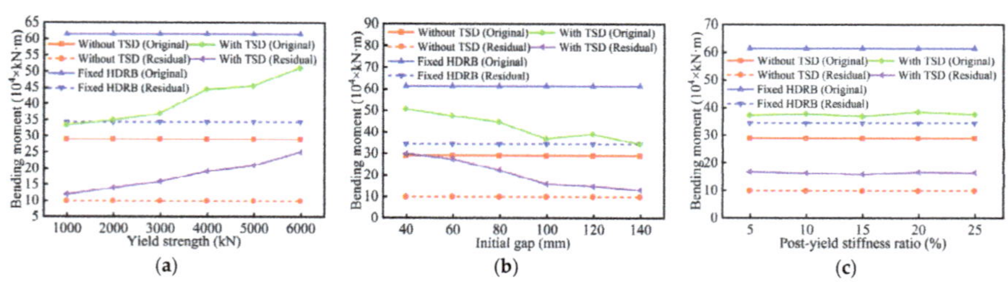

Figure 26. Bottom bending moment: (**a**) yield strength; (**b**) initial gap; (**c**) post-yield stiffness ratio.

5. Conclusions

This study proposes a novel transverse steel damper (TSD) with enhanced lateral strength and energy dissipation capacity. Based on a six-span tall-pier girder bridge, seismic responses under varying TSD parameters are systematically analyzed, while the influence of pulse characteristics is investigated. The key findings are summarized below:

(1) The novel TSD is modeled and validated under cyclic loading using ABAQUS, demonstrating excellent hysteretic behavior and mechanical stability. It provides a viable solution for lateral restraint and energy dissipation in tall-pier girder bridges.

(2) The use of TSDs effectively reduces bearing displacements while increasing pier-top displacements and pier-bottom internal forces as additional inertial forces are transmitted to the pier cap. However, displacements and internal forces in the bridge with TSDs are smaller than those in the bridge with fixed bearings.

(3) As yield strength increases, bearing displacements decrease, while pier-top displacements and pier-bottom internal forces increase. Conversely, larger initial gaps increase bearing displacements but reduce pier-top displacements and pier-bottom internal forces. However, the post-yield stiffness ratio exhibits a negligible influence on the seismic responses of tall-pier bridges. The optimized parameters are a yield strength of 3000 kN, an initial gap of 100 mm, and a post-yield stiffness ratio of 15%. The displacements and internal forces of the pier are greatly reduced using these parameters.

(4) Tall piers develop S-shaped bending moment distributions under short-period and non-pulse motions, while S-shaped shear force distributions manifest under all motions owing to higher-order mode effects. As pier heights decrease, the influence of these modes progressively diminishes, evidenced by bending moment and shear force distributions transitioning toward linear profiles along the pier height.

(5) Medium-period pulse-like motions maximize seismic responses due to resonance (pulse period $\approx$ fundamental structural period). Long-period pulse-like motions also generate large seismic responses owing to the susceptibility of tall-pier girder bridges to high spectral accelerations in the low-frequency range. Under short-period and non-pulse motions, however, seismic responses remain minimal since pulse periods and site characteristic periods are far from the fundamental structural period.

(6) Removing the main velocity pulse significantly reduces pier displacements and internal forces. However, more pronounced S-shaped envelope distributions are found in bending moments and shear forces under residual motions, implying significant higher-order mode effects. This indicates that higher-order modes exhibit greater sensitivity to the residual high-frequency components of pulse-like motions. Furthermore, these effects are masked by dominant responses from large spectral values in medium to long periods.

Despite its promising numerical performance, the TSD faces practical challenges that require attention before its implementation in real bridges. First, TSD installation requires sufficient space and is easy to replace, particularly in retrofits. Second, long-term durability concerns arise due to steel's susceptibility to corrosion. Third, low-cycle fatigue performance requires experimental validation, and hysteretic behavior should be calibrated using measured cyclic responses.

Author Contributions: Conceptualization, Q.Q.; methodology, H.Y.; software, H.Y.; validation, H.Y. and H.C.; investigation, Z.P. and H.C.; data curation, R.Y.; writing—original draft preparation, Z.P.; writing—review and editing, Q.Q.; supervision, C.S.; funding acquisition, Q.Q. and C.S. All authors have read and agreed to the published version of the manuscript.

Funding: This research was funded by the National Key R&D Program of China [No. 2023YFB2604402], the National Natural Science Foundation of China [No. 51978581], and the Sichuan Science and Technology Program [No. 2025ZNSFSC1315].

Data Availability Statement: Data is contained within the article.

Conflicts of Interest: The authors declare no conflict of interest.

References

1. Chen, X.; Li, C. Seismic performance of tall pier bridges retrofitted with lead rubber bearings and rocking foundation. *Eng. Struct.* **2020**, *212*, 110529. [CrossRef]
2. Shao, C.; Qi, Q.; Wang, M.; Xiao, Z.; Wei, W.; Hu, C.; Xiao, L. Experimental study on the seismic performance of round-ended hollow piers. *Eng. Struct.* **2019**, *195*, 309–323. [CrossRef]
3. Chen, X. System fragility assessment of tall-pier bridges subjected to near-fault ground motions. *J. Bridge Eng.* **2020**, *25*, 04019143. [CrossRef]
4. Parsons, T.; Ji, C.; Kirby, E. Stress changes from the 2008 Wenchuan earthquake and increased hazard in the Sichuan basin. *Nature* **2008**, *454*, 509–510. [CrossRef] [PubMed]
5. Ni, S.; Wang, W.; Li, L. The April 14th, 2010 Yushu earthquake, a devastating earthquake with foreshocks. *Sci. China Earth Sci.* **2010**, *53*, 791. [CrossRef]
6. Xu, X.; Wen, X.; Han, Z.; Chen, G.; Li, C.; Zheng, W.; Zhnag, S.; Ren, Z.; Xu, C.; Tan, X. Lushan Ms 7.0 earthquake: A blind reserve-fault event. *Chin. Sci. Bull.* **2013**, *58*, 3437–3443. [CrossRef]
7. Sun, J.; Yu, Y.; Li, Y. Stochastic finite-fault simulation of the 2017 Jiuzhaigou earthquake in China. *Earth Planets Space* **2018**, *70*, 1–12. [CrossRef]
8. Gu, H.; Itoh, Y. Ageing behaviour of natural rubber and high damping rubber materials used in bridge rubber bearings. *Adv. Struct. Eng.* **2010**, *13*, 1105–1113. [CrossRef]
9. Tubaldi, E.; Mitoulis, S.A.; Ahmadi, H. Comparison of different models for high damping rubber bearings in seismically isolated bridges. *Soil. Dyn. Earthq. Eng.* **2018**, *104*, 329–345. [CrossRef]
10. Aghaeidoost, V.; Billah, A.M. An advanced rate-dependent analytical model of lead rubber bearing. *Earthq. Eng. Struct. Dyn.* **2024**, *53*, 1961–1981. [CrossRef]
11. Chen, P.; Wang, B.; Zhang, Z.; Li, T.; Dai, K. A generalized model of lead rubber bearing considering large strain stiffening and degradation. *Eng. Struct.* **2023**, *275*, 115264. [CrossRef]
12. Chen, X.; Wu, S.; Li, J.; Guan, Z.; Xiang, N. Seismic performance assessment and design procedure of base-isolated bridges with lead-rubber-bearing and negative stiffness springs (LRB-NS). *Eng. Struct.* **2024**, *306*, 117871. [CrossRef]
13. Shang, J.; Tan, P.; Zhang, Y.; Han, J.; Mi, P. Seismic isolation design of structure using variable friction pendulum bearings. *Soil. Dyn. Earthq. Eng.* **2021**, *148*, 106855. [CrossRef]
14. Wei, B.; Yang, Z.; Fu, Y.; Xiao, B.; Jiang, L. Seismic displacement response analysis of Friction Pendulum Bearing under friction coupling and collision effects. *Eng. Struct.* **2024**, *310*, 118128. [CrossRef]
15. Zheng, W.; Tan, P.; Li, J.; Wang, H.; Liu, Y.; Xian, Z. Superelastic conical friction pendulum isolator for seismic isolation of bridges under near-fault ground motions. *Struct. Control Health Monit.* **2023**, *2023*, 5497731. [CrossRef]
16. Qi, Q.; Shao, C.; Cui, H.; Huang, H.; Wei, W.; Wang, C.; Zhuang, W. Shaking table tests and numerical studies on the seismic behaviors of FPB in railway continuous beam bridges. *Eng. Struct.* **2023**, *290*, 116318. [CrossRef]
17. Wu, C.; Jing, H.; Feng, Z.; Song, J.; Wan, T.; Chen, Z. Control of longitudinal movement response of suspension bridges induced by passing trains using low-exponent fluid viscous dampers. *Structures* **2024**, *62*, 106330. [CrossRef]
18. Zhao, Y.; Huang, P.; Long, G.; Yuan, Y.; Sun, Y. Influence of fluid viscous damper on the dynamic response of suspension bridge under random traffic load. *Adv. Civ. Eng.* **2020**, *2020*, 1857378. [CrossRef]
19. Liu, Q.; Zhu, S.; Yu, W.; Wu, X.; Song, F.; Ren, X. Ground motion frequency insensitivity of bearing-supported pedestrian bridge with viscous dampers. *KSCE J. Civ. Eng.* **2021**, *25*, 2662–2673. [CrossRef]
20. Han, Q.; Zhou, Y.; Ou, Y.; Du, X. Seismic behavior of reinforced concrete sacrificial exterior shear keys of highway bridges. *Eng. Struct.* **2017**, *139*, 59–70. [CrossRef]
21. Mei, H.; Guo, A. Experimental investigation and finite element analysis for seismic capacity prediction of RC shear keys with different failure modes. *Soil. Dyn. Earthq. Eng.* **2023**, *175*, 108243. [CrossRef]
22. Zhou, L.; Wang, X.; Ye, A. Low cycle fatigue performance investigation on transverse steel dampers for bridges under ground motion sequences using shake-table tests. *Eng. Struct.* **2019**, *196*, 109328. [CrossRef]

23. Yue, K.; Xu, L.; Liu, J.; Fan, L.; Xu, L. Seismic performance of an energy dissipating shear key for highway bridges using butterfly-shaped steel plates. *Eng. Struct.* **2023**, *295*, 116885. [CrossRef]
24. Huang, L.; Wang, W.; Xu, S.; Wang, B.; Li, Z. Shaking table tests and numerical study on the seismic performance of arc-shaped shear keys in highway continuous-girder bridges. *Buildings* **2024**, *14*, 3060. [CrossRef]
25. Bi, K.; Hao, H. Modelling of shear keys in bridge structures under seismic loads. *Soil. Dyn. Earthq. Eng.* **2015**, *74*, 56–68. [CrossRef]
26. Xiang, N.; Li, J. Effect of exterior concrete shear keys on the seismic performance of laminated rubber bearing-supported highway bridges in China. *Soil. Dyn. Earthq. Eng.* **2018**, *112*, 185–197. [CrossRef]
27. Özşahin, E.; Pekcan, G.; Borekci, M. Design of shear keys for mitigating seismic response due to translational and torsional ground motions. *Eng. Struct.* **2024**, *312*, 118195. [CrossRef]
28. Abbasi, M.; Moustafa, M.A. Effect of shear keys on seismic response of irregular bridge configurations. *Transp. Res. Rec.* **2017**, *2642*, 155–165. [CrossRef]
29. Wu, S. Investigation on the connection forces of shear keys in skewed bridges during earthquakes. *Eng. Struct.* **2019**, *194*, 334–343. [CrossRef]
30. Shen, Y.; Li, J.; Freddi, F.; Igarashi, A.; Zhou, J. Numerical investigation of transverse steel damper (TSD) seismic system for suspension bridges considering pounding between girder and towers. *Soil. Dyn. Earthq. Eng.* **2022**, *155*, 107203. [CrossRef]
31. Tubaldi, E.; Tassotti, L.; Dall'Asta, A.; Dezi, L. Seismic response analysis of slender bridge piers. *Earthq. Eng. Struct. Dyn.* **2014**, *43*, 1503–1519. [CrossRef]
32. Chen, X.; Wu, P.; Li, C. Seismic performance assessment of base-isolated tall pier bridges using friction pendulum bearings achieving resilient design. *Structures* **2022**, *38*, 618–629. [CrossRef]
33. Yang, D.; Liu, J.; Yu, R.; Chen, G. Unified framework for stochastic dynamic responses and system reliability analysis of long-span cable-stayed bridges under near-fault ground motions. *Eng. Struct.* **2025**, *322*, 119061. [CrossRef]
34. Jia, Y.; Xin, L.; Yang, D.; Pei, M.; Zhao, L.; Huang, Z. Seismic behavior analysis of long-span cable-stayed bridge under bi-directional near-fault ground motions. *Structures* **2024**, *64*, 106512. [CrossRef]
35. Ozturk, B.; Cetin, H.; Aydin, E. Optimum vertical location and design of multiple tuned mass dampers under seismic excitations. *Structures* **2022**, *41*, 1141–1163. [CrossRef]
36. Guan, Z.; Li, J.; Xu, Y.; Lu, H. Higher-order mode effects on the seismic performance of tall piers. *Front. Struct. Civ. Eng.* **2011**, *5*, 496–502. [CrossRef]
37. *JT/T 842-2012*; High Damping Seismic Isolation Rubber Bearings for Highway Bridges. Ministry of Transport of the People's Republic of China: Beijing, China, 2012.
38. *JTG/T 2231-01-2020*; Specifications for Seismic Design of Highway Bridges. Ministry of Transport of the People's Republic of China: Beijing, China, 2020.
39. *JTG 3363-2019*; Specifications-for-Design-of-Foundation of Highway Bridges and Culverts. Ministry of Transport of the People's Republic of China: Beijing, China, 2019.
40. Silva, P.F.; Megally, S.; Seible, F. Seismic performance of sacrificial exterior shear keys in bridge abutments. *Earthq. Spectra* **2009**, *25*, 643–664. [CrossRef]
41. Qi, Q.; Shao, C.; Wei, W.; Xiao, Z.; He, J. Seismic performance of railway rounded rectangular hollow tall piers using the shaking table test. *Eng. Struct.* **2020**, *220*, 110968. [CrossRef]
42. Qi, Q.; Shao, C.; Yang, H.; Cui, H.; Chen, Z.; Gong, W. Axial-flexure-shear model for seismic analysis of RC thin-walled hollow piers. *Soil. Dyn. Earthq. Eng.* **2025**, *195*, 109375. [CrossRef]
43. Baker, J.W. Quantitative classification of near-fault ground motions using wavelet analysis. *Bull. Seismol. Soc. Am.* **2007**, *97*, 1486–1501. [CrossRef]
44. Shahi, S.K.; Baker, J.W. An empirically calibrated framework for including the effects of near-fault directivity in probabilistic seismic hazard analysis. *Bull. Seismol. Soc. Am.* **2011**, *101*, 742–755. [CrossRef]

MDPI AG
Grosspeteranlage 5
4052 Basel
Switzerland
Tel.: +41 61 683 77 34

Buildings Editorial Office
E-mail: buildings@mdpi.com
www.mdpi.com/journal/buildings

www.ingramcontent.com/pod-product-compliance
Lightning Source LLC
Chambersburg PA
CBHW061442210926
51850CB00008B/1585
9783725866069